£8.00

A Specialist Periodical Report

Electrochemistry
Volume 3

A Review of the Literature published during 1971

Senior Reporter
G. J. Hills, *Department of Chemistry, University of Southampton*

Reporters
A. K. Covington, *University of Newcastle upon Tyne*
B. Fleet, *Imperial College of Science & Technology, University of London*
N. A. Hampson, *University of Technology, Loughborough*
R. D. Jee, *School of Pharmacy, University of London*
T. H. Lilley, *University of Sheffield*
P. M. Robertson, *Eidgenössische Technische Hochschule, Zurich*
D. J. Schiffrin, *I.N.I.T., Buenos Aires*
M. J. Wootten, *University of Southampton*

© Copyright 1973

The Chemical Society
Burlington House, London, W1V 0BN

ISBN: 0 85186 027 3

Library of Congress Catalog Card No. 72-23822

Printed in Great Britain by
Alden & Mowbray Ltd
at the Alden Press, Oxford

Preface

This volume of Electrochemistry surveys the literature of salient parts of the subject up to the beginning of 1972. Certain sections, such as molten salts and membrane phenomena, have been rested whilst the equilibrium, spectroscopic, and transport properties of electrolytes have been taken up again. A beginning has been made to review one other important field, namely that concerned with voltammetry and polarography, particularly at the dropping mercury electrode.

The field continues to grow in range and more aspects will need to be included before the coverage is comprehensive. This responsible task I pass on to my colleague Professor Thirsk of Newcastle University and I wish him well in it.

March 1973 G. J. Hills

Editor's Note. For this volume, the gap between the end of the period of literature coverage and publication is much larger than is usual for volumes in the SPR series. This is mainly due to delays in the delivery of manuscripts, and it is certainly hoped to revert to a much more favourable schedule for Volume 4.

Contents

Chapter 1 Reversible Electrode Systems and Related Topics
by A. K. Covington

1 Introduction	1
2 Conventional Electrode Systems (Electrodes of First and Second Kinds, Redox Couples)	2
3 Glass Electrodes	5
4 Ion-selective Electrodes	10
5 Liquid-junction Potentials	19

Chapter 2 The Conductance of Electrolyte Solutions
by M. J. Wootten

1 Introduction	20
2 Theoretical Considerations	20
3 Aqueous Systems	22
4 Water–Non-aqueous Solvent Mixtures	26
5 Non-aqueous Solvents	29
6 Extreme Conditions	36

Chapter 3 The Solid Metal Electrode in Aqueous Solution
by N. A. Hampson

1 Introduction	41
2 The Interfacial Structure at Solid Metals	41
3 Charge-transfer Reactions	48
Theory of Charge Transfer	48
Diffusional Processes	50
Optical Processes and Noise	55
Electrocrystallization	56
Stress in Electrodeposits	58

4 Reactions of Individual Metal Systems	58
Aluminium	58
Alloys	59
Bismuth	61
Cadmium	62
Cobalt	63
Copper	64
Chromium	69
Carbon	70
Gallium	70
Germanium	71
Gold	71
Iron	72
Iridium	73
Lead	74
Manganese	74
Molybdenum	74
Nickel	74
Palladium	76
Platinum	77
Ruthenium	78
Silver	78
Sodium	79
Tantalum	79
Thallium	79
Tin	79
Titanium	80
Tungsten	80
Uranium	81
Zinc	82
5 Corrosion Processes	83
Theoretical	83
Corrosion of Individual Metal Systems	86
Aluminium	86
Antimony–Lead Alloy	86
Copper and Related Alloys	86
Iron, Steel, and Associated Alloys	87
Iridium	91
Molybdenum	92
Nickel	92
Zinc	93

Chapter 4 Ionic Double Layers and Adsorption
by D. J. Schiffrin

1 Introduction	94
2 Ionic Adsorption and Double-layer Structure at a Liquid Metal–Solution Interface	94
General Characteristics	94
Ionic Adsorption and Double-Layer Theory	99
Theory of the Diffuse Double Layer	101
Gallium	102
Non-aqeuous Solvents	102
3 The Adsorption of Organic Molecules on to Mercury	104
Aliphatic	104
Aromatic and Heterocyclic	107
4 Adsorption Kinetics	110
5 The Interface between Platinum (and Platinum-group Metals) and Aqueous Solutions	110
Thermodymanics and Double-layer Capacitance	110
Hydrogen and Oxygen Adsorption	111
Ionic Adsorption	113
Organic Adsorption	117
6 Experimental	118

Chapter 5 Organic Electrochemistry – Synthetic Aspects
by P. M. Robertson

1 Introduction	120
2 Reductions	121
Compounds containing the Carbonyl Group	121
Reductive Coupling Reactions	130
Reduction of Halides	132
Aliphatic and Heterocyclic Azomethines	137
Componds containing a Quaternary Ammonium Centre	143
Nitro-compounds	147
Other Nitrogen Compounds	151
Carboxylic Acids and Derivatives	153
Organosulphur Compounds	155
Miscellaneous Reductions	157

3 Oxidations	159
Aromatic Hydrocarbons	159
Alkenes and Alkanes	162
Carboxylic Acids	167
Aromatic Amines	172
Other Nitrogen-containing Compounds	175
Oxygen-containing Compounds	177
Alcohols	177
Phenols	179
Benzylic Compounds	181
Furans	181
Anodic Halogenation and Pseudohalogenation of Organic Compounds	183
Miscellaneous Oxidations	184

Chapter 6 Electrolyte Solutions
by T. H. Lilley

1 Introduction	187
2 Liquid Water	188
3 Thermodynamics of Single Electrolytes	190
Osmotic and Activity Coefficients	191
Enthalpies	193
Volumes	193
4 Thermodynamics of Mixed Electrolytes	194
5 Solutions containing Tetra-alkylammonium Ions	196
6 Interactions between Electrolytes and Non-electrolytes	199
Solvation	201
7 Protolytic Equilibria	205
8 Ion Association	207
9 Dielectric Studies and Ultrasonic Absorption	209

Chapter 7 Electroanalytical Chemistry: Voltammetry
by B. Fleet and R. D. Jee

1 Introduction	210
2 Polarography	211
D.C. Polarography	211
D.C. Polarography: Applications	215
Pulse Polarography	219
A.C. Polarography	221

Contents   ix

    3 Stripping Voltammetry   227
        Applications   231

    4 Coulometry   232

Errata   238

Author Index   239

1
Reversible Electrode Systems and Related Topics

BY A. K. COVINGTON

1 Introduction

The literature surveyed for this Report covered the period mid-1970 to mid-1972. The format of the previous Report[1] has been followed to facilitate reference back to topics discussed there, and to avoid the need for excessive repetition of references. Water should be understood to be the solvent used in the investigations described, unless another solvent system is specifically mentioned.

Highlights in the work to be described below include an increasing awareness of the importance of surface films (gel layers) on glass electrodes and their influence on time-dependent potentials, and the development of 'neutral carrier' complexes of alkali-metal and alkaline-earth ions on which a new range of ion-selective electrodes is based.

Buck has extensively revised and extended the chapter 'Potentiometry' in the new Weissberger series.[2] He has also contributed an extensive review (1000 references) to *Analytical Chemistry, Review of Fundamentals*.[3] Rock[4] has advocated the use of glass or amalgam electrodes in double cells without liquid junction when an electro-active species must be prevented, because of the possibility of chemical reaction, from coming into contact with a reference electrode of the second kind, *e.g.* to avoid a reaction such as:

$$Fe(CN)_6^{4-} + AgCl \rightarrow Ag + Cl^- + Fe(CN)_6^{3-}$$

This is essentially the 'bridging technique' in Covington's survey[5] of methods of using reference electrodes. A good recent example is the use of the lanthanum fluoride ion-selective electrode as a reference electrode in nitrate-ion determination with a liquid ion-exchanger ion-selective electrode.[6] The use of two ion-selective electrodes in a cell may present measuring problems

---

[1] A. K. Covington, in 'Electrochemistry', ed. G. J. Hills (Specialist Periodical Reports), The Chemical Society, London, 1970, vol. 1, p. 56.
[2] R. P. Buck, in 'Techniques of Chemistry', ed. A. Weissberger, Vol. 1, Part IIA, Wiley, New York, 1971, pp. 61—162.
[3] R. P. Buck, *Analyt. Chem., Rev. Fundamentals*, 1972, **44**, 270R.
[4] P. A. Rock, *J. Chem. Educ.*, 1970, **47**, 683.
[5] A. K. Covington in 'Ion Selective Electrodes', ed. R. A. Durst, N.B.S. Special Publication No. 314, Washington, 1969, pp. 107—141.
[6] S. E. Manahan, *Analyt. Chem.*, 1970, **42**, 128.

since both may have high resistances. Brand and Rechnitz[7] have described an integrated-circuit differential amplifier, which enables these problems to be overcome.

2 Conventional Electrode Systems (Electrodes of First and Second Kinds, Redox Couples)

Feltham and Spiro[8] have contributed a valuable review on the platinized platinum electrode, that most widely used of all types of electrode. As the authors point out, it is remarkable how little is known about the deposition of platinum from lead-containing or lead-free solutions despite the use of the process for 75 years. The addition of lead acetate to the chloroplatinic acid must surely be the most famous electrochemical recipe, and some important features of its role emerge from this careful review and the authors' own work.[9, 10] Lead can be leached from platinized platinum electrodes prepared in the presence of lead acetate, but only from the first two or three atomic layers where it is present in the form PbO, both acid and oxygen being necessary for the dissolution to occur. The authors consider that the most widely recommended recipe contains too much lead acetate and recommend the following procedure: 3.5% chloroplatinic acid plus 0.005% lead acetate, at a current density of 30 mA cm^{-2} for up to five minutes for hydrogen–e.m.f. or conductance electrodes. Good stirring is considered essential and no gas should be evolved at the platinum cathode. Chlorine evolved at the anode should be prevented from reaching the cathode by use of an H-type plating cell or similar device.

The Milan electrochemistry group[11] report that hydrogen electrodes function reversibly in up to 99% acetonitrile–water mixtures if the electrodes are 'smooth' platinum in the form of a ribbon wound round a glass frit through which hydrogen is bubbled. It is a pity that details are not given about the method of pretreatment of the electrodes. The same workers' capillary inhibition electrode, which is platinized, and α-palladium electrodes function only up to about 25% acetonitrile, but quinhydrone electrodes can be used in acetonitrile–water mixtures. The possibilities[12] of using the palladium hydride electrode as a reference electrode at high temperatures (up to 195 °C), where the presence of hydrogen gas may be objectionable, have been explored, along with certain other features of its behaviour.[13, 14]

Reports of drifts of e.m.f. with time of cells involving the quinhydrone

---

[7] M. J. Brand and G. A. Rechnitz, *Analyt. Chem.*, 1970, **42**, 616.
[8] A. M. Feltham and M. Spiro, *Chem. Rev.*, 1971, **71**, 177.
[9] A. M. Feltham and M. Spiro, *J. Electroanalyt. Chem. Interfacial Electrochem.*, 1970, **28**, 151.
[10] A. M. Feltham and M. Spiro, *J. Electroanalyt. Chem. Interfacial Electrochem.*, 1972, **35**, 181.
[11] P. Giammario, P. Longhi, and T. Mussini, *Chimica e Industria*, 1971, **53**, 347.
[12] J. V. Dobson, M. N. Dagless, and H. R. Thirsk, *J. C. S. Faraday I*, 1972, **68**, 764.
[13] J. V. Dobson, M. N. Dagless, and H. R. Thirsk, *J. C. S. Faraday I*, 1972, **68**, 749.
[14] J. V. Dobson, *J. Electroanalyt. Chem. Interfacial Electrochem.*, 1972, **35**, 129.

Reversible Electrode Systems and Related Topics 3

electrode have been frequent. A linear e.m.f. drift with time in acetate buffers has been traced to nucleophilic attack by the acetate ion on the p-benzoquinone component of quinhydrone.[15] The reaction was followed by observing changes in the u.v. spectrum of the p-benzoquinone grouping at 246 nm, and correlating these with the e.m.f. changes (0.2 mV h^{-1}). The e.m.f. drift does not preclude the use of the quinhydrone electrode in acetate buffers for, since it is linear and small, its contribution can be eliminated by extrapolation back to zero time. The temperature dependence of the salt error of the quinhydrone electrode in 1 mol kg^{-1} lithium chloride solution has been determined by Struck and Schneider[16] who, to obtain the results, needed to find the mean activity coefficient of hydrochloric acid in this salt solution.[17] From ^{1}H n.m.r. studies, Hepfinger, Tomkins, and Turner[18] conclude that there are no significant interactions which will cause the activity coefficients of the appropriate substituted quinone and hydroquinone forms to change, and hence preclude the use of the chloranil electrode in acetonitrile.

Izatt and co-workers[19] report a formal potential of -979 ± 0.005 mV for the Pd|Pd^{2+} couple in 3.94 mol kg^{-1} perchloric acid medium. The cell contained an internal platinum wire connection[1] to avoid precipitation of potassium perchlorate at the liquid junction with a saturated calomel electrode, though an intermediate sodium chloride bridge was also incorporated. Light[20] suggests a solution of acidic ferrous and ferric ammonium sulphates as a standard ('poised', analogous to 'buffered') redox solution for checking-out systems for measuring redox potentials. The search for suitable reference electrodes in non-aqueous solvents continues. The $I_3^-|I^-$ system has been suggested[21] for propene carbonate and the Cu$^+$|Cu^{2+} for pyridine.[22] The standard potential of the latter couple in acetonitrile has been determined[23] and refined[24] using an Owen-cell extrapolation method for eliminating the liquid-junction potential.

In the last Report[1] the resurgence of interest in amalgam electrodes was welcomed, and a useful review is now available.[25] Attention is drawn[25] to the quaternary ammonium amalgams[26] as providing a 'slight chance' of being useful as electrodes reversible to the popular quaternary ammonium ions.

---

[15] G. Dahlgren and M. J. Goodfriend, *Analyt. Chem.*, 1970, **42**, 111.
[16] B. D. Struck and O. Schneider, *J. Electroanalyt. Chem. Interfacial Electrochem.*, 1972, **36**, 41.
[17] B. D. Struck and O. Schneider, *J. Electroanalyt. Chem. Interfacial Electrochem.*, 1972, **36**, 31.
[18] N. F. Hepfinger, R. P. T. Tomkins, and P. J. Turner, *J. Phys. Chem.*, 1972, **76**, 246.
[19] R. M. Izatt, D. J. Eatough, C. E. Morgan, and J. J. Christensen, *J. Chem. Soc. (A)*, 1970, 2514.
[20] T. S. Light, *Analyt. Chem.*, 1972, **44**, 1038.
[21] E. Sutzkover, Y. Nemirovsky, and M. Ariel, *J. Electroanalyt. Chem. Interfacial Electrochem.*, 1972, **38**, 107.
[22] J. Broadhead and P. J. Elving, *Analyt. Chim. Acta*, 1969, **48**, 433.
[23] J. K. Senne and B. Kratochvil, *Analyt. Chem.*, 1971, **43**, 79.
[24] J. K. Senne and B. Kratochvil, *Analyt. Chem.*, 1972, **44**, 585.
[25] H. P. Bennetto and A. R. Willmott, *Quart. Rev.*, 1971, **25**, 501.
[26] J. D. Littlehailes and B. J. Woodall, *Discuss. Faraday Soc.*, 1968, No. 45, p. 187.

Mussini and Pagella[27] report standard potentials for the calcium amalgam electrode at 25—70 °C. Zinc amalgam- and lead amalgam–lead fluoride electrodes have been used to determine[28] the association constant of ZnF^+ from cell measurements. The Cd(Hg) | $CdSO_4$ | Hg_2SO_4 | Hg cell has been studied[29] in dioxan–water mixtures (up to 60 wt % dioxan) and the reference electrode system Cd(Hg) | $CdCl_2$, NaCl has been suggested for use in dimethylformamide.[30]

Baucke,[31] apparently unaware of a contribution[5] discussed in the last Report,[1] presents a lengthy discussion of the effect of excess solubility of electrode material from electrodes of the second kind on their potentials. He reaches the same conclusion,[5] namely that the effect is principally one of enhanced ionic concentration rather than creation of diffusion potentials. In presumably his last contribution on the subject, since he has now retired from the Chair of Electrochemistry at Birkbeck College, Ives (with Prasad[32]) has described a further 'improved' calomel electrode. The 'banjo' cell has a large ratio of mercury surface to solution volume (5 ml). Equilibrium time is now reduced to 9 h, with 30 min to reach new equilibrium after a 5 K temperature rise. The new design was tested with one molality of aqueous hydrochloric acid and then used for determination of pK_1 and pK_2 values over a temperature range for malonic[32] and some substituted malonic acids.[33] We offer our felicitations for a long and happy retirement after 40 years of meritorious contributions to Electrochemistry.

Leuschke and Schwabe[34] report a redetermination by an Owen-cell method of the standard potential of the mercury–mercurous bromide electrode. The value at 25 °C (139.21 ± 0.04 mV) is in good agreement with work from Ives' laboratory. In the last section of this paper[34] the authors' conclusion, that the variation of diffusion potential cannot be accounted for by activity coefficient considerations, is erroneous, being based on an incorrect equation (6).

The saturated (KCl) calomel electrode continues to be a popular choice as reference electrode but cannot be used in some non-aqueous solvents because of solubility difficulties. The system Hg | Hg_2Cl_2 | $(C_2H_5)_4NCl$ has been suggested for propene carbonate with[35] or without[36] the addition of potassium chloride (sat.). The 'standard potential' $(E^{0\prime} + E_j)$ of the electrode with potassium chloride saturated in methanolic solution has been determined by

[27] T. Mussini and A. Pagella, *J. Chem. and Eng. Data*, 1971, **16**, 49.
[28] R. O. Cook, A. Davies, and L. A. K. Staveley, *J. Chem. Thermodynamics*, 1971, **3**, 907.
[29] S. Chakrabati and S. Aditya, *J. Chem. and Eng. Data*, 1972, **17**, 46.
[30] C. W. Manning and W. C. Purdy, *Analyt. Chim. Acta*, 1970, **51**, 124.
[31] F. G. K. Baucke, *Electrochim. Acta*, 1972, **17**, 845, 851.
[32] D. J. G. Ives and D. Prasad, *J. Chem. Soc. (B)*, 1970, 1649.
[33] D. J. G. Ives and D. Prasad, *J. Chem. Soc. (B)*, 1970, 1652.
[34] W. Leuschke and K. Schwabe, *J. Electroanalyt. Chem. Interfacial Electrochem.*, 1970, **25**, 219.
[35] I. Pilac and R. T. Iwamoto, *J. Electroanalyt. Chem. Interfacial Electrochem.*, 1969, **23**, 484.
[36] I. Fried and H. Barak, *J. Electroanalyt. Chem. Interfacial Electrochem.*, 1970, **27**, 167.

Russian workers[37] over the range 15—40 °C. The standard potential of the mercury–mercury(I) picrate electrode has been reported[38] at 25 °C, and a study by Baldwin[39] suggests that the Ag | Ag$_3$PO$_4$ electrode could be a feasible system for the study of phosphate equilibria. The silver–silver perchlorate electrode has the advantage of quick response, long-term stability, and reasonably small bias potentials (0.5 mV) as a reference electrode system for propene carbonate, and moreover appears to tolerate the addition of small amounts of water.[40]

The hydrogen–silver halide cell remains a valuable route to thermodynamic data. We conclude this section by mentioning recent determinations of standard potentials and related information: Ag | AgBr in methanol–water,[41] Ag | AgCl in ethanol (up to 80%)–water,[42] Ag | AgCl in propan-1-ol,[43,44] propan-2-ol (95%)–water,[45] butan-1-ol,[46] t-butyl alcohol (up to 8%)–water,[47] glycerol,[48] and glycerol (up to 70%)–water;[49] Ag | AgCl and Ag | AgBr in the isodielectric mixture methanol–propene glycol[50] (the following three papers give autoprotolysis constants, amalgam standard potentials, and proton-transfer solvent effects[51]); Ag | AgCl in DMF (5 and 10%)–water,[52] in DMSO (5, 10, 20, and 40%)–water;[53, 54] Ag | AgCl, Ag | AgBr, and Ag | AgI in acetonitrile (up to 20%)–water;[55] and some supplementary data for ethanol–water, acetone–water, and dioxan–water.[56]

3 Glass Electrodes

We continue to distinguish glass electrodes from the new range of ion-selective electrodes, if only for historical reasons. Advances in solid-state elec-

---

[37] V. E. Mironov, I. M. Batyaev, and N. L. Belkova, *Soviet Electrochem.*, 1970, **6**, 346.
[38] A. K. Covington and K. V. Srinivasan, *J. Chem. Thermodynamics*, 1971, **3**, 795.
[39] W. G. Baldwin, *Arkiv Kemi.*, 1970, **31**, 407.
[40] E. Kirowa-Eisner and E. Gileadi, *J. Electroanalyt. Chem. Interfacial Electrochem.*, 1970, **25**, 481.
[41] K. H. Khoo, *J. Chem. and Eng. Data*, 1972, **17**, 82.
[42] P. Sagner and H. Pechova, *Coll. Czech. Chem. Comm.*, 1970, **35**, 334.
[43] R. N. Roy, W. Vernon, J. J. Gibbons, and A. L. M. Bothwell, *J. Chem. Thermodynamics*, 1971, **3**, 883.
[44] R. N. Roy, W. Vernon, and A. L. M. Bothwell, *Electrochim. Acta*, 1972, **17**, 5.
[45] R. N. Roy, W. Vernon, and A. L. M. Bothwell, *J. Chem. Thermodynamics*, 1971, **3**, 769.
[46] R. N. Roy, A. L. M. Bothwell, J. J. Gibbons, and W. Vernon, *J. C. S. Dalton* 1972, 530.
[47] R. N. Roy, W. Vernon, A. L. M. Bothwell, and J. J. Gibbons, *J. Chem. and Eng. Data*, 1972, **17**, 79.
[48] R. N. Roy, W. Vernon, J. J. Gibbons, and A. L. M. Bothwell, *J. Electroanalyt. Chem. Interfacial Electrochem.*, 1972, **34**, 106.
[49] K. H. Khoo, *J. C. S. Faraday I*, 1972, **68**, 554.
[50] K. K. Kundu, A. L. De, and N. N. Das, *J. C. S. Dalton*, 1972, 373.
[51] K. K. Kundu, A. L. De, and N. N. Das, *J. C. S. Dalton*, 1972, 378, 381, 386.
[52] R. N. Roy, W. Vernon, and A. L. M. Bothwell, *J. Chem. Soc. (B)*, 1971, 2320.
[53] R. N. Roy, W. Vernon, A. L. M. Bothwell, and J. J. Gibbons, *J. Electrochem. Soc.*, 1972, **119**, 694.
[54] K. H. Khoo, *J. Chem. Soc. (A)*, 1971, 1177.
[55] T. Mussini, P. Longhi, and P. Giammario, *Chimica e Industria*, 1972, **54**, 3.
[56] D. Bax, C. L. deLigny, M. Alfenaar, and N. J. Mohr, *Rec. Trav. chim.*, 1972, **91**, 601.

tronics have rendered the measurement of small potentials from high-impedance sources no problem, and all-solid-state, digital-read-out pH meters are now available. Groups at Oxford and Reading[57] have demonstrated that an integrated-circuit operational amplifier with digital voltmeter read-out, or a vibrating-condenser electrometer backed-off with a vernier potentiometer, can equally well be used for potentiometric titrations with 0.01 mV discrimination, using commercially available high-resistance glass electrodes. McBryde[58] has pointed out that interpretation of pH as $-\log a_{H^+}$, and a suitable estimate of the activity coefficient in order to get the hydrogen ion concentration, is not always the best method, particularly when a supporting medium of high ionic strength is used, as favoured by so many 'complex-ion chemists'. Calibrations are given for three popular supporting electrolytes at several concentrations, thus enabling the pH meter readings to be converted into hydrogen-ion concentrations. The principles are exemplified by the determination of the concentration quotient of sulphosalicyclic acid.

Deviation from hydrogen-electrode function and time-drifts are often noted in solutions uncongenial to the electrode glass. A full account has appeared of work[59] mentioned in the previous Report,[1] where potential drifts in solutions containing various oxyanions have been attributed to the take-up of the anions into the glass, and such behaviour has been detected radiochemically. Other French workers[60] have failed to be able to interpret errors of Corning 015 composition glass electrodes in HCl + NaCl solutions, using Eisenman's equation for mixed H^+/Na^+ response. This is not surprising since Na^+ response would only be expected at higher pH. Karlberg and Johansson[61] have confirmed that electrodes which show high sodium errors in water show high errors in isopropyl alcohol solutions, and conversely '0—14' electrodes show low errors. Japanese workers[62] challenge the usual statement that alkaline errors are steadier and more reproducible than acidic errors and have followed potentials over many days by direct comparison with hydrogen-gas electrodes. It has been found in the Reporter's laboratory and elsewhere that the extent of the alkaline error can decrease after prolonged soaking in water, and after repeated contact with alkali or alternation between acidic and alkaline solution.

The use of glass electrodes in solvents other than water is frequent, but the demonstration of strict hydrogen-ion response is not always easy. Two papers[63, 64] describe the use of glass electrodes in dimethylformamide. In liquid ammonia at $-38\ °C$, deviations from response to NH_4^+ are attributed

[57] R. P. Henry, J. E. Prue, F. J. C. Rossotti, and R. J. Whelwell, *Chem. Comm.*, 1971, 868.
[58] W. A. E. McBryde, *Analyst*, 1969, **94**, 337; 1971, **96**, 739.
[59] G. Goldstein, C. M. Wolff, and J. P. Schwing, *Bull. Soc. chim. France*, 1971, 1195.
[60] J. Dumas, R. Mauger, and J. C. Pariaud, *Compt. rend.*, 1971, **273**, *C*, 1294.
[61] B. Karlberg and G. Johansson, *Talanta*, 1969, **16**, 1545.
[62] H. Matsushi and S. Furuta, *Kogyo Kagaku Zasshi*, 1970, **73**, 2051 (copies of a translation are available from the Reporter).
[63] J. Juillard, *Bull. Soc. chim. France*, 1970, 2040.
[64] G. Demange-Guerin, *Talanta*, 1970, **17**, 1075.

Reversible Electrode Systems and Related Topics 7

to alkali-metal-ion function, the order of selectivity being very different from that in water.[65] Studies of the lithium-ion response of Beckman cation-responsive electrodes in propene carbonate and the effect of interfering ions are reported.[66] Shults and co-workers have continued their work[67] on the effect of methanol on cation-responsive aluminosilicate glass electrodes.

An understanding of the complex behaviour of glass electrodes in such varied media will only follow a better appreciation of the surface reactions of glass in contact with the solution. It is encouraging that more work is being directed to this end, and that advances are being made. It is now widely believed that many glasses are not homogeneous.[68] Hair[69] has suggested that two OH bands in the i.r. spectra of alkali-metal silicate glasses, *viz*. that at 2.8 μm (found only in pure silica) and another at 3.6 μm, which increases in intensity with Na_2O content of the glass, are diagnostic of a pure silica micro-environment and a sodium silicate micro-environment in these glasses. With aluminosilicate glasses[70] two analogous environments are again envisaged, the water distributing itself between the two phases which give rise to the 2.8 μm and 3.6 μm OH bands. A good linear correlation is found between the percentage OH in the 2.8 μm band and $\log K_{NaK}^{pot}$, the Eisenman selectivity constant, throughout the whole range of glass compositions for which data are available. NAS glasses become potassium-selective above an NA^+/Al^{3+} ratio of 2.5. Potassium selectivity is attributed to the sodium silicate phase which gives rise to the 3.6 μm band. Sodium silicate glasses are known to be leachable to yield porous glasses, and it is suggested that response to K^+ is due to the formation of porous hydrated layers; a suggestion based on similarities to sintered porous glasses, where ionic selectivity appears to be a molecular-sieve effect.[71, 72] The presence of micro-inhomogeneities in glass makes questionable some attempts to correlate electrode response properties with bulk glass composition. Since the extent of phase separation will depend on the heat treatment of the glass, inconsistencies in behaviour of nominally identical electrodes are to be expected. Whitfield,[73] in a careful study of the effect of the membrane geometry of the glass electrode on the asymmetry potential and its variation with pressure and temperature, concluded that only by flame annealing of the bulbs could reproducible results be obtained. This is a departure from commercial manufacturing practice for blown bulbs. The long-accepted statement that strain-free flat glass membranes have small and stable asymmetry potentials was confirmed. The best form of electrode which minimizes the asymmetry potential and its

[65] W. M. Baumann and W. Simon, *Helv. Chim. Acta*, 1969, **52**, 2054.
[66] L. M. Mukherjee and D. P. Boden, *Electrochim. Acta*, 1972, **17**, 965.
[67] I. S. Ivanovskaya, V. I. Govrilova, and M. M. Shults, *Soviet Electrochem.*, 1970, **6**, 975.
[68] E. A. Porai-Koshits and V. I. Averjanov, *J. Non-Crystalline Solids*, 1968, **1**, 29.
[69] M. L. Hair, *J. Amer. Ceram. Soc.*, 1969, **52**, 677.
[70] M. L. Hair, *J. Phys. Chem.*, 1970, **74**, 1145.
[71] I. Altug and M. L. Hair, *J. Electrochem. Soc.*, 1970, **117**, 78.
[72] M. L. Hair, *J. Phys. Chem.*, 1970, **74**, 1290.
[73] M. Whitfield, *Electrochim. Acta.*, 1970, **15**, 83.

variation, and which has low resistance coupled with mechanical robustness, is a double-bulb electrode formed by fusing together two normal bulbs with stems at an angle of about 30°. This type is advocated for deep-sea measurements, where frequent standardization is impossible and conditions are extreme (5 °C, 1000 bar).

In continuation of work[74] described in detail in the last Report,[1] Wikby[75, 76] has applied his constant-current pulse method to determine changes in the surface resistance of glass electrodes when subjected to various solutions, either acidic, neutral, or non-aqueous. The surface resistance is obtained by resolution of the constant-current polarization curve, the contributions from the surface resistance and capacitance having time constants of the order of seconds as opposed to the contributions from the bulk glass, which are in the millisecond region. The relative merits of the constant-current pulse and a.c. methods[77, 78] are not clear, although in principle the same information should be obtainable from either. Surface films can only be detected[77] by the a.c. method at frequencies less than 1 Hz. In the earlier paper[74] it was shown that certain electrodes had high surface resistance, which decreased when the electrode was soaked in water (hydrated or 'conditioned'). Additional experiments[75] now show that this high resistance is located at the inside surface of the electrode and can be removed by etching with hydrofluoric acid. Electrodes treated in this way show an increase in surface resistance on conditioning, and the increase continues even after the e.m.f. becomes steady (*e.g.* after 70 h). In isopropyl alcohol solutions of lithium chloride there was found to be a take-up of Cl^- ions that was dependent on the acidity, with a parallel increase in surface resistance; this suggests that the rise in resistance can be attributed to a blocking of the conduction mechanism by HCl molecules. When a partially hydrated electrode is transferred to isopropyl alcohol,[76] the surface layer stops growing, as shown by controlled etching experiments of the type devised by Hungarian workers,[79, 80] and there is an increase in surface resistance. By etching small layers and remeasuring the surface resistance it was shown that the impediment to the conduction process is situated at the interface between gel layer and bulk glass. The Hungarian group[81] has refined the successive etching treatment and chemical analysis so as to give the increased sensitivity necessary to investigate the much smaller gel layers on lithia electrodes. The layer thickness is pH-dependent, and in ethanol only a thin layer (2×10^{-5} cm) is built up, about ten times smaller than that in water. Dobos[82] has used tritium radio-tracer experiments to

[74] A. Wikby and G. Johansson, *J. Electroanalyt. Chem. Interfacial Electrochem.*, 1969, **23**, 23.
[75] A. Wikby, *J. Electroanalyt. Chem. Interfacial Electrochem.*, 1971, **33**, 145.
[76] A. Wikby, *J. Electroanalyt. Chem. Interfacial Electrochem*, 1972, **38**, 429, 441.
[77] M. J. D. Brand and G. A. Rechnitz, *Analyt. Chem.*, 1969, **41**, 1788; 1970, **42**, 304.
[78] R. P. Buck, *J. Electroanalyt. Chem. Interfacial Electrochem.*, 1968, **18**, 381.
[79] G. Bouquet, S. Dobos, and Z. Boksay, *Ann. Univ. Sci. Budapest, Sect. Chim.*, 1964, **6**, 5.
[80] Z. Boksay, G. Bouquet, and S. Dobos, *Phys. and Chem. Glasses*, 1967, **8**, 140.
[81] B. Csakvari, Z. Boksay, and G. Bouquet, *Analyt. Chim. Acta*, 1971, **56**, 279.
[82] S. Dobos, *Acta. Chim. Acad. Sci. Hung.*, 1971, **69**, 43, 49.

determine the concentration profiles of water as well as those of the alkali-metal and alkaline-earth ions in the gel layer. The water concentration falls as expected, fairly sharply, and then more slowly until the gel layer–bulk glass interface is reached. Wikby's experiments[76] are in general agreement with the Hungarian work (the glasses used are different), the necessary information not always being available to derive the thickness of the gel layer from the number of moles of silicon of the network dissolved by the etchant. In some of Wikby's experiments[76] an inner-filling solution containing 0.25% HF was used to remove continuously the surface resistance contribution from the inside surface of the glass bulb. Presumably this has no effect on the e.m.f. of the system. A new method of analysing the surface layers of leached glass has been evolved by workers at the Jena Glaswerk, Mainz.[83] The surface is sputtered with argon ions of high energy in a vacuum and is successively removed in layers, the average penetration of the ions being 6 nm. The luminescence produced by lithium ions in the glass under the bombardment is detected by a photomultiplier and used to obtain the concentration profile across the layer, its thickness being obtained by comparison with interferometric measurements of the depth of etch pits produced after long bombardment. The method seems promising and has been used to investigate the effect of pH on the lithium concentration profile in the gel layer. It is interesting that the presence of lithium in the conditioning solution resulted in layer concentrations being obtained which were greater by a factor of 2. Further experiments with xenon and krypton bombardment and analysis of the concentration profiles of other ion components are promised.

We conclude this section by mentioning some recent applications of hydrogen- and cation-responsive glass electrodes. Eckfeldt and Ott[84] have pointed out that pOH is a more fundamental quantity than pH for basic solutions, and moreover is less temperature-dependent. Circuits are described for deriving an electrical signal from a temperature-sensing resistor which can be used to convert a pH meter to read pOH directly.[84] Low levels of sodium ion in process water can be monitored with sodium-responsive electrodes down to the 10^{-4} p.p.m. level, with addition of triethanolamine, di-isopropylamine, or ammonia to buffer the sample to high pH.[85, 86] Likewise, with a potassium-responsive electrode, ammonia in boiler-feed water can be monitored,[87] (again with addition of triethanolamine). The response of sodium-glass electrodes to silver ion has been utilized for argentometric halogen micro-determination in non-aqueous media.[88]

Of more academic interest, Padova[89] has studied activity coefficient relationships in $KCl + KNO_3$ aqueous mixtures, and Shults and co-workers[90]

[83] H. Bach and F. G. K. Baucke, *Electrochim. Acta*, 1971, **16**, 1311.
[84] E. L. Eckfeldt and W. T. Ott, *Instr. Soc. Amer., Trans.*, 1970, **9**, 45.
[85] E. L. Eckfeldt and W. E. Proctor, *Analyt. Chem.*, 1971, **43**, 332.
[86] A. A. Diggens, K. Parker, and H. M. Webber, *Analyst.*, 1972, **97**, 198.
[87] G. I. Goodfellow and H. M. Webber, *Analyst*, 1972, **97**, 95.
[88] K. Hozumi and N. Akimoto, *Analyt. Chem.*, 1970, **42**, 1312.
[89] J. Padova, *J. Phys. Chem.*, 1970, **74**, 4587.
[90] N. S. Gokman, G. P. Kepnev, and M. M. Shults, *Zhur. fiz. Khim.*, 1971, **45**, 2951.

have extended to lower concentration the work of Lanier[91] on $NaCl + NaNO_3$ and $NaCl + NaClO_4$ aqueous mixtures. Spink and Schrier[92] describe an interesting new route to salting coefficients (effects of non-electrolyte addition of mean ionic activity coefficients of salts) which is claimed to have advantages over solubility or distribution methods. A cation-responsive glass electrode is transferred between two solutions, MCl + acetone + water, having the same MCl, but different acetone concentrations. The difference in e.m.f. (measured using silver–silver chloride reference electrodes) is relatable to the salting coefficient, although the authors' analysis is more sophisticated.

Finally, since mention was made of the effect last time,[93a] another report notes the effect of light on a pH electrode system. A pH change of 0.05 unit was observed, which was traced to the inner reference silver–silver chloride electrode.[93b]

4 Ion-selective Electrodes

The two years under review have seen some stabilization in this field of rapid advancement but also a number of new ideas emerging. Several commercial firms offer a limited range of electrodes, but the most complete range is still that of Orion Research Inc. Heterogeneous electrodes of the Pungor–Radelkis type appear no longer to be available, at least in the U.K.

Several general reviews are available[94—99] and some on more specific aspects, viz. natural water analysis,[100] monitoring and control,[101] and chemical studies.[102] Moody and Thomas[103] have drawn attention to the confusion that can arise from quoting selectivity constants in various forms and without reference to the solution conditions and method used. Often the performances of electrodes of similar response but differing type or origin are compared solely through published selectivity constants instead of by direct experiment. Selectivity constants are only a broad guide to behaviour and should be accorded no more respect than, say, manufacturers' figures for the performance of HiFi equipment.

Classification[1] of ion-selective electrodes into (i) solid-state, (ii) heterogeneous, (iii) liquid ion-exchange, no longer seems appropriate as there is often

---

[91] R. D. Lanier, *J. Phys. Chem.*, 1965, **69**, 3992.
[92] M. Y. Spink and E. E. Schrier, *J. Chem. Thermodynamics*, 1970, **2**, 821.
[93] (a) Ref. 1, p. 64; (b) R. A. McAllister and R. Campbell, *Analyt. Biochem.*, 1970, **33**, 200.
[94] E. Pungor and K. Toth, *Analyst*, 1970, **95**, 625.
[95] R. A. Durst, *Amer. Scientist*, 1971, **59**, 353.
[96] W. Simon, H. R. Wuhrmann, M. Vasak, L. A. R. Pioda, R. Dohner, and Z. Stefanac, *Angew. Chem. Internat. Edn.*, 1970, **9**, 445.
[97] C. Gavach, *Bull. Soc. chim. France*, 1971, 3395.
[98] K. Gammann, *Naturwiss.*, 1970, **57**, 298.
[99] J. T. Clerc, G. Kahr, E. Pretsch, R. P. Scholer, and H. R. Wuhrmann, *Chimia (Switz.)*, 1972, **26**, 287.
[100] M. Whitfield, 'Ion Selective Electrodes for the Analysis of Natural Waters', Australian Marine Sciences Association, Handbook, No. 2, Sydney, 1971.
[101] D. C. Cornish and R. J. Simpson, *Meas. Control*, 1971, **4**, 303.
[102] G. A. Rechnitz, *Accounts Chem. Res.*, 1970, **3**, 69.
[103] G. J. Moody and J. D. R. Thomas, *Talanta*, 1971, **18**, 1251.

Reversible Electrode Systems and Related Topics 11

now no clear indication into which class certain forms fall. For instance, liquid ion-exchanger materials have been incorporated into polymer matrices and hence can presumably be classed as heterogeneous membranes. Support materials which are conducting, and hence dispense with inner filling solutions, further confuse the issue, and in what follows we shall emphasize new forms of fabrication and new materials.

The form of liquid ion-exchange electrode with the active material supported on a thin disc of 'Millipore' filter separating two aqueous solutions is not entirely convenient for some applications, and it is not easy to assemble. Attempts have been made to form solid electrodes by incorporation of the active material in a polymeric matrix. An early calcium electrode employed collodion for this purpose[1] and was marketed, but reports on it were not favourable. Thomas and co-workers[104, 105] have successfully incorporated the Orion calcium- and nitrate-active materials in PVC membranes, and discuss the optimum composition ratio.[106] The plastic is cast in a tray and discs are cut from the sheet with a cork borer and then sealed into tubes, like solid-state electrodes. The antibiotic materials discussed below have similarly been incorporated in plastic.

In some applications (*e.g.* use of the electrode in an inverted position), inner filling solutions are inconvenient, and solid contacts have been developed, a coating of silver chloride for instance in the case of the lanthanum fluoride electrode. Some recent papers[107—109] report the use of simple 'coated-wire' electrodes, for example the Orion calcium exchanger in PVC, coated directly on to a platinum wire; a 31 mV per decade response between 10^{-1} and 10^{-4} mol^{-1} $CaCl_2$ was claimed.[108] Similarly, sulphides in epoxy resin or silicone rubber have been coated on to wires. Greater selectivity is claimed for this form of electrode over interfering ions than for the Millipore-filter Orion-type construction.[109]

Instead of using a non-conducting porous medium such as cellulose acetate, one can dispense with the need for an inner solution by use of a porous conductor such as carbon or graphite. This is the basis of Ruzicka's 'universal "Selectrode"'.[110—115] It is necessary to prevent the aqueous test solution from penetrating the graphite, so its surface is rendered hydrophobic with Teflon. The active material is then introduced on to the surface by one of a variety of methods as appropriate, *viz.* dipping in molten halide, rubbing in an aqueous suspension, dry rubbing, or simply dipping in an organic liquid

[104] G. J. Moody and J. D. R Thomas, *Analyst*, 1970, **95**, 910.
[105] J. E. W. Davies, G. J. Moody, and J. D. R. Thomas, *Analyst*, 1972, **97**, 87.
[106] G. M. Griffiths, G. J. Moody, and J. D. R. Thomas, *Analyst*, 1972, **97**, 420.
[107] H. Hirata and K. Date, *Talanta*, 1970, **17**, 883.
[108] R. W. Cattrall and H. Freiser, *Analyt. Chem.*, 1971, **43**, 1905.
[109] H. J. James, G. P. Carmack, and H. Freiser, *Analyt. Chem.*, 1972, **44**, 856.
[110] J. Ruzicka and C. J. Tjell, *Analyt. Chim. Acta*, 1970, **49**, 346.
[111] J. Ruzicka and C. J. Tjell, *Analyt. Chim. Acta*, 1970, **51**, 1.
[112] J. Ruzicka and K. Rald, *Analyt. Chim. Acta*, 1970, **53**, 1.
[113] J. Ruzicka and C. G. Lamm, *Analyt. Chim. Acta*, 1971, **53**, 206.
[114] J. Ruzicka and C. G. Lamm, *Analyt. Chim. Acta*, 1971, **54**, 1.
[115] E. H. Hansen, C. G. Lamm, and J. Ruzicka, *Analyt. Chim. Acta*, 1972, **59**, 403.

containing ion-exchanger material. The coating is only superficial and it is claimed that it can readily be renewed or even changed for another type. Electrical connection to the graphite is by way of a threaded stainless-steel rod. One would imagine that various contact potentials are not particularly well defined, and frequent restandardization of the electrodes may be necessary. The application of such electrodes is probably restricted to titrimetry. A combined electrode assembly–titration cell for liquid ion-exchange materials has been described[116] which disperses the active material in a glass sintered disc. In view of the expense of electrode assemblies but the relative cheapness of active materials, this suggestion could be useful for student instruction.

In the last Report[1] mention was made of enzyme electrodes, in which a layer of enzyme immobilized in a polyacrylamide matrix causes hydrolysis of a compound to ions which are detected by a cation-selective glass electrode, on to which the polymer layer is coated. Guilbault[117] has reviewed these solid enzyme probes, which can also take the form of an enzyme in liquid solution inside a plastic bag covering the sensor electrode. The latter are similar to the Severinghaus CO_2 electrode, in which CO_2 diffuses through a membrane, dissolves in a bicarbonate solution, and displaces the hydrogen-ion activity, the change being detected by a glass electrode. A variant of this has been described using instead the quinhydrone electrode.[118] In the first use of an ion-selective electrode as sensor, Rechnitz and Llenado[119, 120] have coated the Orion cyanide electrode (silver iodide + sulphide) with a polymer containing the enzyme β-glucosidase, which with amygdalin gives glucose, benzaldehyde, and cyanide, the latter being detected by the sensor.

Materials for ion-selective electrodes can be broadly classified into:

(a) insoluble salts, such as lanthanum fluoride and silver sulphide;
(b) long-chain ion-exchange materials, such as tetra-alkylammonium salts and alkyl phosphate salts;
(c) complexing agents, often those used in solvent extraction, such as 2-thenoyltrifluoroacetone and dithizone. These are called 'neutral carriers' (they provide neutral sites for exchange), but charged sites have been employed, such as the o-phenanthroline chelating group (1) used for the Orion nitrate electrode.

$$\left[Ni \left(\underset{N}{\overset{N}{\diagdown}} \right)_3 \right]^{2+}$$

(1)

[116] A. K. Covington and J. M. Thain, *Analyt. Chim. Acta*, 1971, **55**, 453.
[117] G. G. Guilbault, *Pure. Appl. Chem.*, 1971, **25**, 727.
[118] L. H. Van Kempen, H. Deurenberg, and F. Kreuzer, *Resp. Physiol.*, 1972, **14**, 366.
[119] G. A. Rechnitz and R. Llenado, *Analyt. Chem.*, 1971, **43**, 283.
R. A. Llenado and G. A. Rechnitz, *Analyt. Chem.*, 1971, **43**, 1457.

The lanthanum fluoride single-crystal electrode still reigns supreme in importance, selectivity, and application. Using fluoride-complexing cations, Baumann[121] has shown that, theoretically, it responds to free fluoride down to 10^{-9} mol l^{-1}. The limit, of course, is the fluoride solubility from the membrane, and she gives 10^{-24} (mol $1^{-1})^4$ for the solubility product associated with a single crystal. Russian workers[122] have prepared and studied a lanthanum fluoride electrode which, however, showed Nernstian response only over the limited range of 10^{-3} to 1 mol 1^{-1}. From a.c. impedance measurements, Brand and Rechnitz[123] conclude that 'it is not improbable' that the fluoride electrode has a surface film resembling that on the glass electrode, because the impedance behaviour is the same. Potassium fluoride solutions have been proposed[124] as pF standards, because, unlike sodium fluoride, the potassium salt is free of ion-association and more soluble. Arguments based on the Robinson and Stokes hydration theory are given for equating the single ion-activity coefficients of K^+ and F^-.

Incorporation of a seond material with slightly greater solubility along with the first produces electrodes analogous to those of the third kind. The use of CuS with Ag_2S to form a Cu^{II} solid-state electrode was discussed in the last Report,[1] and a heterogeneous form has been described.[125] Mixing of CaF_2 with LaF_3 is the basis for a patent on a calcium electrode.[126] Lead(II) sulphide with silver sulphide is used for the Orion solid-state lead electrode, and this has been studied in non-aqueous media by Rechnitz and Kenny.[127] Sintered (so-called ceramic) and heterogeneous forms $PbS + Ag_2S$, $PbS + Cu_2S + Ag_2S$, and also with PbSe and PbTe replacing PbS, have been described.[128—131] Hirata has reviewed his work on these chalcogenide-containing electrodes.[132] The use of the semiconducting Cu_2S alone as a Cu^{II} sensor has been described[133—135] (actually $Cu_{1.79}S$ is said to be the optimum composition[134]). Hirata has also mentioned electrodes for Zn^{II}, Mn^{II}, Ni^{II}, and Cd^{II}.[132, 136] As well as the solubility criterion, the question of conductivity is of critical importance, and the role of Ag_2S is probably to provide ionic conductivity. However, Brand and Rechnitz[123] draw a different conclusion

---

[121] E. W. Baumann, *Analyt. Chim. Acta*, 1971, **54**, 189.
[122] R. R. Tarasyants, R. N. Potsepkina, V. P. Roze, and E. A. Bondarevskaya, *Zhur analit. Khim.*, 1972, **27**, 808.
[123] M. J. Brand and G. A. Rechnitz, *Analyt. Chem.* 1970, **42**, 478.
[124] R. A. Robinson, W. C. Duer, and R. G. Bates, *Analyt. Chem.*, 1971, **43**, 1862.
[125] M. Mascini and A. Liberti, *Analyt. Chim. Acta*, 1971, **53**, 202.
[126] G. M. Farren, Ger. P. 2101 339/1971.
[127] G. A. Rechnitz and N. C. Kenny, *Analyt. Letters*, 1970, **3**, 259.
[128] H. Hirata and K. Date, *Analyt. Chem.*, 1971, **42**, 279.
[129] H. Hirata and K. Higashiyama, *Analyt. Chim. Acta*, 1971, **54**, 415.
[130] M. Mascini and A. Liberti, *Analyt. Chim. Acta*, 1972, **60**, 405.
[131] H. Hirata and K. Higashiyama, *Analyt. Chim. Acta*, 1971, **57**, 476.
[132] H. Hirata and K. Higashiyama, *Talanta*, 1972, **19**, 391.
[133] H. Hirata and K. Date, *Talanta*, 1970, **17**, 883.
[134] H. Hirata, K. Higashiyama, and K. Date, *Analyt. Chim. Acta*, 1970, **51**, 209.
[135] N. I. Savvin, V. S. Shterman, A. V. Gordievskii, and A. Y. Syrchenkov, *Zavodskaya Lab.*, 1971, **37**, 1015.
[136] H. Hirata and K. Higashiyama, *Z. analyt. Chem.*, 1971, **257**, 104.

from comparison of a.c. impedance behaviour. Silver sulphide is present is an important new sulphate sensor.[137] A mixture of 32 mole % Ag_2S, 31% PbS, 32% $PbSO_4$, and 5% Cu_2S is hot pressed at above 170 °C and 7000 atm. The role of the Cu_2S is to improve response time. Alternate soaking in dilute sodium sulphate followed by silver nitrate + lead(II) nitrate for several hours is the recommended conditioning procedure. A Nernstian slope in the pH range 3—10 is claimed, with a fast (1 minute) response time.

What may prove to be an important new development is described by Baker and Trachtenburg.[138] A chalcogenide-glass semiconducting electrode is claimed to give Nernstian response to Cu^{II} and Fe^{III} in the range 10^{-5} to 10^{-1} mol l^{-1}. It is composed of 60% Se, 28% Ge, and 12% Sb, doped with Fe^0(2%) Co^0, or Ni^0 to lower the resistivity. Discs are sealed into plastic tubes with inner-filling elements or gold-plated contacts. The authors consider that electronic conduction and a redox potential mechanism are responsible for the electrode response. The sensing element is considered to act as a reducing agent which generates the lower oxidation state of the redox couple and maintains it at constant activity, hence the electrode responds to Fe^{3+} and not to Fe^{2+}. The sensors do not respond to Fe^{2+} in solution, and the presence of Fe^0 in the glass is not a prerequisite for response to Fe^{3+}. Also in a difficult-to-classify category comes the report of a silicone-rubber-based caesium-responsive electrode which employs caesium 12-molybdophosphate.[139] A response of less than half theoretical in the pH range 4—6 is observed, but the electrode is useful for titrimetric determinations.[140]

Most important amongst the liquid ion-exchanger electrodes are still the calcium, nitrate, and perchlorate electrodes, although the potassium electrode to be discussed below merits inclusion. Russian workers, in a belated entry into the field, have studied the calcium salt of ethylhexyl hydrogen phosphate in toluene as a calcium-selective material,[141] and salts of di-2-ethylhexyl phosphate in chlorobenzene for strontium and barium.[142] Tetraoctylammonium perchlorate and nitrate and tetradecylammonium iodide have been studied for perchlorate-,[143] nitrate-,[144] and iodide-selective[145] electrodes, respectively. From a systematic study of long-chain alkylammonium salts in ethyl bromide,[146] tetra-n-heptylammonium perchlorate was singled out as a promising perchlorate-responsive electrode for use in a concentration cell

[137] G. A. Rechnitz, G. H. Fricke, and M. S. Mohan, *Analyt. Chem.*, 1972, **44**, 1098.
[138] C. T. Baker and I. Trachtenburg, *J. Electrochem. Soc.*, 1971, **118**, 571.
[139] C. J. Coetzee and A. J. Basson, *Analyt. Chim. Acta*, 1971, **57**, 478.
[140] C. J. Coetzee and A. J. Basson, *Analyt. Chim. Acta*, 1971, **56**, 321.
[141] A. L. Grekovich, E. A. Materova, and F. A. Belinskaya, *Soviet Electrochem.*, 1970, **6**, 1004.
[142] E. A. Materova and V. V. Mukhovikov, *Soviet Electrochem.*, 1971, **7**, 1684.
[143] A. L. Grekovich, E. A. Materova, and F. A. Belinskaya, *Soviet Electrochem.*, 1971, **7**, 1227.
[144] A. V. Gordievskii, A. Y. Syrchenov, V. V. Sergeivskii, and N. I. Savvin, *Elecktrokhimiya*, 1972, **8**, 520.
[145] A. L. Grekovich, E. A. Materova, and I. I. Pronkina, *Soviet Electrochem.*, 1971, **7**, 421.
[146] E. Dubini-Paglia, T. Mussini, and R. Galli, *Z. Naturforsch.*, 1971, **26a**, 154.

without transport to determine the activity coefficients of perchloric acid:
Pt, H_2 | $HClO_4$ (fixed) | ClO_4^- membrane | $HClO_4$ (variable) | H_2, Pt.
Results[147] show a scatter and reproducibility of about ±0.1 or 0.2 mV and furnish activity coefficient values in reasonable agreement with other e.m.f. and isopiestic measurements. An alternative perchlorate-responsive system, and that used in the Orion electrode, is that based on o-phenanthroline chelates. Ishibashi and Kohara[148] have compared a number of related compounds and concluded that tris-(o-phenanthroline)ferrous perchlorate in nitrobenzene has excellent selectivity for ClO_4^- over NO_3^-, but quaternary ammonium perchlorates would seem to be slightly superior. Systems for sulphonate[149] (Crystal Violet benzenesulphonate in nitrobenzene) and salicylate[150] (tetra-n-heptyl ammonium salicylate in n-decanol) have been described. Zn^{II}- and Pd^{II}-selective electrodes based on tetrachloro-zinc (or -palladous) tetralaurylammonium chlorides in benzene have been shown to give linear response to free zinc or palladous ions in aqueous solutions.[151] A quaternary ammonium compound (Aliquat 336S), when shaken with the sodium or potassium salt of an amino-acid such as tryptophan or phenylalanine, is the basis for amino-acid-anion-responsive electrodes.[152] The first electrode designed for measurement of an organic cation is claimed[153] by Corning Scientific Instruments for acetylcholine [$CH_3CO\cdot O\cdot(CH_2)_2N(CH_3)_3^+$]. A selectivity of 10^3 over K^+ and 15 over choline is claimed, the electrode having applications to organopesticide analysis[154] and determination of acetylcholinesterase[155] (by enzyme-catalysed hydrolysis). The active material is acetylcholine tetra-(p-chlorophenyl)borate in, for example, 3-o-nitroxylene. Organic radical-ion salts, for example silver and copper(II) salts of 7,7,8,8-tetracyanoquinodimethane (2), are semiconducting and can be used in the form of

$$NC\diagdown C=\diagup\diagdown=C\diagup CN$$
$$NC\diagup \qquad \diagdown CN$$
(2)

pelletized discs.[156] Electrodes responsive to Ag^+, Cu^{2+}, and $(C_2H_5)_4N^+$, amongst others, have been investigated.[156, 157]

Undoubtedly the most important development in ion-selective electrodes in the two year period under review has been the introduction of so-called neutral carrier complexes to make electrodes selective to alkali-metal ions,

[147] T. Mussini, R. Galli, and E. Dubini-Paglia, *J. C. S. Faraday I*, 1972, **68**, 1322.
[148] N. Ishibashi and H. Kohara, *Analyt Letters*, 1971, **4**, 785.
[149] N. Ishibashi and H. Kohara, *Jap. Analyst.*, 1972, **21**, 100.
[150] W. M. Haynes and J. H. Wagenknecht, *Analyt. Letters*, 1971, **4**, 491.
[151] G. Scibona, L. Mantella, and P. R. Danesi, *Analyt. Chem.*, 1970, **42**, 844.
[152] M. Matsui and H. Freiser, *Analyt. Letters*, 1970, **3**, 161.
[153] G. Baum, *Analyt. Letters*, 1970, **3**, 105; *J. Phys. Chem.*, 1972, **76**, 1872.
[154] G. Baum, *Analyt. Biochem.*, 1970, **39**, 65.
[155] G. Baum and F. B. Ward, *Analyt. Chem.*, 1971, **43**, 947.
[156] M. Sharp and G. Johansson, *Analyt. Chim. Acta*, 1971, **54**, 13.
[157] M. Sharp, *Analyt. Chim. Acta*, 1972, **59**, 137; **61**, 99.

principally potassium but also ammonium. The simplest neutral-carrier electrode is probably the iodide electrode of Ruzicka,[112] who impregnated graphite with a solution of iodine in carbon tetrachloride and found that the electrode showed a selectivity for iodide over bromide of 10^3. The reason must surely be the formation of the $I_3^-|I_2$ couple. However, the materials used for the new range of sensors are large organic ligands with high selectivity for alkali-metal ions, which they almost completely envelop. These developments originate from an observation of Moore in 1964 that neutral macrocyclic antibiotics induced ion permeation in mitochondria. Electrodes based on antibiotics were first developed by Simon in Zurich, who studied a number of antibiotics, including nonactin, monactin, and valinomycin. Pioda, Stankova, and Simon[158] reported that a 0.009 mol l^{-1} solution of valinomycin in diphenyl ether gave a 58.3 mV per decade slope for potassium, with a sensitivity over Na^+ of 4000, this being a great advance over that of potassium-responsive glass electrodes (about 10). An ammonium-ion-selective electrode based on a mixture of nonactin and monactin in tris-(2-ethylhexyl) phosphate has been described by the same group.[159] In 1970, Frant and Ross[160] reported on an Orion experimental electrode, using a 5—10% valinomycin solution in suitable organic solvent, and this was appraised by Lal and Christian.[161] A 52 ± 1 mV per decade response to K^+ and a selectivity for K^+ over Na^+ of 11 were found. Butler and Huston[162] attempted to use the Orion electrode to determine activity coefficients of potassium chloride solutions, but noted deviations from expected behaviour of up to 14 mV, probably because of decomposition of the active material. The electrode lifetime claimed by Simon[158] was only 2 weeks.

Beckman workers announced 'solid organic' membrane electrodes selective

(3)

[158] L. A. R. Pioda, V. Stankova, and W. Simon, *Analyt. Letters*, 1969, **2**, 665.
[159] R. P. Scholer and W. Simon, *Chimia (Switz.)*, 1970, **24**, 372.
[160] M. S. Frant and J. W. Ross, *Science*, 1970, **167**, 987.
[161] S. Lal and G. D. Christian, *Analyt. Letters*, 1970, **3**, 11.
[162] J. N. Butler and R. Huston, *Analyt. Chem.*, 1970, **42**, 676.

to potassium[163] and ammonium,[164] without stating the nature of the active materials, a practice which no Editor of a reputable scientific journal should permit, otherwise it becomes just another outlet for manufacturers' publicity. The ability of macrocyclic antibiotics to complex K^+ so selectively lies in their ability to provide a hole of the right size. Nonactin (3) has been described as wrapping itself round a potassium ion in a configuration like the seams on a tennis ball.

The solution conformation of the K^+ complex of valinomycin is reported[165] to have a pore structure with a diameter of 4.5 Å, which is considered approximately correct for the hydrated K^+ ion. A second conformation, termed a 'core structure', with the acyl oxygens pointing outwards instead of inwards, leaves a small cavity for an unhydrated ion.

Entirely comparable selectivities can be obtained with the more stable synthetic 'crown' compounds, first synthesized by Pedersen in 1967. So great is the affinity of 'crowns' for alkali-metal cations that they cannot be handled in ordinary glassware. Typical 'crowns' are (4)—(8).

A number of excellent reviews on these compounds and the related 'cryptates' are available.[166-169] A potassium-responsive electrode may be based on dicyclohexyl-18-crown-6 (the first figure refers to the ring size, the second to the number of hetero-atoms). The possibility of tailoring ligands to provide highly specific electrodes for a whole range of ions is an exciting prospect, and

[163] I. H. Krull, C. A. Mask, and R. E. Cosgrove, *Analyt. Letters*, 1970, **3**, 43.
[164] R. E. Cosgrove, C. A. Mask, and I. H. Krull, *Analyt. Letters*, 1970, **3**, 457.
[165] M. Ohnishi, *Science*, 1970, **168**, 1091.
[166] Article in *Chem Eng. News*, 1970, **48**, No. 9, p. 26.
[167] C. J. Pedersen and H. K. Frensdorff, *Angew. Chem. Internat. Edn.*, 1972, **11**, 16.
[168] M. R. Truter and C. J. Pedersen, *Endeavour*, 1971, **30**, 142.
[169] J. J. Christensen, J. O. Hill, and R. M. Izatt, *Science*, 1971, **174**, 459.

requires real interdisciplinary collaboration, with important implications for the use of these model systems as an aid in attempting to understand ion transport in biological membranes.

Great advances are therefore confidently expected in this area. Levins[170] has described a barium-ion-selective electrode, which uses as active material the barium complex of Igepal Co 880, a nonylphenoxypolyoxyethylene-ethanol with twelve ethylene oxide units per mole of Ba^{2+}. A further paper correcting values for selectivity constants has just appeared at the time of writing.[171] Simon and co-workers[172] have synthesized a ligand (9) which is selective to Ca^{2+}, providing a cavity of about 1 Å, which they consider is necessary to discriminate between bivalent and univalent ions. Such a ligand should contain no acid–base functional groups, in order to avoid H^+ selectivity. The skeletal formula is as shown.

Mechanistic studies of valinomycin systems have been undertaken by Eyal and Rechnitz[173] and by Simon's group.[174] Transport of the K^+ ion is by a carrier relay process, rather like the Grotthus chain mechanism for proton conduction in acid solutions. Rechnitz's group[175] have recently reported determinations of selectivity constants for electrodes utilizing cyclic polyethers ('crowns') and have attempted to relate these to equilibrium constants for the formation of the appropriate complexes. They conclude that it is possible to predict selectivity properties of electrode systems on this basis as a first approximation, but other workers do not agree[176] that the selectivity ratio is simply the quotient of the formation constants of the respective complexes.[173]

In the foregoing we have emphasized electrochemical rather than purely analytical aspects of ion-selective electrodes. The number of applications of

[170] R. J. Levins, *Analyt. Chem.*, 1971, **43**, 1045.
[171] R. J. Levins, *Analyt. Chem.*, 1972, **44**, 1544.
[172] D. Ammann, E. Pretsch, and W. Simon, *Tetrahedron Letters*, 1972, 2473.
[173] E. Eyal and G. A. Rechnitz, *Analyt. Chem.*, 1971, **43**, 1090.
[174] H. K. Wipf, A. Olivier, and W. Simon, *Helv. Chim. Acta*, 1970, **53**, 1605.
[175] G. A. Rechnitz and E. Eyal, *Analyt. Chem.*, 1972, **44**, 370.
[176] H. J. James, G. P. Carnack, and H. Freiser, *Analyt. Chem.*, 1972, **44**, 853.

these electrodes continues to increase, and interested readers are directed to Buck's summary tables[3] for recent developments.

5 Liquid-junction Potentials

There is little to report under this heading. Førland, Thulin, and Østvold[177] have advocated the discussion of concentration cells with liquid junctions, using the methods of irreversible thermodynamics. Such derivations in terms of gradients of chemical potential of neutral species, will, they argue, make it easier for undergraduates to grasp the subject. Whilst agreeing with them, one must ask who teaches irreversible thermodynamics to undergraduates?

Hickman[178] has calculated the l.j.p. for free diffusion junctions by perturbation techniques. Henderson's equation appears as the leading term of an expansion. Goldberg and Frank,[179] using computer simulation, calculate the l.j.p. for concentration cells with transference (homoionic junctions) and reach the entirely predictable conclusion that no information about single ion activities can come from such cells. Treatment of three ion junctions is promised.

Whilst measurement of the l.j.p. is not possible, changes can be measured, when the composition of the solutions forming the junction is changed in a regular fashion, as in the Biedermann–Sillèn treatment. Baumann[180] has suggested how changes in the junction

$$\begin{array}{c|c} \text{NaCl} & (2-x) \text{ mol l}^{-1} \text{ NaClO}_4 \\ 2 \text{ mol l}^{-1} & x \text{ mol l}^{-1} \text{ HClO}_4 \end{array}$$

can be measured with the addition to the right-hand side of 10^{-4} mol l^{-1} NaI so that an iodide-ion-selective electrode can be used as a reference electrode. A correction for change in l.j.p. was required for the study of tantalum fluoride complexes in media of varied acidity.

L.j.p., single ion activities, and single-ion free energies of transfer between solvents are, of course, intimately related, and we conclude by drawing attention to the excellent and clear review by Popovych[181] on medium effects, and some recent related papers.[182—184]

The literature survey on which this Report is based was derived from experimental computer information-retrieval experiments carried out by the Oxford Experimental Information Unit, and from the current-awareness publication which was the outcome of these experiments, the 'Electrolyte Solution Bulletin', published monthly by the University of Newcastle Library Information Section. Thanks are due to Dr. P. Leggate, Dr. G. Corfield, and Mrs. M. A. Watson for their co-operation and enthusiasm.

[177] T. Førland, L. V. Thulin, and T. Østwald, *J. Chem. Educ.*, 1971, **48**, 741.
[178] H. J. Hickman, *Chem. Eng. Sci*, 1970, **25**, 381.
[179] R. N. Goldberg and H. S. Frank, *J. Phys. Chem.*, 1972, **76**, 1758.
[180] E. W. Beumann, *J. Electroanalyt. Chem. Interfacial Electrochem.*, 1972, **34**, 238.
[181] O. Popovych, *Crit. Rev. Analyt. Chem.*, 1970, **1**, 73.
[182] O. Popovych, A. Gibofsky, and D. H. Berne, *Analyt. Chem.*, 1972, **44**, 811.
[183] C. L. deLigny, H. J. M. Denessen, and M. Alfenaar, *Rec. Trav. chim.*, 1971, **90**, 1265.
[184] D. Bax, C. L. deLigny, and M. Alfenaar, *Rec. Trav. chim.*, 1972, **91**, 452.

2
The Conductance of Electrolyte Solutions

BY M. J. WOOTTEN*

1 Introduction

This Report covers the period from the middle of 1969 to the end of 1971 and is thus a continuation of the first Report.[1] A small degree of overlap in references cited has been unavoidable in order to make the continuation smooth.

No major review articles on conductance of electrolyte solutions have appeared during the period studied, although several reviews on closely alligned topics are worth mentioning. Thus Hinton and Amis[2] review solvation numbers of ions and devote a section to conductimetric techniques. A review on transport processes in liquids has appeared although conductance studies are omitted.[3] Blandamer has reviewed the structure and properties of aqueous salt solutions[4] and Wood and Reilly[5] have given a general review of electrolyte solutions. The principles and concepts of ion mobility has been reported[6] and a general review of electrochemistry in non-aqueous solvents has been given by Kolthoff.[7] The principles of electrolyte conductance measurements have been reviewed citing applications to molten salts and to solid electrolytes as well as to aqueous electrolytes.[8] Papers presented at the nineteenth meeting of the French Society of Physical Chemistry on 'Problems of Structure and of Transport Processes in Concentrated Electrolyte Solutions and in Fused Salts' have been published.[9] As in the previous Report, certain aspects such as the conductance of ionic and semiconductor solids together with molten salt systems are not considered.

2 Theoretical Considerations

The previous Report on the conductance of electrolyte solutions brought up

---

[1] G. J. Hills, in 'Electrochemistry', ed. G. J. Hills, (Specialist Periodical Reports), The Chemical Society, London, 1970, vol. 1, p. 73.
[2] J. F. Hinton and E. S. Amis, *Chem. Rev.*, 1971, **71**, 627.
[3] H. J. V. Tyrrell and E. G. Neal, *Ann. Reports (A)*, 1969, **66**, 3.
[4] M. J. Blandamer, *Quart. Rev.*, 1970, **24**, 169.
[5] R. H. Wood and P. J. Reilly, *Ann. Rev. Phys. Chem.*, 1970, **21**, 387.
[6] L. Onsager, *Angew. Chem.*, 1969, **81**, 1009.
[7] I. M. Kolthoff, *Pure Appl. Chem.*, 1971, **25**, 305.
[8] J. Braunstein and G. D. Robbins, *J. Chem. Educ.*, 1971, **48**, 52.
[9] Papers in *J. Chim. phys.*, special issue, October, 1969.

* *Present address*: Westinghouse Research Laboratory, Rue Gatti de Gamond 95, B-1180 Brussels, Belgium.

to date the theoretical treatment for the dependence of conductance on the composition of the system. During the period of review no new theoretical treatments have been brought to notice although several revisions, criticisms, and comparisons of earlier theories have been attempted. Pitts and co-workers[10] have made a small correction to the earlier Pitts theory and they then compared it numerically with the Fuoss–Onsager approach. Five main differences between the two models were pointed out, concerning the electric force on the ion, the contact distance, the contribution of ionic Brownian motion to the electrophoretic contribution, and the exact electrostatic boundary condition to be used with the equations. In a subsequent paper[11] a comparison of experimental data was made with the two treatments and although the three-parameter version of Fuoss involving Λ_0, a, and an association constant K_a could be fitted closely to experimental data, the values required were unreasonably small and K_a found to be negative in nearly all cases. The revised Pitts theory which involves only Λ_0 and a was found to give a more satisfactory fit.[11] Garman[12] has criticized both the Pitts and Fuoss models in that they both give an incorrect logarithmic term and that they are not mutually exclusive since they are based on the same model. Justice[13] has reported a modified Fuoss–Onsager equation based on Bjerrum's critical distance q. This equation has been commented on by Fuoss[14] as drawing incorrect conclusions concerning the distance parameter, a, because the calculation of the $J_{3/2}$ coefficient of the $c^{3/2}$ concentration term was based on the 1957 Fuoss–Onsager equation which neglected $c^{3/2}$ terms. The criticisms were not accepted by Justice[15] and a further examination[16,17] led to the conclusion that the generalization of the original Fuoss–Onsager conductance equations to the case of association was identical to that proposed by Bjerrum in 1926. Fernandez-Prini[18] has also made modifications to the Fuoss–Hsia conductance equations to calculate the coefficients J_1 and J_2 of the equation

$$\Lambda = \Lambda_0 - S\sqrt{c} + Ec \ln c + J_1 c - J_2 c^{3/2}$$

When the J_1 and J_2 coefficients were used, the discrepancies between the Fuoss and Pitts theories were found to be reduced. Other mathematical corrections have been made to the theories of Fuoss and Onsager and extensions made to cover the case of asymmetric binary electrolytes.[19] Results for 2–2 salts in water have been analysed using simplified versions of the Pitts and Fuoss–Hsia equation.[20]

[10] E. Pitts, B. E. Tabor, and J. Daly, *Trans. Faraday Soc.*, 1969, **65**, 849.
[11] E. Pitts, B. E. Tabor, and J. Daly, *Trans. Faraday Soc.*, 1970, **66**, 693.
[12] P. Carman, *J. Phys. Chem.*, 1970, **74**, 1653.
[13] R. Bury, M. C. Justice, and J. C. Justice, *Compt. rend.*, 1969, **268**, C, 670.
[14] R. M. Fuoss, *J. Chim. phys.*, 1969, **66**, 1191.
[15] J. C. Justice, *J. Chim. phys.*, 1969, **66**, 1193.
[16] M. C. Justice, R. Bury, and J. C. Justice, *Electrochim. Acta*, 1971, **16**, 701.
[17] J. C. Justice, *Electrochim. Acta*, 1971, **16**, 701.
[18] R. Fernandez-Prini, *Trans Faraday Soc.*, 1969, **65**, 3311.
[19] J. J. Murphy and E. G. Cohen, *J. Chem. Phys.*, 1970, **53**, 2173.
[20] E. M. Hanna, A. D. Pethybridge, and J. E. Prue, *Electrochim. Acta*, 1971, **16**, 677.

An alternative approach proposed earlier by Falkenhagen and co-workers has received much attention from the Falkenhagen school,[21—30] particularly to the problem of the theory of conductivity of associating electrolytes. Conductance formulae were established for systems with large Bjerrum parameters,[24] for ions with a square-well potential, and for the Wien effect. With the introduction of the mass action law and the degree of dissociation, the range of validity was extended to strongly associating electrolytes.[21] The results were compared to the Fuoss–Onsager treatment for KCl in water–dioxan and NaCl in water–propan-1-ol mixtures.[28] The theory was also extended and discussed for solutions of tetra-alkylammonium ions.[26,27] Other authors[31,32] have based their calculations on the three-parameter Fuoss–Onsager equation for calculating ion association constants. Guggenheim[33] suggested a simple straightforward method for analysing conductance measurements in aqueous solutions of 1:1 electrolytes which does not need a computer but utilizes a method of corresponding conductances.

The Born–Fuoss–Boyd–Zwanzig calculation of the dielectric friction coefficient on a moving ion has been revised by Zwanzig.[34] The relative motion of the ion and its surrounding fluid is taken into account using both sticking and slipping boundary conditions. The electrostatic problem of finding the electric field due to a change in a moving medium was formulated and solved exactly. Using the slipping boundary condition the maximum conductance predicted for a singly charged ion in water was about 46 conductance units. This revised theory has been examined for representative univalent ions in a number of solvents.[35] The theory was shown to predict quantitatively the dependence of mobility on the ion radius in dipolar aprotic solvents, but the absolute magnitude in protic solvents show deviations from the predictions. During the period under review many other papers have been published which use the above mentioned methods to evaluate Λ_0, K_a, and other parameters. These have been split into three sections, aqueous systems, non-aqueous media, and mixed solvent systems where one solvent is water.

3 Aqueous Systems

The conductances of several alkali-metal perchlorates in water at 25 °C was

[21] W. D. Kraeft and W. Ebeling, *Z. phys. Chem. (Leipzig)*, 1969, **240**, 141.
[22] D. Kremp, H. Ulbricht, and G. Kelbg, *Z. phys. Chem. (Leipzig)*, 1969, **240**, 65.
[23] D. Kremp, H. Ulbricht, and G. Kelbg, *Z. phys. Chem. (Leipzig)*, 1969, **240**, 80.
[24] G. Kelbg, ref. 9, p. 83.
[25] H. Ulbricht, *Z. phys. Chem. (Leipzig)*, 1970, **243**, 255.
[26] H. Ulbricht and H. Falkenhagen, *Z. phys. Chem. (Leipzig)*, 1970, **243**, 305.
[27] H. Ulbricht and H. Falkenhagen. *Z. phys. Chem. (Leipzig)*, 1970, **243**, 313.
[28] W. D. Kraeft and R. Sandig, *J. Chim. phys.*, 1970, **67**, 1265.
[29] G. Kelbg and H. Ulbricht, *Z. phys. Chem. (Leipzig)*, 1970, **244**, 125.
[30] W. D. Kraeft and R. Sandig, *Z. phys. Chem. (Leipzig)*, 1971, **247**, 343.
[31] V. M. Tsentovskii, *Electrokhimiya*, 1971, **7**, 258.
[32] B. S. Krumgalz, P. Y. Staroshitskii, and G. G. Dorfmann, *Electrokhimiya*, 1971, **7**, 1748.
[33] E. A. Guggenheim, *Trans. Faraday Soc.*, 1969, **65**, 2474.
[34] R. Zwanzig, *J. Chem. Phys.*, 1970, **52**, 3625.
[35] R. Fernandez-Prini and G. Atkinson, *J. Phys. Chem.*, 1971, **75**, 239.

studied by D'Aprano[36] in a concentration range up to 0.06 mol l^{-1}. Association constants were obtained using the Fuoss–Hsia conductance equation and it was found that association increased with atomic number, i.e. it was negligible for lithium, marginal for sodium. Pair-wise association constants calculated for the others were: $K_a(K^+) = 1.00$; $K_a(Rb^+) = 1.35$; and $K_a(Cs^+) = 1.70$. A similar variation of conductance with the position of the element in the Periodic Table was found by Artamonov and Grilikhes.[37] Bury[38] has also studied the association of alkali-metal chlorates and perchlorates in water at 25 °C by conductimetric techniques. In a study of the conductance of dilute aqueous solutions of hexafluorophosphoric acid at 25 °C,[39] the results were presented by plots of Stokes' radii of AsF$_6$ $^-$, PF$_6$ $^-$, and various other spherically symmetric anions against their crystallographic radii. The points for the fluorinated anions did not fall on the smooth curve obtained with the other anions, which the authors suggest is indicative of a different kind of solvent–anion interaction. An analysis of the conductances of some aqueous acid salts has been carried out by Spiro[40] and conductance data have been used to estimate the hydration numbers of lanthanide ions in solution.[41] The equivalent conductances of CsCl solutions from 0.01 to 0.1 mol l^{-1} have been analysed by the Fuoss and Guggenheim methods.[42] Agreement between the two methods was found to be excellent. Kay and workers,[43—45] using a potentiometric method for detecting moving boundaries, have evaluated cation transference numbers for K$^+$ at 10 and 1 °C and using the Kay–Dye equation obtain single ion conductances and the temperature coefficient.

Much interest has been placed in the study of the transport properties of large ions in water. Thus Broadwater and Evans have measured conductances of large bolaform ions in water at 10 and 25 °C.[46] From a comparison of limiting ionic mobilities and concentration dependence of the tetra-alkylammonium and bolaform ions, they concluded that cation–cation pairing does not make a significant contribution to the concentration dependence of R$_4$N$^+$ ions in dilute aqueous solutions. Kay et al.[47] have also reported conductances of large hydrophobic ions in water at 25 °C and they postulated a two-step association process involving solvent-separated and contact ion pairs. The electrical conductance of tetrabutylammonium nitrate has been reported at temperatures of 10 and 50 °C.[48] Desnoyers and co-

[36] A. D'Aprano, J. Phys. Chem., 1971, 75, 3290.
[37] B. P. Artamonov and M. S. Grilikhes, Russ. J. Phys. Chem., 1970, 44, 172.
[38] R. Bury, J. Chim. phys., 1970, 67, 2045.
[39] E. Baumgartner, M. Busch, and R. Fernandez-Prini, J. Phys. Chem., 1970, 74, 1821.
[40] M. Spiro, Rev. Port. Quím., 1970, 12, 1.
[41] S. L. Bertha and G. R. Choppin, Inorg. Chem., 1969, 8, 613.
[42] P. Paterson, S. K. Jalota, and H. S. Dunsmore, J. Chem. Soc. (A), 1971, 2116.
[43] R. L. Kay, G. A. Vidulich, and A. Fratiello, Chem. Instr., 1969, 1, 161.
[44] R. L. Kay, K. Pribadi, and B. Watson, J. Phys. Chem., 1970, 74, 2724.
[45] R. L. Kay and G. A. Vidulich, J. Phys. Chem., 1970, 74, 2718.
[46] T. L. Broadwater and D. F. Evans, J. Phys. Chem., 1969, 73, 3985.
[47] R. L. Kay, D. F. Evans, and G. P. Cunningham, J. Phys. Chem., 1969, 73, 3322.
[48] I. V. Kuduyavtseva, Zhur. priklad. Khim., 1971, 44, 355.

workers[49] have determined the conductances of a series of n-alkyl hydrobromides over the range ammonium to n-octylammonium in water at 25 °C. The n-octyl salt showed clear evidence of micelle formation. In a temperature study of the conductance of dilute solutions of N-alkypyridinium halides,[50] deviations from theoretical behaviour for 1 : 1 electrolytes were apparent below the critical micelle concentration. The deviations were more pronounced at low temperatures whereas at higher temperatures the results closely approached the predictions of the Onsager limiting law. The deviations were ascribed to dimerization of the long-chain cations. Conductances of aqueous solutions of [v-phenenyltris(oxyethylene)]tris[triethylammonium] iodide and trimetaphosphate ($SynI_3$ and $SynP_3O_9$) have been reported at 25 °C.[51] $SynI_3$ showed larger negative deviations from the Onsager limiting law than $LaCl_3$ or $La(NO_3)_3$, whereas $SynP_3O_9$ showed much smaller negative deviations. These results were suggested as being indicative of an interaction between the Syn^{3+} cation and the I^- anion that is not entirely electrostatic in nature and which is not present in the Syn^{3+} and $P_3O_9{}^{3-}$ case. Donation of π electrons by the benzene ring to the I^- or a hydrophobic bond between Syn^{3+} and I^- were suggested as possible causes. Conductivity data have been reported for aqueous $[Co(NH_3)_5NO_2]SO_4$ over the concentration range 10^{-4} to 10^{-3} mol l^{-1}.[52] The results were analysed by the Shedlovsky method and ion-pair dissociation constants calculated. A comparison with the Fuoss theory of ion-pair formation suggested that non-electrostatic forces are of major importance in 2:1 and 1:1 co-ordination compounds.

During the period under review it is evident that more interest has been placed on mixed electrolyte systems. In this way the effect of the second electrolyte on the other electrolyte and solvent can be examined. A theoretical model for thermodynamic coefficients of aqueous LiCl and CsCl mixtures has been given[53] but no model is available for the conductance of mixed electrolytes. An empirical expression for 1:1 electrolyte mixtures has been put forward by Quint and Viallard[54] in terms of mole fraction and a single parameter A, and in a subsequent paper they presented experimental data for several mixed electrolyte systems and a complete analysis by this method.[55] The same authors have calculated the relaxation coefficients for the conductivity of an ion at infinite dilution in an electrolyte mixture.[56] Various other authors have examined mixed electrolyte systems by conductimetric techniques.[57—59] A

[49] J. E. Desnoyers, M. Arel, and P. A. Leduc, *Canad. J. Chem.*, 1969, **47**, 547.
[50] C. G. Butler, J. A. Stead, and H. Taylor, *J. Colloid. Interface Sci.*, 1969, **30**, 489.
[51] G. Rizzardi and A. Indelli, *Electrochim. Acta*, 1969, **14**, 845.
[52] W. L. Masterton and T. Bierly, *J. Phys. Chem.*, 1970, **74**, 139.
[53] H. L. Friedman and P. S. Ramanathan, *J. Phys. Chem.*, 1970, **74**, 3756.
[54] J. Quint and A. Viallard, *Compt. rend.*, 1969, **268**, C, 913.
[55] J. Quint and A. Viallard, *Compt. rend.*, 1969. **268**, C, 2153.
[56] J. Quint and A. Viallard, *Compt. rend.*, 1971, **273**, C, 1223.
[57] A. G. Kulkarni, *J. Indian Chem. Soc.*, 1971, **48**, 401.
[58] V. G. Gleim, A. N. Vishnevetskaya, and V. Ya. Khentov., *Russ. J. Phys. Chem.*, 1969, **43**, 278.
[59] R. A. Crawford, W. B. Darlington, and L. B. Kliever, *J. Electrochem. Soc.*, 1970, **117**, 2791.

temperature study was carried out on aqueous solutions of mixtures of sodium chloride, phosphate, and silicate from 25 to 75 °C,[58] and the sodium chloride–chlorate system was reported in aqueous solutions at temperatures above 80 °C.[59] Numerous researchers have maintained interest in the conductance of solutions outside of the region where the usual theoretical limiting laws apply. A short review on problems encountered in more concentrated solutions has been given by Hasse,[60] together with experimental data for the aqueous perchloric acid system at 25 °C. A study of all the alkali-metal halides (except fluorides) at 25 °C from 0.1 mol l^{-1} up to saturation was reported by Molenat.[61] A continuous evolution in the shape of the equivalent conductivity curves from LiCl to CsI was observed. The conductance of concentrated KCl solutions was studied by Hine and co-workers[62] and Lind and Sageman[63] have examined their data for the variation of conductance with concentration near the limit of the fused salt. Angell and co-workers have continued their work in the electrical conductance of concentrated solutions of electrolytes with LiCl–H$_2$O[64] and the alkali-metal nitrate+nitrite system with water concentrations ranging from 0 to 82 mol%.[65] The composition of the specific conductance maximum was interpreted in terms of the characteristic behaviour of the 'ideal glass' transition temperature in aqueous solutions at concentrations of 4—6 mol l^{-1}.[66, 67] A similar interpretation has been placed on experimental data for concentrated aqueous KOH solutions over the concentration range 0—10 mol l^{-1} and temperature range of -70 to $+50$ °C.[68] An extensive study of the transport properties of NaNO$_3$, NaClO$_4$, and NaCNS in concentrated aqueous solution has been made by Janz and co-workers.[69] The data were analysed and compared with that for NaCl and examined as an indication of ion–ion and solvation type interactions and structural transitions in very concentrated salt solutions. The possibility of collaboration of conductance data with Raman spectroscopy was also examined for nitrate solutions.[70, 71] Roessler and Schweider have reported transport properties of concentrated aqueous solutions of silver nitrate up to a concentration of 14 mol kg^{-1}.[72]

[60] R. Hasse, ref. 9, p. 100.
[61] J. Molenat, *J. Chim. Phys.*, 1969, **66**, 825.
[62] F. Hine, M. Yasuda, and S. Inuta, *Denki Kagaku*, 1971, **39**, 934.
[63] J. E. Lind and D. R. Sageman, *J. Phys. Chem.*, 1970, **74**, 3269.
[64] C. T. Moynihan, R. D. Bressel, and C. A. Angell, *J. Chem. Phys.*, 1971, **55**, 4414.
[65] C. A. Angell, *Austral. J. Chem.*, 1970, **23**, 929.
[66] C. A. Angell and E. J. Sare, *J. Chem. Phys.*, 1969, **52**, 1058.
[67] C. A. Angell, *J. Chem. Educ.*, 1970, **47**, 583.
[68] J. C. Jarrousseau and G. Valensi, ref. 9, p. 33.
[69] G. J. Janz, B. G. Oliver, G. R. Lakshminarayan, and G. E. Mayer, *J. Phys. Chem.*, 1970, **74**, 1286.
[70] G. J. Janz, K. Balasubrahamanyan, and B. G. Oliver, *J. Chem. Phys.*, 1969, **51**, 5723.
[71] B. G. Oliver and G. J. Janz, *J. Phys. Chem.*, 1970, **74**, 3819.
[72] N. Roessler and H. Schweider, *Ber. Bunsengesellschaft. phys. Chem.*, 1970, **74**, 1225.

4 Water–Non-aqueous Solvent Mixtures

The research reviewed in this section can be roughly divided into two classes. On one hand these systems have been examined by workers interested in structural properties of aqueous solutions and the data are analysed in terms of the structural effects the non-aqueous component has on water, whereas on the other hand these systems have received much interest because of the wide range of dielectric constant available by addition of the second component. An excellent example of the latter is dioxan–water mixtures, which have been a favourite solvent system for electrochemical measurements because of the large variation of dielectric constant. In his plenary lecture to the Second Australian Conference on Electrochemistry,[73] Fuoss presented a wide range of experimental data for KCl and NaCl in this mixed solvent system. Conductance curves of Λ/Λ_0 against the Debye–Hückel kappa showed large negative deviations from the limiting slope with decreasing the dielectric constant. Thallous chloride solutions were also investigated in dioxan–water mixtures[74] and the K_a values calculated were found to be always an order of magnitude higher at any given value of the dielectric constant than in the alkali-metal halides even though the rate of increase of K_a with decreasing ε was approximately the same. Panckhurst[75] has pointed out that the D'Aprano and Fuoss continuum treatment does not satisfactorily account for the observed trends in association constants for the thallous halides in water at 25 °C (the trend being $K_{TlCl} < K_{TlBr} < K_{TlI}$), although the K_{TlCl} value of 5.2 ± 0.5 l mol^{-1} in water at 25 °C was in good agreement with the value obtained from solubility and spectrophotometric measurements.[76] Kay and Broadwater,[77] as part of an extensive study of solvent structure in aqueous mixtures, have made precise conductance measurements on KCl, KBr, and Bu$_4$NBr in dioxan–water mixtures at 25 °C. The data were examined in terms of the effect of dioxan on the structure of water and it was concluded that dioxan does not enhance long-range order in water. In contrast to ethanol–water[78] and t-butanol–water,[79] the small initial increase in the Walden product with added dioxan was attributed to a dehydration effect rather than to better structure breaking properties of the ions. As was shown with the alcohol–water mixtures, the Bu$_4$N$^+$ ion loses its structure-making properties continuously as dioxan is added to water. Dunn and Marshall[80] have measured conductances and ionization behaviour of sodium chloride in dioxan–water mixtures at 100 °C. Measurements for the potassium salts of several rigid bolaform disulphonic

---

[73] R. M. Fouss, presented at the Second Australian Conference on Electrochemistry, Butterworths, London, 1970, p. 9.
[74] A. D'Aprano and R. M. Fuoss, *J. Amer. Chem. Soc.*, 1969, **91**, 279.
[75] M. H. Panckhurst, *J. Phys. Chem.*, 1969, **73**, 2097.
[76] J. B. Macaskill and M. H. Packhurst, *Austral. J. Chem.*, 1969, **22**, 317.
[77] R. L. Kay and T. L. Broadwater, *Electrochim. Acta*, 1971, **16**, 667.
[78] R. L. Kay, G. P. Cunningham, and D. F. Evans, in 'Hydrogen Bonded Solvent Systems', ed. A. Covington and P. Jones, Taylor and Francis, London, 1968, p. 249.
[79] T. L. Broadwater and R. L. Kay, *J. Phys. Chem.*, 1970, **74**, 3802.
[80] L. A. Dunn and W. L. Marshall, *J. Phys. Chem.*, 1969, **73**, 2619.

acids in dioxan–water mixtures at 25 °C have been reported by Staples and Atkinson.[81] The distance of closest approach in solution between the cation and anion as calculated by thermodynamics of association, conductance J parameter and dielectric relaxation drag effect, and hydrodynamic methods was compared. All methods agreed fairly well and the trend observed was generally in increasing distance of closest approach with increasing charge separation. The conductances of uranyl acetate in dioxan–water mixture were measured by Khairy and co-workers.[82] K_1 for the ion-pair and K_3 for the triple-ion dissociation were evaluated using Λ_0 values of the method of Fuoss–Kraus–Bray. From these calculations, interionic distances a_1 (ion-pair) and a_3 (triple ion) were eliminated. The conductances of Alizarine Saphirol S.E. in 70 and 82% dioxan–water mixtures at various temperatures have been reported.[83] Ramananurty and Yadav[84] have applied the Fuoss–Onsager–Skinner conductance–concentration relationship to $LiNO_3$ in dioxan–water systems at 25 °C. A study of lithium and sodium chlorates in the same system at 25 °C was carried out by Oliver and Campbell[85] from dilute solutions up to saturation. Λ_0 of both electrolytes was greatly reduced as the dioxan concentration was increased. The conductances of potassium and silver nitrates in water–dioxan solutions were measured and Λ_0 and K_a values evaluated[86] from the Fuoss–Onsager equations. Erdey-Gruz[87—91] and co-workers have extended their studies of transport processes in mixed aqueous mixtures to include the water–dioxan system. Singh and Mishra[92] investigated lithium bromide in water–dioxan at different temperatures and in a study of electrolyte–solvent interactions the same authors reported conductance measurements of Cu, Zn, and Mg sulphates in water–tetrahydrofuan solutions at 25 °C.[93] Tetrabutylammonium bromides and perchlorates[94] and cesium bromide[95] have been studied conductimetrically in water–THF mixtures by Treiner and Justice. Mixtures of low dielectric constants were utilized and the data were analysed by the Justice–Bjerrum model and interpreted by water–THF interactions.

Conductance measurements in mixtures of water with various alcohols

[81] B. R. Staples and G. Atkinson, *J. Phys. Chem.*, 1969, **73**, 520.
[82] F. M. Khairy, A. El-Said Mahgaub, and A. I. Mosaad, *J. Electroanalyt. Chem. Interfacial Electrochem.*, 1969, **23**, 115.
[83] M. Mitsuish, M. Haihizume, A. Katayama, and N. Kuroki, *Bull. Chem. Soc., Japan*, 1970, **43**, 955.
[84] M. V. Ramananurty and R. C. Yadav, *Indian J. Chem.*, 1971, **9**, 1003.
[85] B. G. Oliver and A. N. Campbell, *Canad. J. Chem.*, 1969, **47**, 4207.
[86] I. D. McKenzie and R. M. Fuoss, *J. Phys. Chem.*, 1969, **73**, 1501.
[87] T. Erdey-Gruz and K. Kugler, *Magyar Kém. Folyóirat*, 1970, **76**, 583.
[88] T. Erdey-Gruz, *Magyar Kém. Folyóirat*, 1971, **77**, 145.
[89] T. Erdey-Gruz, *Acta Chim. Acad. Sci. Hung.*, 1971, **67**, 283.
[90] T. Erdey-Gruz and B. Levay, *Acta Chim. Acad. Sci. Hung.*, 1971, **69**, 215.
[91] T. Erdey-Gruz, B. Levay, P. F. Csanyi, and E. G. Szilagyi, *Acta Chim. Acad. Sci. Hung.*, 1971, **69**, 423.
[92] D. Singh and A. Mishra, *Indian J. Chem.*, 1969, **7**, 86.
[93] D. Singh and A. Mishra, *Indian J. Chem.*, 1969, **7**, 1219.
[94] C. Treiner and J. C. Justice, *Compt. rend.*, 1969, **269**, C, 1364.
[95] C. Treiner, *J. Chim. phys.*, 1971, **68**, 56.

have been studied by many authors. Thus Amis and co-workers have extended their investigation into these systems.[96, 97] The conductance of magnesium sulphate at three different temperatures and several weight % ethanol compositions were reported.[96] Activation energies were calculated for the conductance process and compared with the theoretical requirements of the Onsager equation. Accascina and co-workers have continued their studies on the behaviour of the Walden product in mixed aqueous solutions with a complete investigation of the t-butanol–water system.[98] Alkaline perchlorate conductances were analysed by the Fuoss–Onsager theory. A comparison with water–dioxan and water–methanol systems was discussed and other authors[77, 79] have carried this comparison out also. Figure 1 shows the Walden products as a function of dielectric constant for sodium perchlorate in the three solvent mixtures at 25 °C.[98] In a study of the effect of

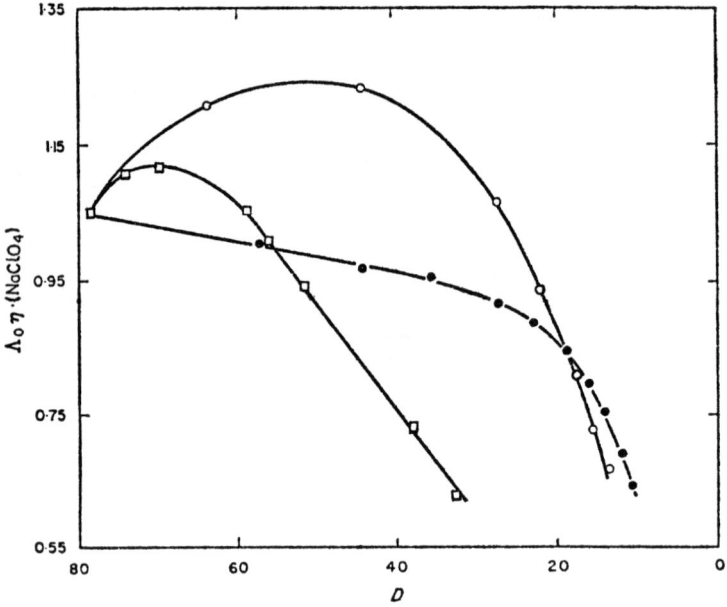

Figure 1 *Walden products as a function of dielectric constant for sodium perchlorate in water–t-butanol* ○, *in water–methanol* □, *and in water–dioxan* ●, *mixtures at 25 °C*
(Reproduced by permission from *Electrochim. Acta*, 1970, **15**, 1209)

water on proton migration in alcohols, the limiting equivalent conductance of HCl was found to decrease when H_2O was added to alcohols and then to

---

[96] E. S. Amis and J. F. Castell, *J. Electrochem. Soc.*, 1970, **117**, 213.
[97] C. Hibbs, E. S. Amis, and J. O. Wear, *J. Inorg. Nuclear Chem.*, 1971, **33**, 1659.
[98] F. Accascina, R. De Lisi, and M. Goffredi, *Electrochim. Acta*, 1970, **15**, 1209.

increase again.[99] The conductance data were analysed by the Fuoss–Onsager equations and the equilibrium constant and standard free energy change for the reaction:

$$ROH_2{}^+ + H_2O \rightleftharpoons ROH + H_3O^+$$

were determined. Values obtained are shown in Table 1.

Table 1

Solvent	MeOH	EtOH	Pr^{n}OH	Bu^{n}OH	Bu^{i}OH
ΔG°_{298}(ROH)/kcal mol^{-1}	253.48	253.13	252.86	252.47	252.24

Demey and co-workers[100, 101] have made precise measurements of the conductance of NaI[100] and tetrabutylammonium iodide[101] in ethanol–water mixtures. A temperature study over the range 25—230 °C was reported for the conductance of NaI in methanol–water solutions.[102] From these measurements it was found that a solution containing approximately 17 mole % methanol could be considered maximally ordered in comparison with other compositions. Kotaka[103] has reported conductances of ammonia, NaOH, and NH$_4$Cl in propan-1-ol–water mixtures. An extensive analysis of the conductance of KIO$_3$ in glycerol–water mixtures using the Fuoss–Onsager theory was carried out by Sadek *et al.*[104] Over the dielectric constant range studied, 55.6 < ε < 78.5, K_a and the a_0 parameter decreased as ε decreased but a plot of log K_a against $1/\varepsilon$ was not linear. These facts were explained by solvent-separated ion pairs found in a system characterized by strong hydrogen bonding. Hemmes and Petrucci[105] have made measurements on various bolaform electrolytes in glycerol–water mixtures.

5 Non-aqueous Solvents

During the period under review many extensions of earlier work have been carried out in these solvent systems by exploring the degree of association and other effects of dielectric constant and by direct comparison of different solvent systems using various theoretical models. Evans and co-workers have continued their studies of transport processes in hydrogen-bonding solvents.[106—109] As in earlier papers,[106, 110] the results were critically analysed by the Fuoss–Onsager equation and also with a modified equation proposed

[99] R. De Lisi and M. Goffredi, *Electrochim. Acta*, 1971, **16**, 2181.
[100] J. P. Demey, G. Delesalle, J. M. Hochary, and P. Deurainne, *Compt. rend.*, 1971, **273**, C, 935.
[101] J. P. Demey, G. Delesalle, and P. Deurainne, *Compt. rend.*, 1971, **273**, C, 1677.
[102] U. I. Korobkov and A. D. Mikhilev, *Soviet Electrochem.*, 1970, **6**, 969.
[103] M. Kotaka, *J. Chem. Soc. Japan*, 1971, **92**, 18.
[104] H. Sadek, A. M. Hafez, and F. Y. Khalil, *Electrochim. Acta*, 1969, **14**, 1089.
[105] P. Hemmes and S. Petrucci, *J. Amer. Chem. Soc.*, 1969, **91**, 275.
[106] D. F. Evans and P. Gardam, *J. Phys. Chem.*, 1969, **73**, 158.
[107] J. Thomas and D. F. Evans, *J. Phys. Chem.*, 1970, **74**, 3812.
[108] M. A. Matesich, J. A. Nadas, and D. F. Evans, *J. Phys. Chem.*, 1970, **74**, 4568.
[109] D. F. Evans, J. A. Nadas, and M. A. Matesich, *J. Phys. Chem.*, 1971, **75**, 1708.
[110] D. F. Evans and P. Gardam, *J. Phys. Chem.*, 1968, **72**, 3281.

by Justice based on Bjerrum's critical distance q.[109] The concentration dependence of several salts in formamide was compared to that observed in alcohols and water.[107] A multiple step association process was suggested as being a feature common to all hydrogen-bonding solvents. A series of tetra-alkylammonium salts were studied in propan-2-ol[108] and the association constants calculated were found to increase with increasing anion size and to be larger by a factor of 2 to 3 more than predicted by the behaviour in the normal alcohols.[106, 110] An interesting series of measurements were carried out on several electrolytes in 2,2,2-trifluoroethanol.[109] On comparison with the nearly isodielectric solvent ethanol, the association constants calculated were found to differ in magnitude, cation dependence, and anion dependence. The general conclusion from the above series of papers was that there seems to be a scale for solvating characteristics of ionizing solvents. One extreme where cation solvation has preference over anion solvation is for example in the aprotic solvents. At the other end of the scale, anion solvation is enhanced at the expense of cation solvation, as in 2,2,2-trifluorethanol. In between these two are solvents such as water, the amides, and the hydrocarbon alcohols, where both cation and anion solvation are important and complex ionic association patterns result. Various other workers have maintained their interest in alcoholic solutions. Shkodin and co-workers[111] measured the electrical conductance of potassium iodide in seven aliphatic alcohols (propyl, butyl, amyl, hexyl, heptyl, octyl, and nonyl). Limiting equivalent conductances and dissociation constants were calculated and the variation of $\lambda_0 \eta_0$ and pK on 1/dielectric permittivity was determined. As in an earlier study of lithium chloride in these alcohols, the Walden product varies in a non-systematic way.[112] Macua and Lamberts[113] have examined ion-pairing of n-butylammonium picrate in aliphatic alcohols by conductance measurements. Desieno et al.[114] analysed the conductance of several tetra-alkylammonium halides in ethylene glycol by the Fuoss–Onsager equation and found no evidence of association of the halide and tetra-alkylammonium ions. A comparison between spectroscopic and conductimetric measurements for alcoholic solutions of cobalt(II) chloride has been given by Osugi.[115] The conductance of HCl in glycol solution has been reported over the temperature range 20—50 °C.[116] Beronius[117] et al. have made a study of the conductance and association of alkali-metal halides in methanol and ethanol. Brookes and co-workers,[118, 119] using a four-electrode alternating current potentiometer,

---

[111] A. M. Shkodin, L. P. Sadovnichaya, and S. G. Rosenko, *Soviet Electrochem.*, 1971, **7**, 46.
[112] A. M. Shkodin, L. P. Sadovnichaya, and V. A. Podolyanko, *Ukrain. Khim. Zhur.*, 1969, **35**, 144.
[113] J. Macau and L. Lamberts, *Chem. Comm.*, 1970, 724.
[114] R. P. Desiento, P. W. Greco, and R. C. Mamajek, *J. Phys. Chem.*, 1971, **75**, 1722.
[115] J. Osugi, *J. Chem. Soc. Japan*, 1969, **90**, 640.
[116] V. I. Vigdorovich and I. J. Pchel'nikov, *Soviet Electrochem*, 1969, **5**, 660.
[117] P. Beronius, G. Wikander, and A. Nilsson, *Z. Phys. Chem. (Frankfurt)*, 1970, **70**, 52.
[118] H. C. Brookes, M. C. B. Hotz, and A. H. Spong, *J. Chem. Soc. (A)*, 1971, 2410.
[119] H. C Brookes, M. C. B. Hotz, and A. H. Spong, *J. Chem. Soc. (A)*, 1971, 2415.

measured the conductances of some perchlorates, perchloric acid, and lithium chloride in acetone at 25 °C. Limiting ionic conductances were calculated and the results indicated that although anion solvation effects were apparent, cation solvation was generally greater. The conductances of several tetraphenylborates in tetrahydrofuran were reported by Tersoc and Boileau.[120] A detailed examination of transport behaviour in dimethyl sulphoxide has been presented by Yao and Bennion.[121—123] Conductance measurements were made at 25 °C for $NaClO_4$, NaSCN, CF_3SO_3Na, $NaBPh_4$, $(1\text{-amyl})_3BuNBPh_4$, and CH_3SO_3Na. The first three of these salts were found to be completely dissociated and the $NaBPh_4$ slightly dissociated. Association constants of 12 and 100 l mol^{-1} were found for $(1\text{-amyl})_3BuNBPh_4$ and CH_3SO_3Na, respectively. Corrected Stokes radii in Å calculated for the ions at 25 °C were:

Ion	$(1\text{-amyl})_3 BuN^+$	BPh_4^-	Na^+	$CF_3SO_3^-$	ClO_4^-	SCN^-
Stokes radius	5.12	5.12	4.32	3.03	2.82	2.37

The effective ionic radii of these ions were found to be insensitive to temperature over the range 25—55 °C. The conductance of tetra-N-amylammonium thiocyanate was also studied in this solvent from infinite dilution to molten salt at 55 °C.[123] The limiting conductance at that temperature was found to be 64.16 cm^2 ohm^{-1} mol^{-1} and an association constant of 16.2 l mol^{-1} was calculated. The cation was postulated as being unsolvated with an effective radius of 5.21 Å. Monica[124] et al. have measured conductances in dimethyl sulphoxide and treated them with a Fuoss–Onsager type analysis. From transference number of ClO_4^- in $AgClO_4$ in the same solvent (measured by the Hittorf technique) they obtained single ion conductances. Bolzan and Arvia[125—127] have also studied dimethyl sulphoxide solutions but with their main interest in the temperature dependence of the conductances of HCl and HBr. At 25 °C the limiting molar conductance of HCl was 38.7 mho cm^2 mol^{-1}. The data were interpreted using the Fuoss theory for associated 1–1 type electrolytes and an association constant related to pair formation of 115.7 l mol^{-1} was calculated for HCl,[125] whereas a value of 10 l mol^{-1} was obtained for HBr.[127] The mechanism for HCl resembled that for other 1–1 electrolytes in dimethyl sulphoxide in which charges are transported by simple entities in a viscous medium. A conductance study of several quarternary ammonium halides in dimethyl sulphoxide was reported by Arrington and Griswold.[128] Single ion conductivities were calculated using tetraisoamyl-

---

[120] G. Tersoc and S. Boileau, *J. Chim. phys.*, 1971, **68**, 903.
[121] N. P. Yao and D. N. Bennion, *J. Electrochem. Soc.*, 1971, **118**, 1097.
[122] N. P. Yao and D. N. Bennion, *J. Phys. Chem.*, 1971, **75**, 1727.
[123] N. P. Yao and D. N. Bennion, *J. Phys. Chem.*, 1971, **75**, 3586.
[124] M. D. Monica, D. Masciopinto, and G. Tessari, *Trans. Faraday Soc.*, 1970, **66**, 2872.
[125] J. A. Bolzan and A. J. Arvia, *Electrochim. Acta*, 1970, **15**, 39.
[126] J. A. Bolzan and A. J. Arvia, *Electrochim. Acta*, 1970, **15**, 827.
[127] J. A. Bolzan and A. J. Arvia, *Electrochim. Acta*, 1971, **16**, 531.
[128] D. E. Arrington and E. Griswold, *J. Phys. Chem.*, 1970, **74**, 123.

ammonium tetraphenylborate and tetraisoamylammonium tetraisoamylborate as reference electrolytes. The halide order was found to be $Cl^- > Br^- > I^-$, which was expected for unsolvated ions. Slight association to ion pairs was found for Me_4NBr and Me_4NI. Several alkali-metal halide solutions in dimethyl sulphoxide were studied at 25 °C at concentrations from 0.004 to 0.028 mol l^{-1}.[129] The results were analysed by the Fuoss–Onsager theory and explained in terms of solvation of the cations and the dipolar structure of the solvent.

A continued interest has been maintained in electrochemical measurements in acetonitrile. Kay and co-workers[47, 130] measured the conductances of the halides of the $(isoamyl)_3BuN^+$ and $(isoamyl)_4N^+$ ions in acetonitrile and compared results with those in water and methanol. The transference number of $(CH_3)_4N^+$ ion in $(CH_3)_4NClO_4$ in acetonitrile was obtained using a rising boundary sheared cell with tetraphenylarsonium perchlorate as indicator and with electrical monitoring of the boundary.[130] A scale of single-ion conductivities based on this transference number ($t_+ = 0.4768 \pm 0.0002$) differed by 0.35 conductivity unit from a scale based on the assumption that the tetraisoamylammonium and tetraisoamylboride ions have equal mobilities, *i.e.* a Caplan–Fuoss split. Libus and Strzelecki[131] compared conductance measurements of Ni, Cu, and Zn perchlorates with visible spectra of the same salts in acetonitrile. The co-ordination form indicated by both measurements was $[ML_6]^{2+}2ClO_4^-$ where L = MeCN. The conductances of some halogenoborate and boronium salts in acetonitrile have been measured by Ahmed and Schmulbach.[132] They found that tetraethylammonium phenyltrichloroborate and tetramethylammonium tetrabromoborate both behaved as strong 1–1 electrolytes with no detectable ion pairing. The transport properties of sodium iodide in acetonitrile have been studied from infinite dilution to saturation at 25 °C.[133] K_a was obtained from a Fuoss–Onsager treatment and for concentrated solutions the data were accounted for by the Wishaw–Stokes equation with 4.5 Å as the ion size parameter over the entire concentration range. The value obtained for Λ_0 from the Wishaw–Stokes analysis was in exact agreement with that from the Fuoss–Onsager treatment of the dilution data, *i.e.* 179.4 ohm^{-1} cm^2 mol^{-1}. Forcier and Olver[134] studied trifluoroacetic acid and its salts in acetonitrile by conductimetric methods. Limiting conductances were evaluated for the tetraethylammonium and sodium salts and found to be 183.5 and 176.1 ohm^{-1} cm^2 mol^{-1}, respectively. Precise measurements were carried out on the conductances of copper and silver salts with symmetrical anions in acetonitrile at 25 °C,[135] *e.g.* $CuClO_4$, $CuBF_6$, $CuPF_6$, $AgClO_4$, $AgBF_4$, $AgPF_6$, and $AgNO_4$. An analysis by a Fuoss–

[129] F. Calmes-Perraud and Y. Doucet, *Compt. rend.*, 1970, **271**, *C*, 780.
[130] C. H. Springer, J. F. Coetzee, and R. L. Kay, *J. Phys. Chem.*, 1969, **73**, 471.
[131] W. Libus and H. Strzelecki, *Electrochim. Acta*, 1970, **15**, 703.
[132] I. Y. Ahmed and C. D. Schmulbach, *Inorg. Chem.*, 1969, **8**, 1411.
[133] R. P. T. Tomkins, E. Andalaft, and G. J. Janz, *Trans. Faraday Soc.*, 1969, **65**, 1906.
[134] G. A. Forcier and J. W. Olver, *Electrochim. Acta*, 1970, **15**, 1609.
[135] H. L. Yeager and B. Kratochvil, *J. Phys. Chem.*, 1969, **73**, 1963.

Onsager treatment indicated association for $AgNO_3$, $CuBF_4$, and $CuPF_6$. Single ion conductivities for all the salts were calculated. The anion values were discussed on a crystallographic size basis while those of copper and silver were interpreted in terms of specific solvation by the solvent. The same authors also report conductances of thallium perchlorate and fluoroborate in acetonitrile at 25 °C.[136]

Mukherjee and co-workers[137, 138] have continued their studies into the transport properties of electrolytes in propene carbonate. All conductance data were treated by the Fuoss–Accascina equation. Lithium chloride and bromide were found to be associated with K_a values of 557 and 19 l mol^{-1}, respectively. Λ_0 Values of (1-amyl)$_4$N(1-amyl)$_4$B and transference numbers of ClO_4^- and Li^+ (determined from concentration cells) have been utilized in obtaining single ionic conductances of Li^+, K^+, Et_4N^+, $Bu_4^nN^+$, and (1-amyl)$_4N^+$, as well as Cl^-, Br^-, I^-, ClO_4^-, and (1-amyl)$_4B^-$ at 25 °C.

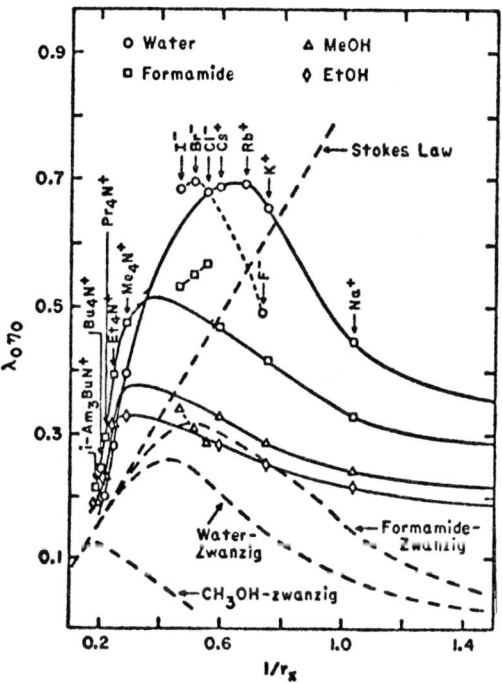

Figure 2 *The limiting Walden product for the halide, alkali-metal and tetra-alkylammonium ions as a function of crystallographic size at 25 °C* (Reproduced by permission from *J. Phys. Chem.*, 1970, **74**, 3812)

[136] H. L. Yeager and B. Kratochvil, *J. Phys. Chem.*, 1970, **74**, 963.
[137] L. M. Mukherjee, D. P. Boden, and R. Lindauer, *J. Phys. Chem.*, 1970, **74**, 1942.
[138] L. M. Mukherjee and D. P. Boden, *J. Phys. Chem.*, 1969, **73**, 3965.

The conductances of the alkaline perchlorates, AgClO$_4$, and Tl(ClO$_4$)$_3$ in propene carbonate at 25 °C were reported by Courtot–Coupez and L'Her.[139]

The amides are another group of non-aqueous solvents that have been extensively studied during the period under review. As reported above, Evans and Thomas[107] compared formamide with other solvents. Figure 2 shows their measurements for the dependence of $\Lambda_0\eta$ upon ionic size in formamide at 25 °C for a large number of ions. Extensive data for conductance, viscosity and density of NaCl and CsCl solutions in N-methylformamide at 5 and 25 °C over the concentration range 0.004—0.2 mol l^{-1} have been reported by Mostkova and co-workers.[140] It was shown that the solvation of the free ions in these solutions was no less than in water and increased with increasing temperature. Paul and co-workers[141–145] have studied some substituted ammonium perchlorates conductimetrically in dimethylformamide[143] and formamide.[144] Lithium chloride was studied in dimethylformamide[141] and the single ion conductances were calculated. A modified Hittorf treatment of the transport number of ClO$_4$$^-$ in AgClO$_4$ in NN-dimethylformamide[145] was reported and, from the limiting conductance of AgClO$_4$, single ionic conductances of other salts were calculated. In a study of large ions in solvents of high dielectric constant, Gopal[146] reported electrical conductances for several ions with long alkyl chains in formamide together with apparent molal volume data for common electrolytes[147] to study theoretical aspects of formamide electrolyte solutions. An investigation into ion–solvent interactions of tetra-alkylammonium and other common ions in N-methylformamide has been reported by Rastogi.[148]

Shkodin and co-workers[149–153] have made several systematic studies of the conductance of electrolytes in mixed non-aqueous solvents. For example, they have continued their work on methanol–dioxan[154, 155] with a study of NaCl and CsCl in this solvent mixture.[151] A transport study of dimethylformamide–dioxan mixtures was carried out and found to be in contrast to

[139] J. Courtot-Coupez and M. L'Her, *Compt. rend.*, 1970, **271**, C, 357.
[140] R. I. Mostkova, Ya. M. Kessler, and V. N. Semenova, *Soviet Electrochem.*, 1971, **7**, 620.
[141] R. C. Paul, J. P. Singla, and S. P. Narula, *J. Phys. Chem.*, 1969, **73**, 741.
[142] R. C. Paul, *Indian J. Chem.*, 1970, **8**, 63.
[143] R. C. Paul, *Indian J. Chem.*, 1970, **8**, 936.
[144] R. C. Paul, J. P. Singla, D. S. Gill, and S. P. Narula, *Indian J. Chem.*, 1971, **9**, 981.
[145] R. C. Paul, J. P. Singla, D. S. Gill, and S. P. Narula, *J. Inorg. Nuclear Chem.*, 1971, **33**, 2953.
[146] R. Gopal, *Kolloid-Z.*, 1970, **239**, 699.
[147] R. Gopal and K. Singh, *Z. Phys. Chem. (Frankfurt)*, 1971, **75**, 219.
[148] P. P. Rastogi, *Bull. Chem. Soc. Japan*, 1970, **43**, 2442.
[149] A. M. Shkodin, N. K. Levitskaya, and E. P. Nikitskaya, *Soviet Electrochem.*, 1969, **5**, 654.
[150] A. M. Shkodin, *Izvest V.U.Z. Khim. i khim. Tekhnol.*, 1969, **12**, 1013.
[151] A. M. Shkodin and N. K. Levitskaya, *Ukrain. Khim. Zhur.*, 1970, **36**, 454.
[152] A. M. Shkodin and I. A. Sergeeva, *Soviet Electrochem.*, 1971, **7**, 532.
[153] A. M. Shkodin, N. K. Levitskaya, and L. I. Tkachenko, *Soviet Electrochem.*, 1971, **7**, 846.
[154] A. M. Shkodin and N. K. Levitskaya, *Ukrain Khim. Zhur*, 1968, **34**, 330.
[155] A. M. Shkodin and N. K. Levitskaya, *Soviet Electrochem.*, 1968, **4**, 544.

the methanol–dioxan system. In dimethylformamide–dioxan, log $\lambda_0\eta_0$ and pK have a linear relationship with $1/\varepsilon$. The authors[149] suggested that this result was expected since no material structural changes occur in the mixed solvent itself as the composition was varied. The electrical conductivities of several alkali-metal perchlorates were measured in mixtures of acetone and chloroform[152] and the conductance of NaI was reported in isodielectric mixtures of alcohols containing non-polar components.[153] For example, a range of ε from 6.6 to about 30.0 was obtained using the three systems methanol–carbon tetrachloride, ethanol–carbon tetrachloride, and ethanol–dioxan. In these three systems no systematic variation of $\lambda_0\eta_0$ with dielectric properties was observed. The conductance of tetraethylammonium bromide in methanol–dioxan solutions was studied by Singh and Aggarwal.[156] Johari[157] has investigated the conductance of lanthanum hexacyanoferrate(III) tetrahydrate in dioxan–formamide and acetone–formamide mixtures at 25 °C. The data were analysed by Shedlovsky's method for Λ_0 and K_a. K_a Values at a comparable dielectric constant were higher in the dioxan mixtures than in the acetone solutions. The Walden product was found to decrease with decreasing dielectric constant in both systems but relatively more so in the acetone mixture. Pistoia and Pecci[158] found a complex Walden product relationship with solvent composition in acetone–ethanol mixtures, but a limited solubility prevented the investigation being carried out over the entire composition range. An extensive study of propan-1-ol–acetone mixtures at 25 °C and in acetone itself has been reported.[159] The conductance data were analysed by the Fuoss–Onsager equation and limiting conductances and ionic association were compared with pure solvent behaviour. The interpretation given to the Walden product and association constant behaviour with solvent composition was that there is preferential anion solvation by propan-1-ol throughout most of the composition range. Conductance measurements of various alkali-metal alcoholates MOR (where M = Li, Na, K, Rb, or Cs and R = Me, Et, Pr^i, or Pr^n) in pure alcohols have been made with high precision.[160-163] The limiting conductance Λ_0 and K_a values were calculated. It was shown that the addition of a non-polar component, for example cyclohexane or benzene, did not increase significantly the association constant with decreasing dielectric constant of the solution.[161, 163] In fact the more or less complete dissociation of potassium methoxide remained unchanged even in solvents of low dielectric constant. These results were explained in terms of specific solvation of the ions with the host alcohol solution.[163]

[156] D. Singh and I. P. Aggarwal, *Z. Phys. Chem. (Frankfurt)*, 1971, **76**, 50.
[157] G. P. Johari, *J. Phys. Chem.*, 1970, **74**, 934.
[158] G. Pistoia and G. Pecci, *J. Phys. Chem.*, 1970, **74**, 1450.
[159] D. F. Evans, J. Thomas, J. A. Nadas, and M. A. Matesich, *J. Phys. Chem.*, 1971, **75**, 1714.
[160] J. Barthel, M. Knerr, G. Schwitzgebel, and R. Wachter, *Z. Phys. Chem. (Frankfurt)*, 1970, **72**, 222.
[161] J. Barthel, M. Knerr, and G. Engel, *Z. Phys. Chem. (Frankfurt)*, 1970, **69**, 283.
[162] R. Wachter and J. Barthel, *Electrochim. Acta*, 1971, **16**, 713.
[163] R. Wachter, J. Barthel, and M. Knerr, *Electrochim. Acta*, 1971, **16**, 723.

6 Extreme Conditions

Several review articles have appeared, during the period of this Report, concerned with the effect of pressure and high-temperature on the conductance of electrolyte solutions. Hills has reviewed the techniques of high-pressure measurements,[164, 165] and a survey of recent results on the properties of water and aqueous solutions at high pressures and temperatures with emphasis on supercritical conditions has been given separately by Franck[166, 167] and Marshall.[168] The latter author and co-workers[169—171] have extended their investigations into the effect of pressure and temperature on aqueous solutions by the examination of mixed solvent systems, and in particular dioxan–water mixtures.[80, 172] Their reported measurements were up to 800 °C and 4000 bar for sodium halide solutions. At various temperatures, pressures, and solvent compositions, net changes (k) in waters of solvation upon ionization and K_0 values (the complete constant of dissociation) were obtained. The general conclusion from the whole series of measurements was that there is a generalization for simple salts, over a wide range of temperature, that varying the concentration of water either by hydrostatic pressure or by dilution with dioxan does not change the isothermal value of the complete constant. The complete constant only depended on temperature. This idea that ion-pair formation of electrolytes in water does not depend on changes in the solvent dielectric constant but only on changes in the solvent density and temperature was examined and analysed in terms of a modification of the Fuoss equation for the ion-pair dissociation constant by Gilkerson,[173] and rejected. This author[173] maintained that the data reported by Marshall and co-workers were uniquely suited to experimental separation of the effects of specific ion–solvent and ion-pair–solvent interaction and the effects of changes in the solvent dielectric constant. Experimental data have also been reported for aqueous KNO_3 and tetramethylammonium bromide in water up to 800 °C and 4000 bar.[174] Franck and co-workers continue to excel in their experimental techniques to achieve high pressure conductance data. An internally heated autoclave for conductivity measurements at 300—1000 °C and pressures up to 12 kbar was reported,[175] together with precautions taken to prevent cor-

---

[164] G. J. Hills, in 'Advances in High Pressure Research', ed. R. S. Bradley, Academic Press, London and New York, 1969.
[165] G. J. Hills, presented at the Second Australian Conference on Electrochemistry, Butterworths, London, 1970, p. 39.
[166] E. U. Franck, ref. 9, p. 9.
[167] E. U. Franck, *Pure Appl. Chem.*, 1970, **24**, 13.
[168] W. L. Marshall, presented at the Second Australian Conference on Electrochemistry, Butterworths, London, 1970, p. 62.
[169] L. A. Dunn and W. L. Marshall, *J. Phys. Chem.*, 1969, **73**, 723.
[170] A. S. Quist and W. L. Marshall, *J. Phys. Chem.*, 1969, **73**, 978.
[171] W. L. Marshall, *Rec. Chem. Progr.*, 1969, **30**, 61.
[172] L. B. Yeatts, L. A. Dunn, and W. L. Marshall, *J. Phys. Chem.*, 1971, **75**, 1099.
[173] W. R. Gilkerson, *J. Phys. Chem.*, 1970, **74**, 746.
[174] A. S. Quist, *J. Chem. Eng. Data*, 1970, **15**, 375.
[175] K. Mangold and E. U. Franck, *Ber. Bunsengesellschaft phys. Chem.*, 1969, **73**, 21.

rosion of the vessel.[176] Franck and workers have used argon as a diluent and by this method they obtained dielectric constant variations from 20 to 4.[177] Association constants and degrees of dissociation were obtained for LiCl, KCl, and CsCl by application of the Fuoss theory. It was shown that, in dilute aqueous solutions, dissociation of the alkali-metal halides was incomplete even at water densities equal to 1.0 g cm^{-3} and higher at temperatures above 300 °C. The conductance of pure water up to 1000 °C and 100 kbar has been analysed extensively by Holzaple[178] and at the higher temperatures and pressures the dissociation was found to increase markedly. The attainment of very high pressures was the subject of a paper by Hamann and Linton.[179] Over a range of 70—133 kbar, by shock methods, they examined the conductivities of aqueous solutions of KCl, KOH, and HCl.[180] They found that the molar conductivity of KCl was not measurably different from its normal value and this was interpreted as implying that although the temperature jump in the shock front tends to increase the mobilities of the K^+ and Cl^-, the simultaneous jump in density had an opposite and cancelling effect. On the other hand, the conductivities of HCl and KOH increased several fold, indicating that the excess mobilities of the H ions and OH ions increased by a factor of 3—6 under shock conditions. The K_ω of water was found to increase to about 0.09 mol^2 kg^{-2} at 133 kbar shock pressure and it is suggested that dissociation of water into H_3O^+ and OH^- is nearly complete at 200 kbar.[180] A valuable note to workers in the high-pressure field has been published jointly by the Franck group and Marshall and co-workers.[181] This report suggested that aqueous 0.01 mol kg^{-1} KCl solution could be used as a reference solution for electrical conductance measurements up to 800 °C and 12 kbar. Table 2 and Figure 3 have been taken from this paper. The values are computed averages of published values up to the present time and will be useful for determining cell constants and checking reliability of apparatus used at these critical conditions. The error between the two groups was estimated at less than 1% at lower temperatures and pressures, rising to 3—4% at the higher conditions.

Not all researchers in the field of electrochemistry under pressure go to the extreme conditions that have been reported above. Especially in the field of aqueous electrolytes much work has been done at pressures up to 3 kbar and temperatures below 100 °C. At these lower temperatures and pressures water still shows some of the abnormalities that are associated with this highly H-bonded structured solvent. Gancy and Brummer[182] investigated the effect of solution concentration on ionic conductance for LiCl, KCl, RbCl, and

[176] H. Renkert and E. U. Franck, *Ber. Bunsengesellschaft phys. Chem.*, 1970, **74**, 40.
[177] D. Hartman and E. U. Franck, *Ber. Bunsengesellschaft phys. Chem.*, 1969, **73**, 514.
[178] W. B. Holzaple, *J. Chem. Phys.*, 1969, **50**, 4424.
[179] S. D. Hamann and M. Linton, presented at the Second Australian Conference on Electrochemistry, Butterworths, London, 1970, p. 82.
[180] S. D. Hamann and M. Linton, *Trans. Faraday Soc.*, 1969, **65**, 2186.
[181] A. S. Quist, W. L. Marshall, E. U. Franck, and W. Osten, *J. Phys. Chem.*, 1970, **74**, 2241.
[182] A. B. Gancy and S. B. Brummer, *J. Phys. Chem.*, 1969, **73**, 2429.

Table 2 Specific conductances/ohm^{-1} cm^{-1} × 10^5 of 0.01 mol kg^{-1} KCl solution at integral temperatures and pressures

Pressure/bar	Temperature/°C													
	100	150	200	250	300	350	400	450	500	600	700	800
1000	368	498	599	663	689	685	650	578	460	125	—	—
2000	364	491	593	659	698	707	701	679	643	516	336	116
3000	358	483	585	651	695	715	718	709	689	625	528	405
4000	351	481	574	640	687	712	723	724	714	670	610	526
6000	338	466	555	617	665	699	722	733	732	707	672	628
8000	326	446	536	598	645	679	703	716	721	711	692	664
10000	—	—	—	—	625	665	690	707	711	709	694	675
12000	—	—	—	—	601	644	674	694	704	703	688	673

NH$_4$Cl at 25 °C and pressures up to 2 kbar. Their results were analysed by the Debye–Hückel–Onsager limiting law and they showed that $\Lambda_p/\Lambda_{p=1}$ fitted the law up to to 20 mol l^{-1} within 0.1%. These authors suggested that the best method to extrapolate $\Lambda_p/\Lambda_{p=1}$ to infinite dilution was to use one set of data at about 3 mmol l^{-1} and the limiting law slope.[183] Utilizing this method, the same authors have reported data for eleven 1:1 electrolyte systems[183, 184] over a temperature range of 3—55 °C and pressures of 1—2300 atm. The infinite

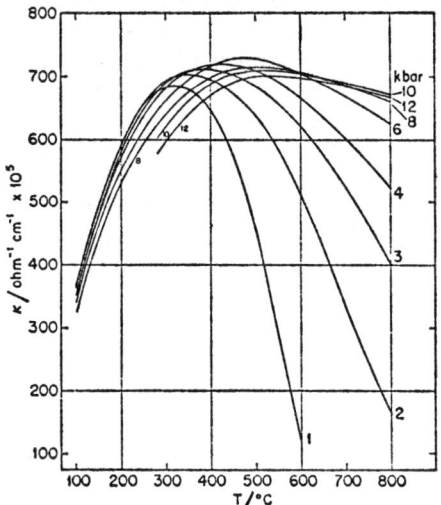

Figure 3 *Specific conductances of* 0.01 mol kg^{-1} KCl *solutions at integral temperatures and pressures*
(Reproduced by permission from *J. Phys. Chem.*, 1970, **74**, 2241)

dilution ratios were fitted to a third-order polynomial in pressure and were internally consistent to 0.05% and accurate to <0.2%. Pearce and Strauss[185] have reported conductances for KCl and tetrabutylammonium picrate in dioxan–water mixtures over a dielectric range of 78.3 (pure H$_2$O) to 11.98 (80% dioxan) and pressures up to 2.5 kbar. The association of KCl in low dielectric constants was found to be reduced by increasing pressure so that the conductances of the solutions of finite concentration were not reduced as much by pressure as at infinite dilution. In contrast to this, the Bu$_4$N picrate solutions were wholly dissociated even in very low dielectric constant solvents, as shown by the limited concentration dependence of the conductance pressure characteristics. Lown and co-workers[186—188] have continued their studies in

[183] A. B. Gancy and S. B. Brummer, *J. Chem. Eng. Data*, 1971, **16**, 385.
[184] A. B. Gancy and S. B. Brummer, Final Report Contract No. 14-01-0001-966, Dept. of Interior, Office of Saline Water, Washington D.C., March 1971.
[185] P. J. Pearce and W. Strauss, *Austral. J. Chem.*, 1970, **23**, 905.
[186] D. A. Lown, H. R. Thirsk, and Lord Wynne-Jones, *Trans. Faraday Soc.*, 1970, **66**, 51.
[187] D. A. Lown and H. R. Thirsk, *Trans. Faraday Soc.*, 1971, **67**, 132.
[188] D. A. Lown and H. R. Thirsk, *Trans. Faraday Soc.*, 1971, **67**, 149.

the effect of high pressure on proton transfer in aqueous solutions by investigation of HCl, CH_3CO_2H, alkali-metal hydroxides, and orthophosphoric acid to pressures of 3 kbar and temperatures to 200 °C. The suggestion from the set of papers was that the proton-transfer mechanism of conductance in water is obliterated as the solute concentration was increased and, at higher concentrations, hydrogen and hydroxyl ionic conductance occurs by a hydrodynamic mechanism. Molenat[189] has reported conductance measurements of saline solutions as a function of pressure and in a study of ionic solutions under pressure[190] the $Co(NH_3)_6^{3+}$ and SO_4^{2-} mobilities were investigated. The effect of pressure on the conductance of 4-methoxycarbonyl-N-methylpyridinium iodide and NaI in acetone and in isobutyl alcohol has been described by Ewald and Scudder.[191] Limiting conductances and ion-pair association constants were derived at pressures up to 3 kbar. The results showed incomplete dissociation of the salts at all concentrations. An extensive investigation of concentrated alkali-metal halide solutions has been carried out by Salvinien and co-workers[192] from 0.1 mol l^{-1} up to saturation, and pressures up to 1800 kg cm^{-2}. The results were qualitatively interpreted and quantitatively compared with theoretical formulae.

[189] J. Molenat, *J. Chim. phys.*, 1970, **67**, 368.
[190] M. Nakahara, *Rev. Phys. Chem. Japan*, 1970, **40**, 1.
[191] A. H. Ewald and J. A. Scudder, *Austral. J. Chem.*, 1970, **23**, 1939.
[192] J. Salvinien, B. Brun, and J. Molenat, ref. 9, p. 19.

3
The Solid Metal Electrode in Aqueous Solution

BY N. A. HAMPSON

1 Introduction

This article continues the review of solid metal electrodes that commenced in the second volume of this Specialist Periodical Reports series to cover the period up to the end of 1971. The papers reviewed are mainly those which have appeared in 1971. However, also included are a number of articles published in earlier years, which for one reason or another were not available to the reviewer when the previous Report was written. As far as possible the Report is again confined to the solid metal–aqueous solution interphase in the absence and presence of a faradaic reaction, although this scope has had to be broadened, especially in dealing with corrosion, where some aspects of film growth are touched on.

2 The Interfacial Structure at Solid Metals

A critical examination of values of work function, ϕ, and potential of zero charge, E_z, for metals has been made by Trasatti.[1] This shows that the linear relationships with unit slope between ϕ and E_z that are widely quoted in the literature are generally incorrect and to be expected to hold only for transition metals, where H_2O is forced to the position of maximum orientation by additional forces arising from chemisorption phenomena. It has accordingly been proposed that successive approximations may be used to estimate ϕ values. The best relationship (through three successive approximations) between the two quantities is expressed by the equation:

$$E_z = \phi - 4.61 - 0.40\alpha$$

where $\alpha = (2.1 - x_m)/0.6$ is defined as the degree of orientation of H_2O at the interface and x_m is the Pauling electronegativity. Two groups of metals were identified. These were (i) transitional metals, for which $E_z = \phi - 5.01$, *i.e.* $x_m = 0.5\phi - 0.29$ (exceptions were Ga, Zn, and Al, which gave $x_m = 0.5\phi - 0.55$), and (ii) non-transitional metals, for which $E_z = \phi - 4.69$, *i.e.* $x_m = 1.5$. For transitional metals $\alpha = 1$, showing complete orientation of water molecules, and $\alpha = 0$ for Au, corresponding to no preferred orientation of H_2O on the surface. The orientation of H_2O at interfaces was discussed in some

---

[1] S. Trasatti, *J. Electroanalyt. Chem. Interfacial Electrochem.*, 1971, **33**, 351.

detail. Lists of calculated ϕ, E_z, and electronegativity values were given for certain metals for which experimental data were lacking. It will be interesting to see how far these predicted values are borne out by experiment. In this area, determination of the work function is difficult, particularly when it is remembered that the ease of loss of electrons may differ for different crystallographic planes making up a metal surface. Other quantities characteristic of a metal surface can be measured much more accurately than the work function. Gokhshtein[2] has considered the surface tension γ of solid metal electrodes in relation to the potential of zero charge (p.z.c.). Using the nomenclature $\gamma_q = d\gamma/dq$ (q = surface charge density), γ_q was termed the 'stance'. A potential derivative, the 'E-stance', $\gamma_E = d\gamma/dE$, is related to the stance by:

$$\frac{d\gamma}{dq} = \frac{d\gamma}{dE} \cdot \frac{dE}{dq} = \frac{1}{C}\frac{d\gamma}{dE}$$

where C is the electrode capacitance. The slope of the stance–potential curve at zero stance is a measure of the effect of an elastic deformation on the charge density of the metal at constant potential. The slope $d\gamma_q/dE$ was shown to be equal to unity if elastic deformation has no effect on the charge density at constant potential, and this relationship indicates the extent to which elastic deformation affects charge density. For lead, bismuth, cadmium, and thallium, zero stance coincided with the p.z.c. and indicated that the effect of elastic deformation on the surface charge density was not significant. In the case of dilute solutions, it was shown that the second harmonic of the potential of the response of the interphase to a.c. stimulus passes through zero at the p.z.c. (minimum in differential capacitance), and the p.z.c. obtained from this method was compared with zero stance. For cadmium, near the p.z.c. (zero stance) in 0.01 M-NaF solution, the slope of the stance–potential curve was greater than 3, which indicated the formation of bonds between H_2O molecules in the inner double-layer and individual electrode atoms. The effect of an elastic deformation on the charge density was considered to be the result of the interaction of geometrical, dielectric, dipole, and various energetic factors. Thus the stance of a number of oxidized metal surfaces, including platinum, was positive, owing to the deformation of the dipoles in the oxide monolayer. Morcos[3] has determined double-layer parameters from electrocapillary curves of partially immersed electrodes. The modern theories of wetting and electrocapillary phenomena were used in the interpretation of the variation of liquid meniscus height and potential. The results obtained using an amalgamated gold electrode in KI solutions agreed with the theory developed. A comparison between the charge–potential relationship obtained from experimental data using the meniscus-rise (h) method and the equation:

$$q = 265\,(1+8.5\,h^2)\left(\frac{\partial h}{\partial E}\right)_\mu$$

[2] A. Ya. Gokhshtein, *Elektrokhimiya*, 1971, **7**, 3.
[3] I. Morcos, *Coll. Czech. Chem. Comm.*, 1971, **36**, 689.

The Solid Metal Electrode in Aqueous Solution 43

and electrocapillary measurements on liquid mercury exhibited significant differences. It was argued that these were due to two main causes, the influence of the electric field on the water orientation and on the dispersion interaction at the electrode–solution interface. Brzostowska et al.[4] described a method of double-layer capacitance measurements at solid electrodes using a pulse potentiostatic method, and a number of results of studies of the interphase that is established between large solid surfaces and electrolyte solutions are reported. Tschernikovski and Gileadi[5] also report on new methods for determining the double-layer capacitance at solid metals. Ershler et al.[6] have reviewed and evaluated the methods used for measuring the time-lag on a potentiogram as a measure of the amount of reacting substances present, applied to the case of a reversible reduction of adsorbed particles. The methods selected as the best were based on extrapolations of the purely capacitative portions of the time-lag, the slope of these portions being independent of the mechanisms of the electrode processes. Russian workers[7] have considered the form of the isotherm for the case of adsorption of a capillary-active dipolar material on a solid electrode. The electrostatic repulsion of the adsorbed dipoles has an insignificant effect on the form of the adsorption isotherms when the coverage $\theta \leqslant 0.5$, and consequently the differences between the Frumkin isotherm and that proposed by the authors in this region do not exceed the possible experimental errors. On the other hand, when $\theta \geqslant 0.5$, the repulsion between the dipoles of the organic material results in a decrease of θ by comparison with Frumkin's isotherm, particularly when there are negative surface charges. Since a decrease in θ at identical values of $y\ (=C/C_{\theta=0.5}) > 1$ corresponds to a decrease in the effective value of the attraction constant a_{eff} in Frumkin's isotherm, the extension of the theory is in qualitative agreement with the experimentally established principle of a decrease in a_{eff} as $q > 0$ becomes $q < 0$.

Armstrong and Race[8] have more clearly identified a principal component of the dispersion of capacitance at a metal–electrolyte interphase in the case of a dilute electrolyte (0.01—0.0002 mol l^{-1}) in the range 100—500 kHz. The series capacitance and resistance values showed a considerable frequency dependence which could be almost entirely accounted for by considering the geometrical capacitance between the electrode (in this case a mercury drop plus Hg thread) and the counter-electrode. Campanella[9] has reviewed a part of the data that have been reported for the p.z.c. on a number of solid metals. Kheifets et al.[10] also review a large number of p.z.c. determinations for many

[4] M. Brzostowska, J. Dabkowski, and S. Minc, *Roczniki Chem.*, 1969, **43**, 2171.
[5] N. Tshernikovski and E. Gileadi, *Electrochim. Acta*, 1971, **16**, 579.
[6] A. B. Ershler, N. N. Ovsyannikov, and K. S. Korshunova, *Elektrokhimiya*, 1970, **6**, 876.
[7] V. A. Kir'yanov, V. S. Krylov, B. B. Damaskin, and A. V. Chizhov, *Elektrokhimiya*, 1970, **6**, 1020.
[8] R. D. Armstrong and W. P. Race, *J. Electroanalyt. Chem. Interfacial Electrochem.*, 1971, **33**, 285.
[9] L. Campanella, *J. Electroanalyt. Chem. Interfacial Electrochem.*, 1970, **28**, 228.
[10] V. L. Kheifets, B. S. Krastskov, and A. L. Rotinyan, *Elektrokhimiya*, 1970, **6**, 916.

metal–solution interphases. On the basis of an analysis of the literature data on the constants 'a' of the Tafel equation for the cathodic discharge of the hydroxonium ions and on the p.z.c. values of solid metals, it was shown that all experimental data were embraced by the equation:

$$i = i_{p.z.c.} \exp[-\alpha F(E-E_z)/RT]$$

where i is the current density at E and $i_{p.z.c.}$ is current density at p.z.c. E_z, which therefore relates hydroxonium ion discharge rate on the reduced or rational potential scale. The collected data could be divided into two groups, those with $i_{p.z.c.} \approx (2.1 \pm 0.5) \times 10^{-6}$ A cm^{-2} and those with $i_{p.z.c.} \approx (1.0 \pm 0.5) \times 10^{-3}$ A cm^{-2}. The former group comprised those metals on which the h.e.r. proceeds by the slow discharge mechanism and the latter probably comprised those metals for which the slow stage of the h.e.r. is the electrochemical desorption.[10]

A number of investigations of the interphase at solid metals in aqueous solutions have been reported during the past year, although few new experimental procedures have been described. A useful contribution by Devanathan and Tilak[11] described a simple transformer bridge for differential capacitance measurements. Applied to a growing mercury drop, measurements could be made to an accuracy of ±0.2%.

Bismuth has been studied by Palm, Past, and Pullerits.[12] The adsorption of surfactants was investigated using spherical electrodes of very pure bismuth ($< 5 \times 10^{-4}$ % impurity). The electrodes had easily measurable areas and were subject to a uniform electric field on their surface. The differential capacitances in aqueous solutions of K$_2$SO$_4$, KCl, and (Me$_4$N)$_2$SO$_4$ were measured at 23 °C, and from these data the p.z.c. could be estimated as −0.4 V. In the frequency range 200—3000 Hz the dispersion of capacitance was ≤2%.

Copper has been the subject of three recent papers. In the first two of these Egorov and Novosel'skii[13] have considered the differential capacitance of a polycrystalline copper electrode in K$_2$SO$_4$ and NaCl solutions. In 0.8—0.001 M-NaF solutions the behaviour of the copper electrode is in agreement with the diffuse-double-layer theory. The minimum in the capacitance–bias potential curve in the case of 0.00125 M-K$_2$SO$_4$ occurs at a potential of −0.04 ± 0.015 V (n.h.e. is normal hydrogen electrode) and for 0.003 M-NaCl at −0.03 ± 0.015 V. From the concentration dependence of the measured values of the capacitance minimum, the capacitance measurements were considered to represent contributions of the diffuse region of the electrical double layer. Thus the minimum in the capacitance–potential curves for dilute K$_2$SO$_4$ and NaCl solutions occurs at the p.z.c. In this case the p.z.c. of copper in 0.00125 M-K$_2$SO$_4$ (after allowing for a correction for the asymmetry of the electrolyte) is −0.01 ± 0.015 V (n.h.e.). By comparing the values of the p.z.c. of copper in 0.001 M-NaF (0.09 V), K$_2$SO$_4$ (−0.01 V), and NaCl (−0.03 V), the

[11] M. A. V. Devanathan and B. V. K. S. A. Tilak, *Electrochim. Acta*, 1971, **16**, 2121.
[12] U. Palm, V. Past, and R. Pullerits, *Tartu Riikliku Ulikooli Toim.*, 1968, **219**, 63.
[13] L. Ya. Egorov and I. M. Novosel'skii, *Elektrokhimiya*, 1970, **6**, 521, 869.

dependence of these on the anion can be traced throughout the series by increasing adsorption. In the third of these recent papers on copper, Champion, Crespy, and Royon[14] have determined the p.z.c. of copper in aqueous K_2SO_4 using 0.002 $M-K_2SO_4$ and a.c. of 40 Hz. The p.z.c. reported (−0.35 V) is somewhat at variance with that reported in the first two papers.

Graphite is attracting attention in the electrochemical field because of its inertness. Randin and Yeager[15] have reported the differential capacitance of the basal plane of stress-annealed pyrolytic graphite. The electrolyte interface was found to be quite abnormal and it was concluded that the space-charge characteristics of graphite must be taken into account in examining the electrochemical properties of this material.

Gold has been examined in some detail by French workers. Hamelin and Lecoeur[16] have reported data characteristic of a gold monocrystal in dilute solutions of NaF. A capacitance minimum was obtained with the three crystallographic orientations studied; the potentials at which these occurred (s.c.e. is saturated calomel eletrode) were −0.05 V (110); 0.14 V (100), and 0.26 V (111); the minimum obtained with the (110) face was more marked than either of the others. The ions of NaF solution have been shown recently by Huong and Clavilier[17] to be unadsorbed (specific) at the electrode. The structure of the gold–aqueous NaF interphase was shown to be relatively anisotropic. For the orientations (110), (100), and (111) the corresponding p.z.c's were reported to be −0.05 V, 0.14 V, and 0.26 V (s.c.e.), respectively. For the interphase established between gold and KBr solution, Hamelin and Dechy[18] reported that with a neutral solution of 0.01 M-KBr, Br⁻ was adsorbed differently on different planes of the crystal lattice, and that a relationship exists between the atomic structure of the crystalline faces and the adsorption peaks observed in the differential capacitance curves.

Levin and Rotinyan[19] have determined the p.z.c. of the indium electrode by measuring the differential capacitance in dilute Na_2SO_4 solution at pH 3. Potentials were measured against the hydrogen electrode in the same solutions (0.005 and 0.0025 $M-Na_2SO_4$). A capacitance minimum was observed at -0.70 ± 0.01 V (n.h.e.). Assuming a 30 mV shift due to electrolyte asymmetry by analogy with the mercury electrode, the p.z.c. was therefore confirmed as -0.67 ± 0.01 V (n.h.e.), agreeing with earlier work.

Hackerman continues his interest in the iron electrode. In collaboration with Viadyanathan[20] he measured the differential capacitance of high-purity zone-refined iron, using a single-pulse technique. The capacitance–bias potential relationship in dilute aqueous solutions of $NaClO_4$ was reported, and the results showed that the assignment of an unambiguous value of potential

[14] P. Champion, G. Crespy, and J. Royon, *Compt. rend.*, 1971, **272**, *C*, 1541.
[15] J-P. Randin and E. Yeager, *J. Electrochem. Soc.*, 1971, **118**, 711.
[16] A. Hamelin and J. Lecoeur, *Coll. Czech. Chem. Comm.*, 1971, **36**, 714.
[17] N. V. Huong and J. Clavilier, *Compt. rend.*, 1971, **272**, *C*, 1404.
[18] A. Hamelin and P. Dechy, *Compt. rend.*, 1971, **272**, *C*, 1450.
[19] E. D. Levin and A. L. Rotinyan, *Elektrokhimiya*, 1971, **7**, 372.
[20] H. Vaidyanathan and N. Hackerman, *Electrochim. Acta*, 1971, **16**, 2193.

to the p.z.c. was not possible from capacitance minima. Measurements in NaX solution, where X = F, Cl, or Br, showed a behaviour different from that in NaI solution. Electrode capacitances in an aqueous solution of the furoic acid–furoate buffer were determined over a ±300 mV range either side of the open-circuit potential. The effect of specifically adsorbed furoate ion on the morphology of the capacitance curves was reportedly very significant. Szklarska-Smialowska and Wieczorek[21] have studied the form of adsorption isotherms on mild steel in 0.25 M-H_2SO_4 solution for primary aliphatic compounds (amines, alcohols, and acids) differing in chain length (6—12 carbon atoms). It was found that the compounds considered were adsorbed on the iron surface according to the Frumkin isotherm used in the form: $\theta/(1-\theta)$ exp $(f\theta) = kC$. The degree of coverage of the iron surface by adsorbed molecules is influenced by the factors (a) solubility of the given compounds, (b) presence of functional groups, (c) interaction between adsorbed molecules. Within the given homologous series the contribution of the functional group to the adsorption was found to decrease with the length of the chain of the aliphatic compound. Conversely, the contribution of the interaction between adsorbed molecules increases with their length.

The temperature dependence of the capacitance of the electrical double-layer on lead indicated[22] that the differential capacitance curves corresponding to the compact part of the double layer on lead resemble those obtained for mercury at higher temperatures. Recent studies of the differential capacitance of lead in a number of aqueous electrolyte solutions have been reported.[23,24]

Burshtein et al.[25] have determined the specific surface of nickel from charging curves.

The interphase at a platinum electrode in aqueous solution has received considerable attention. During the review period four papers[26—29] have appeared. The double-layer capacitance at a smooth platinum electrode in 0.5 M-H_2SO_4 has been studied by the impedance technique.[26] Using a charge-step decay method, Flinn, Rosen, and Schuldiner[27] have confirmed the results of an earlier study of the same system using the galvanostatic method. It was substantiated that in passing from 0.5 M- to 1.8×10^{-4} M-H_2SO_4 the 'hump' virtually disappeared, and a distinct minimum appeared in the potential region 0.22—0.33 V (n.h.e.) which apparently was a diffuse-layer minimum. The p.z.c. was considered to be 0.23 V. An alternative approach is

[21] Z. Szklarska-Smialowska and G. Wieczorek, *Corrosion Sci.*, 1971, **11**, 843.
[22] K. V. Rybalka, *Elektrokhimiya*, 1971, **7**, 242.
[23] J. P. Carr, N. A. Hampson, S. N. Holly, and R. Taylor, *J. Electroanalyt. Chem. Interfacial Electrochem.*, 1971, **32**, 345.
[24] J. P. Carr and N. A. Hampson, *J. Electrochem. Soc.*, 1971, **118**, 1262.
[25] R. Kh. Burshtein, A. G. Pshenechnikov, V. D. Kovalskaya, and M. E. Belyaeva, *Elektrokhimiya*, 1970, **6**, 1756.
[26] V. I. Luk'yanycheva, E. M. Strochkova, V. S. Bagotskii, and L. L. Knots, *Elektrokhimiya*, 1971, **7**, 267.
[27] D. R. Flinn, M. Rosen, and S. Schuldiner, *Coll. Czech. Chem. Comm.*, 1971, **36**, 454.
[28] A. Ya. Gokhshtein, *Elektrokhimiya*, 1970, **6**, 979.
[29] V. V. Polev and O. A. Petrii, *Elektrokhimiya*, 1970, **6**, 1726.

reported by Gokhshtein[28] in a brief communication of a possible method of measuring $d\gamma/dE$ as a function of bias potential. The adsorption properties of rhodium and palladium in acid solutions have been reported by Polev and Petrii.[29] Silver has also received a great deal of attention during the past 12 months. Valette and Hamelin[30] have determined the p.z.c. on the (111) face of silver in NaF as -0.69 ± 0.01 V (s.c.e.). It was reported that there was no evidence for the adsorption of F^-, contrary to that which was found for SO_4^{2-}. The Stern theory, which considers the interphase as two capacitances in series, is verified for the case of a solid monocrystalline electrode. Valette[31] has measured and analysed differential capacitance curves on (110) monocrystalline silver in NaF. The p.z.c. of the (110) face was placed at -1.01 V (s.c.e.). By integration of the differential capacitance–potential curves, the differential capacitance–charge curve was obtained. The roughness factor of the electrode was calculated and hence the differential capacitance of the compact double-layer was studied. The system conformed to a model in which specific adsorption was absent. Andrusev et al.[32] have reported differential capacitance curves on an electrolytically polished electrode. Using 0.5 and 0.005 M-Na_2SO_4 solutions, the p.z.c. was identified at -0.95 V (s.c.e.). It was interesting to note that the results of the experiments indicated that the roughness factor of the silver electrode was ~ 1.5. BuOH was adsorbed at the electrode, showing characteristic peaks and minima; 0.6 and -1.3 V (s.c.e.) were the limits of the adsorption region. It was noted that curves with clearly developed minima could be obtained only after long contact of the electrode with the solution of organic substance. This investigation was extended[32] to study the adsorption of n-amyl alcohol on the silver electrode. With 0.5 M-Na_2SO_4 as the background electrolyte, the differential capacitance curves were back-integrated from the p.z.c., which was determined from the position of the minimum which developed in dilute solution. The degree of coverage, θ, of the electrode was calculated from $\theta = (C_0 - C)/(C_0 - C')$, where C_0 and C' are capacitances for $\theta = 0$ and $\theta = 1$ respectively. A logarithmic dependence of θ on time was observed. The adsorption complies with the Temkin isotherm. The linear dependence of θ on log t and log (concentration) corresponds to the case where the rate of adsorption, v_a, of n-amyl alcohol on silver is described by the Roginskii–Zel'dovich equation for slow adsorption on an energetically heterogeneous surface:

$$v_a = \frac{d\theta}{dt} = k_a c_{BuOH} \exp[-f\theta]$$

where k_a is the adsorption constant. Integration gives:

$$\theta = 1/f \ln k_a c_{BuOH} + 1/f \ln t$$

where $v_a \approx 10^{-4}$–10^{-9} s^{-1} and $k_a = 1.6$ s^{-1}

[30] G. Valette and A. Hamelin, *Compt. rend.*, 1971, **272**, C, 602.
[31] G. Valette, *Compt. rend.*, 1971, **273**, C, 320.
[32] M. M. Andrusev, A. B. Érshler, and G. A. Tedoradze, *Elektrokhimiya*, 1970, **6**, 1163; 1971, **7**, 1159.

The structure of the electrical double layer on tellurium has been studied by Rybakov et al.[33]

Bartenev, Sevast'yanov, and Leikis[34] have investigated the differential capacitance of the tin–dilute aqueous solution interphase. For Na_2SO_4 electrolytes it was observed that differential capacitance curves measured in solutions of 0.0125—0.05 M-Na_2SO_4 showed a clearly defined capacitance minimum which deepened with dilution of the solution. Taking into account the effect of the asymmetrical valence nature of the electrolyte, which results in the minimum of the differential capacitance curves occurring at potentials more negative than the p.z.c., it was reported that the p.z.c. occurred at −0.43 V (s.c.e.), in agreement with other data in the literature. The effect of surface-active substance (*e.g.* n-hexanol) was to cause a significant decrease in capacitance at small surface charges and a clearly defined desorption peak at a negative surface charge. The authors concluded that the ideas concerning the structure and properties of the electrical double-layer which have been checked experimentally at the mercury–aqueous solution interphase apply quantitatively to the tin–aqueous solution interphase.

Bartenev et al.[35] also reported the results of an investigation of the differential capacitance of zinc in dilute aqueous solution. The p.z.c. was identified by both diffuse-layer minimum and the results of n-hexanol adsorption, and was considered to be at −0.7 V (n.h.e.) from the position of the desorption peak. It was argued that since this potential was more positive than its equilibrium potential in the same solutions, the rapid dissolution of zinc would prevent the direct measurement of the differential capacitance at low charge densities.

3 Charge-transfer Reactions

Theory of Charge Transfer.—Enyo and Yokoyama[36] have explored the reaction-order method of Vetter and have described extensions to make it applicable to reactions for which a chemical step controls the overall rate. The stoicheiometric number of the rate-determining step is taken into consideration in order to make the method more general. In their paper the authors derived their result starting from a general rate equation rather than the charge-transfer equation, which clearly limited the application to processes controlled by charge transfer. The diffusion theory of electron-transfer reactions that was originally formulated by Marcus for the rate of electron-transfer processes has been developed by Šolc.[37] The stationary diffusion equationn ad reaction-rate equation were derived on the basis of a statistical model of electron-transfer reactions in solution. The diffusion equation was

---

[33] B. N. Rybakov, G. V. Maslova, and L. A. Sinyagovskaya, *Elektrokhimiya*, 1970, **6**, 1237.
[34] V. Ya. Bartenev, É. S. Sevast'yanov, and D. I. Leikis, *Elektrokhimiya*, 1971, **7**, 1868.
[35] V. Ya. Bartenev, É. S. Sevast'yanov, and D. I. Leikis, *Elektrokhimiya*, 1970, **6**, 1197.
[36] M. Enyo and T. Yokoyama, *Electrochim. Acta*, 1971, **16**, 223.
[37] M. Šolc, *Ber. Bunsengesellschaft phys. Chem.*, 1970, **74**, 1244.

solved for the limiting cases of rapid and slow electron transfer and for a truncated dependence of electron-transfer probability $K_e(r)$ on the reactant distance r. For a reaction rate controlled by electron transfer the rate integral was solved for truncated and smooth functions. The temperature dependence of second-order rate constants and the relationship to the form of the function $K_e(r)$ was discussed in detail.

Dogonadze and co-workers have continued their interest in the kinetics of electrochemical processes. A semi-phenomenological theory is given[38] in terms of potential–current density characteristics of the continuous electron spectrum in the electrode using the Brönsted relation. In a second paper[39] the qualitative form of the potential curve is derived and some of its general properties were investigated on the basis of the Arrhenius equation for the probability of the elementary event of an electrochemical reaction, with no assumptions made concerning the model. The existence of regions of normal, non-barrier, and non-activation discharge was demonstrated with highly general assumptions as to the electronic subsystem in a metallic electrode. The potential curves for semiconductor electrodes were examined. Dogonadze[40] also considered charge-transfer processes in polar media. It was emphasized that the influence of the solvent cannot be described by the dielectric constant only, as implicit in the transition-state theory, but the polarization fluctuations of the solvent have to be taken into account. A theoretical treatment is given for the reactivity of solvated electrons on the basis of quantum mechanics. With Urushadze, Dogonadze[41] calculated the rate of chemical reactions taking place in a polar medium under kinetic conditions. An analytic expression was derived for the rate constant, transfer coefficient, and entropy of activation. Krishtalik,[42] in two consecutive papers, investigated ψ_1 effects on electrode processes in relation to Parsons' assumption that the effect of iodide and other anions was not associated with a ψ_1 effect but with an abrupt change in γ for the activated complex, although neither the ψ_1 effect nor changes in γ can explain the abrupt increase in the reaction rate in the absence of significant adsorption of hydrogen ions. The author was able to explain the fundamental facts associated with the ψ_1 effects both in the absence and presence of specifically adsorbed ions by separating the 'double-layer' cations into those found in its external shell and those in direct contact with the electrode. In the subsequent paper, dealing with charge transfer in relation to a heavy particle, the various effects related to the quantum character of the vibrations of the heavy particle were examined. The case where after charge transfer the co-ordinate of the heavy particle is far from equilibrium is considered, that is, where a greatly elongated bond is formed as a result of charge

---

[38] R. R. Dogonadze and A. M. Kuznetsov, *Elektrokhimiya*, 1971, **7**, 172.
[39] M. A. Vorotyntsev, R. R. Dogonadze, and A. M. Kuznetsov, *Elektrokhimiya*, 1971, **7**, 306.
[40] R. R. Dogonadze, *Ber. Bunsengesellschaft phys. Chem.*, 1971, **75**, 628.
[41] R. R. Dogonadze and Z. D. Urushadze, *J. Electroanalyt. Chem. Interfacial Electrochem.*, 1971, **32**, 235.
[42] L. I. Krishtalik, *Elektrokhimiya*, 1970, **6**, 1165, 1168.

transfer. From an approximate calculation it was clear that the total probability curve was practically constant and corresponded to $\alpha \sim 0.46$. The fact that α is practically constant in spite of an increase in ΔU_0 (the difference in initial and final states of the system when the magnitude of the vibrational quantum for the intermolecular vibrations of the reacting particles has been taken into account) is related to two compensating effects; the increase in the contribution from excited levels and the approach of the discharging particles to the electrode. The transfer coefficient has been discussed by Parsons,[43] who shows that the usual formal definition of α is satisfactory for the unambiguous description of an experiment. More detailed knowledge of the mechanism of an electrode reaction requires distinction between the intrinsic (true) α_I and the environmental part α_E of the overall coefficient α.

Diffusional Processes.—Diffusion in solution and associated phenomena have been the subject of a number of theoretical studies. Cheytanov[44] proposed new definitions of diffusion current and kinetic current which show that the kinetic current could not be explained by increasing the concentration gradient of the electroactive substance and that the latter was not modified by the superposition of a coupled chemical reaction on an electrochemical one. Hsueh and Newman[45] calculated the limiting current and concentration profile in a stagnant diffusion cell and compared it with that observed optically. Juodkazis and Visomirskis[46] have reported an electrochemical study of mass transfer to a vertical plate electrode. Some selected flowing systems were investigated in which limiting currents corresponding to laminar flow differed from those calculated using the Levich equation. An empirical correction factor was therefore introduced into the Levich equation. Roušar et al.[47] derived equations for calculating the local and average limiting current densities for an electrode located on the walls of a rectangular channel. The equation is complicated; however, tables of value are given. The validity of the equation for the average limiting current density was checked experimentally. The distribution of potential in a porous electrode under conditions of flow electrolysis has been treated by Sioda,[48] using a simplified model of the potential distribution in a porous electrode under limiting current conditions and with flowing solution. The model was based on several approximations, including the uniform flow velocity in the porous electrode. Equations were derived for the potential distributions in the solution filling the porous electrode, and for the difference of the potentials in the solution between the front and rear ends of the electrodes. Theoretical predictions agreed with experiment within a limited velocity range. Using numerical methods, Dunning

---

[43] R. Parsons, *Croat. Chem. Acta*, 1970, **42**, 281.
[44] H. Cheytanov, *Bull. Soc. chim. France*, 1970, 3777.
[45] L. Hsueh and J. Newman, *Electrochim. Acta*, 1971, **16**, 479.
[46] K. Juodkazis and R. Visomirskis, *Liet. T.S.R. Mokslu Akad. Darb: Ser. B*, 1968, **4**, 33.
[47] I. Rousar, J. Hostomsky, V. Cezner, and B. Stverak, *J. Electrochem. Soc.*, 1971, **118**, 881.
[48] R. E. Sioda, *Electrochim. Acta*, 1971, **16**, 1569.

The Solid Metal Electrode in Aqueous Solution 51

et al.[49] developed a model for the working of a porous electrode in which sparingly soluble reactants are present. Theoretical developments of the results of experimental techniques are of general application to the study of electrode reactions at solid metals. Rangarajan[50] proved the generalized form of the double-pulse galvanostatic method. Essentially this was shown to be equivalent to the reduction of the 'dual-pulse' problem to that of a single pulse. The measurement and analysis of impedance–admittance have been considered by a number of authors; Birke[51] deals with the operational admittance of an electrode process where an *a priori* dependence of rate constant on potential was not assumed. Using a linearized current–potential expression and considering double-layer charging and faradaic processes to be unseparated, simplification of the operational admittance yielded a term $\left(\frac{\partial q}{\partial E}\right)_{\Gamma_0, \Gamma_R}$ in parallel with the remaining admittance. The expressions for diffusional impedance and adsorption capacitance resulting from this treatment contain terms relating to the electrode reaction rates. MacDonald[52] gives expressions for the equivalent circuits for various situations of a galvanic cell with two plane parallel electrodes a short distance apart. A Laplace plane analysis of the impedance of faradaic and non-faradaic electrode processes, in which a unified theory of the impedance of electrode processes was used, has been given by Pilla.[53]

The potentiostatic method has been studied by a number of workers. Birke[54] has described the study of electrode processes in relation to the operational immittance and presented a treatment of the relationships between time domain and operational domain functions in both the Laplace and Fourier (frequency) domains. Response functions generated by the application of a potential impulse to various equivalent circuits for electrode processes are derived. In addition, a computerized system for the on-line calculation of operational impedances and admittances was discussed. Niki *et al.*[55] have described the determination of kinetic parameters from potential-step measurements with a digital computer. Johnson, Barnartt, and Glasser[56] continue a series of considerations of potentiostatic kinetics. In a contribution concerning potentiostatic kinetics at spherical electrodes, the i–t relation for an electrode reaction of any order was given to a good approximation by an equation of the form (δ and λ arise from the Laplace method):

$$i/i_{t=0} = \delta(1+\delta)^{-1} + (1+\delta)^{-1} \exp[\lambda^2(1+\delta)^2] \exp[\lambda(1+\delta)t^{\frac{1}{2}}]$$

[49] J. S. Dunning, D. N. Bennion, and J. Newman, *J. Electrochem. Soc.*, 1971, **118**, 1251.
[50] S. K. Rangarajan, *J. Electroanalyt. Chem. Interfacial Electrochem.*, 1971, **32**, 329.
[51] R. L. Birke, *J. Electroanalyt. Chem. Interfacial Electrochem.*, 1971, **33**, 201.
[52] J. R. MacDonald, *J. Electroanalyt. Chem. Interfacial Electrochem.*, 1971, **32**, 317.
[53] A. A. Pilla, *J. Electrochem. Soc.*, 1971, **118**, 1295.
[54] R. L. Birke, *Analyt. Chem.*, 1971, **43**, 1253.
[55] K. Niki, Y. Okuda, T. Tomonari, E. Buck, and N. Hackerman, *Electrochim. Acta*, 1971, **16**, 487.
[56] C. A. Johnson, S. Barnartt, and F. D. Glasser, *J. Electroanalyt. Chem. Interfacial Electrochem.*, 1970, **28**, 1; *J. Electrochem. Soc.*, 1971, **118**, 576.

This equation coincided closely with actual i–t curves over much of the reaction time, which is useful for the determination of the charge-transfer current density $i_{t=0}$. For a given electrode reaction the coincidence was extended by a judicious choice of the experimental variables. It was shown that the authors 'difference-ratio' method developed for first-order reactions applied equally well to reactions of higher order. In a later paper Johnson, Barnartt, and Glasser[56] have also shown that, in electrode reactions under potentiostatic control, the coupled concentration changes of two substances involved in the reaction can give rise to concentration reversion, when the absolute value of one substance passes through a maximum at some finite reaction time. It was shown that this behaviour could occur at spherical and cylindrical electrodes for both first- and higher-order reactions but that it cannot occur at planar electrodes. For the case of first-order reactions at spherical electrodes the experimental conditions which would give the reversion phenomenon were predicted by using closed-form solutions for the concentration–time behaviour.

The electrical field of a circular cathode and a system of plane anodes has been discussed by Ivanov and Yaroshko.[57] The method of conformal lattices, a combination of the methods of linear and conformal mappings, was used, and it was shown that a solution can also be obtained for the case in which a system of plane anodes is arranged on both sides of the cathode.

The correction for the ohmic resistance in potentiostatic experiments has been treated by a number of authors.[58–61] In the first of these papers Herrmann et al.[58] have shown that a correction can be made at low frequency when the resistance correction is approximately equal to the resistance of the electrolyte. Pilla[59] has concluded that ohmic compensation could be employed unambiguously in the time and as well as in classical a.c. frequency domain if the critical damping frequency is accurately known and the frequencies employed are sufficiently lower than this frequency; ohmic compensation could be advantageously employed in combination with the transient impedance technique over the entire accessible frequency range. For the elimination of resistance polarization and its compensation in potentiostatic studies,[60, 61] an a.c. having an amplitude proportional to the intensity of the d.c. flowing through the cell was superimposed on cell current; hence measured a.c. voltage, measured between the reference electrode and the working electrode, is proportional to the d.c. potential arising from the resistance of the solution. The a.c. voltage was amplified and added, after rectification, to the d.c. potential controlling the potentiostat. The potential of the working electrode was always maintained at a desired value independent of the resistance of the cell.

The potentiostatic technique has been applied by Backmann[62] to the

[57] V. T. Ivanov and N. M. Yaroshko, *Elektrokhimiya*, 1970, **6**, 1111.
[58] C. C. Herrmann, C. Lamy, and P. Malaterre, *Compt. rend.*, 1971, **273**, C, 1593.
[59] A. A. Pilla, *J. Electrochem. Soc.*, 1971, **118**, 702.
[60] J. Dévay, B. Lengyel, and L. Mészaros, *Magyar Kém. Folyóirat*, 1970, **76**, 209, 212.
[61] B. Lengyel and J. Dévay, *Magyar Kém. Folyóirat* 1970, **76**, 321.
[62] K. J. Backmann, *J. Electrochem. Soc.*, 1971, **118**, 226.

metal|metal ion electrode with two charge-transfer steps: Metal-alloy|M^{z+} and $M^{z+}|M^{(z+1)+}$ under charge-transfer and diffusion control. The theoretical treatment has been developed for the concentration of electrochemically active component in the alloy and of the corresponding metal ions of different valence states which form the redox couple in the electrolyte, respectively, as a function of the distance from the interphase and the duration of the potential step. Expressions for the partial current densities carried by the two consecutive charge-transfer reactions, $M = M^{z+} + ze^-$ and $M^{z+} = M^{(z+1)+} + e^-$, and for the total current density flowing through the cell have been evaluated. Analysis of the current density transients on the basis of those expressions allowed interpretation of experimental curves in terms of the kinetic parameters of the electrochemical system.

The advantages of charge measurements for the determination of kinetic parameters were the subject of a publication by Osteryoung and Osteryoung.[63] It was argued that chronocoulometry is a better technique than chronoamperometry for determining kinetic parameters because the charge measurements contain information at very short times, where the amount of kinetic information in the data is greatest. It was shown that in certain respects charge is a better experimental quantity than current for determining kinetic parameters, *i.e.* in relation to (i) the information content of data, (ii) the provision of derived kinetic parameters, and (iii) the correction of data for double-layer charging.

Voltammetry at solid electrodes may be briefly mentioned. Farsang *et al.*[64] have considered theoretical current–voltage curves. Valcher[65] described an apparatus for solid electrode voltammetry with periodic renewal of the diffusion layer. This involved measurements at null faradaic current coupled with controlled-current voltammetry. Alder, Fleet, and Kane[66] have reviewed and evaluated a wide range of substances as potential electrodes for use in anodic voltammetry.

Vondrák and Špalek[66a] described an electrolytic cell and the electronic circuitry for chronopotentiometric measurements using solid electrodes. The maximum current output was 2 A, with rise time in the range up to 1 µs; the circuit was claimed to be relatively free from oscillations, so that the potential of working electrodes may be obtained within one µs after the pulse has developed.

The effect of adsorption on the characteristics of potentiodynamic curves has been re-examined by Nesterov,[67] for it was pointed out that the previous treatment did not take into account the degree of electrode surface coverage. An equation was proposed having the form: $i_p = (A/\chi_T) \cdot C^0 v^{\frac{1}{2}}$, where A is the

---

[63] J. Osteryoung and R. A. Osteryoung, *Electrochim. Acta*, 1971, **16**, 525.
[64] Gy. Farsang, B. Rozsondai, and T. Tomcsànyi, *Magyar Kém. Folyóirat*, 1970, **76**, 223.
[65] S. Valcher, *J. Electroanalyt. Chem. Interfacial Electrochem.*, 1971, **31**, 349.
[66] J. F. Alder, B. Fleet, and P. O. Kane, *J. Electroanalyt. Chem. Interfacial Electrochem.*, 1971, **30**, 427.
[66a] J. Vondràk and O. Špalek, *Chem. listy*, 1970, **64**, 609.
[67] B. P. Nesterov, *Elektrokhimiya*, 1970, **6**, 881.

constant in the Delahay equation $i_p = A\, C^0\, v^{\frac{1}{2}}$ and χ_T is approximately constant.

Several papers have appeared during the review period dealing with the rotating disc and ring-disc electrode, and mainly concerned with the extension of theoretical principles without specific application to solid metal reactions.[68—75] The extensions include those relating to preceding and proceeding reactions,[68, 69] eccentricity,[70] hydrodynamic voltammetry,[71] surface roughness,[72] adsorption,[73] and current distribution in solutions of low conductance.[74] A further interesting extension relates to convective diffusion at a spherical electrode.[75] Kiss and Farkas[76, 77] have investigated the dissolution and deposition of metals by means of a rotating-disc electrode. A relationship was established for electrode processes composed of a number of steps in which various numbers of electrons interchange in single steps having an integral number difference of starting and final products. Equations were derived for the description of the electrode potential and reaction rate, depending upon which step is rate-determining. The theory of ring-disc electrodes has been extended by Albery and his group[78, 79] to cover transient currents and a.c. current measurements; size relationships of ring and disc to rotation velocity have been examined by Fujishima et al.,[80] and Nanis and Kesselman[81] have considered the engineering applications of current and potential distributions in ring-disc electrode systems. An area related to rotating electrodes is vibrating electrodes, treated by Lopatin et al.,[82] who have calculated the magnitude of an alternating diffusion current allowing for the electrode reaction rate.

The applications of fluidized beds in electrochemistry have been reviewed.[83] Fleischmann et al.[84] have presented the results of a study of the electrodeposition of copper in a fluidized bed of copper-coated spheres. The effective resistivity of the discontinuous metal phases was derived from measurements of the potential distribution in solution. The values observed agreed with

[68] V. A. Lapatin, B. M. Grafov, and V. G. Levich, *Elektrokhimiya*, 1971, **7**, 123.
[69] F. Aouanouk and M. Daguenet, *J. Chim. phys.*, 1970, **67**, 1956, 1959.
[70] M. B. Bardin and A. I. Dikusar, *Elektrokhimiya*, 1970, **6**, 1147.
[71] H. Matsuda and J. Yamada, *J. Electroanalyt. Chem. Interfacial Electrochem.*, 1971, **30**, 261, 271.
[72] M. Daguenet, M'H. Meklati, and G. Cognet, *Compt. rend.*, 1971, **272**, C, 1355.
[73] É. M. Podgaetskii and V. Yu. Filinovskii, *Elektrokhimiya*, 1970, **6**, 1178.
[74] W. J. Albery and M. L. Hitchman, *Trans. Faraday Soc.*, 1971, **67**, 2408.
[75] D-T. Chin, *J. Electrochem. Soc.*, 1971, **118**, 1434.
[76] L. Kiss and J. Farkas, *Acta. Chim. Acad. Sci. Hung.*, 1970, **66**, 33, 241.
[77] L. Kiss and J. Farkas, *Magyar Kém. Folyóirat*, 1970, **76**, 186, 287.
[78] W. J. Albery, *Trans. Faraday Soc.*, 1971, **67**, 153.
[79] W. J. Albery, J. S. Drury, and M. L. Hitchman, *Trans. Faraday Soc.*, 1971, **67**, 161, 166, 2162, 2169.
[80] A. Fujishima, H. Iketani, and K. Honda, *Bull. Chem. Soc. Japan*, 1970, **43**, 3949.
[81] L. Nanis and W. Kesselman, *J. Electrochem. Soc.*, 1971, **118**, 454.
[82] V. A. Lopatin, B. M. Brafov, and V. G. Levich, *Elektrokhimiya*, 1971, **7**, 120.
[83] P. Le Goff, F. Coeuret, and J. Bordet, *Ind. and Eng. Chem.*, 1969, **61**, 8.
[84] M. Fleischmann, J. W. Oldfield, and L. Tennakoon, *J. Appl. Electrochem.*, 1971, **1**, 103.

previously reported data for silver-coated particles. It was shown, by the comparison of the experimental data and theoretical predictions derived for the postulated system in which charge sharing occurred during single-particle elastic collisions, that a stagnation zone close to the feeder electrode was likely. Further evidence for this was obtained by the comparison of experimental and theoretical distributions of potential in the metal phase. Using aqueous 10^{-4} M-CuSO$_4$, the faradaic removal of copper under diffusion control followed the expected behaviour, although the mass-transfer coefficient indicated a high degree of turbulence within the bed. Scale-up factors of about 300 were considered possible in the processing of such dilute solutions. A final suggestion by the authors was that, to offset the high resistivity of the metal phase, practical systems should arrange for a mutually orthogonal current and fluid flow. The reduction of O_2 at a fluidized bed of solid silver spheres[85] may be noted here. It was shown by Janssen[85] that the current at which the overpotential is minimal depends on the bed expansion, and that it occurred at a bed expansion of 20 and 24% respectively for spheres of 1.0 and 0.61 mm diameter. The η–i relationship was linear over a large range of η and the results suggested a similarity between the fluidized bed and the suspension electrode. From the data provided, the active fluidized-bed electrode effectively consisted of a current feeder and those spheres less than 6 mm from the current feeder.

Optical Processes and Noise.—Optical methods of investigating interphases have been reviewed by Bockris, Damjanovic, and O'Grady.[86] It was pointed out forcibly by the authors that Mössbauer spectroscopy in electrochemistry was capable of making a contribution equal to that of the other established optical techniques. Watkins and Tvarusko[87] describe a Lloyd mirror interferometer which determines the refractive index gradient in a diffusion layer. Various modes of measurement were discussed which might be suitable for transparent or opaque electrodes. The time dependence of fringe movemen during electrodeposition from acid and neutral CuSO$_4$ solution was discussed, in addition to the effect of incidence angle and cathode displacement. For both solutions the concentration decreased linearly with $it^{\frac{1}{2}}$. Pleskov et al.[88] have studied the photoemissive effect at a Hg–electrolyte interphase.

Brodsky et al.[89] have related the photocurrent due to electron photoemission from metallic electrodes irradiated in the visible and near-u.v. to the structure of the double layer. From the results a method was proposed for the determination of the double-layer parameters from the deviations of the

---

[85] L. J. J. Janssen, *Electrochim. Acta*, 1971, **16**, 151.
[86] J. O'M. Bockris, A. Damjanovic, and W. E. O'Grady, *J. Colloid Interface Sci.*, 1970, **34**, 387.
[87] L. S. Watkins and A. Tvarusko, *Rev. Sci. Instr.*, 1970, **41**, 1860; *J. Electrochem. Soc.*, 1971, **118**, 580.
[88] Yu. V. Pleskov, Z. A. Rotenberg, and V. I. Lakomov, *Elektrokhimiya*, 1970, **6**, 1787.
[89] A. M. Brodsky, Yu. Ya. Gurevich, and S. V. Sheberstov, *J. Electroanalyt. Chem. Interfacial Electrochem.*, 1971, **32**, 353.

I–E characteristics of the irradiated electrode from the 'law of five-halves' for sufficiently concentrated solutions:

$$I = A\,[\hbar\omega - \hbar\omega_0(0) - eE]^{5/2}$$

where I is the photoemission current density, A is a constant, and $\hbar\omega_0(0)$ is the ϕ value at an arbitrary zero electrode potential. For dilute solutions it was shown that E should be replaced by $(E - \psi_1)$. A comparison of theoretical and experimental data was made for the adsorption of aliphatic alcohols, and from this it was found possible to determine the thickness of the adsorption layer and the electronic work-function for emission into the hydrocarbon layer. The effect of the adsorption layer on photoemission at small surface coverage was analysed using a zero range potential model, and a theoretical expression relating the photocurrent to the absorbate concentration was established.

Electrochemical noise analysis appears to be a valuable supplement to existing electrochemical kinetic studies and yields interesting information about electrode mechanisms. Tyagai[90] analysed faradaic noise, taking into account adsorption, diffusion, and homogeneous first-order chemical reactions with an electrochemical rate-determining step. It was evident from the treatment that the frequency dependence of the faradaic noise as well as its dependence on current and reactant concentration enabled the parameters of the various component steps in the electrode reaction to be calculated.

Electrocrystallization.—The theories of the electrocrystallization of metals have been discussed in a number of papers during the period under review. Despić[91] has treated problems associated with the electrodeposition of metals, with special interest being devoted to (i) location of discharge of metal ions onto a solid electrode, and (ii) the effect of diffusion on the morphology of metal deposits.

Experimental results quoted by Despić indicate that discharge on step edges is preferred due to much lower energies of activation. Other results confirmed the theory of the Bockris school, according to which in diffusion-controlled processes an exponential increase in surface roughness with time is to be expected. The crystalline roughness of electrodeposited metals was also considered by Gnusin,[92] who showed that the roughness could be treated as the result of an unordered accumulation of crystals growing during the deposition and differing in size. Pangarov[93] has predicted the orientation of silver crystallites on a platinum base from theoretical principles and verified the theory in practice. The dependence of orientation and formation rate on the type of two-dimensional nuclei, crystal structure, and crystallization overpotential was established. Mamontov[94] discussed the effect of

[90] V. A. Tyagai, *Electrochim. Acta*, 1971, **16**, 1647; *Elektrokhimiya*, 1971, **7**, 69.
[91] A. R. Despić, *Croat. Chem. Acta*, 1970, **42**, 265.
[92] N. P. Gnusin, *Elektrokhimiya*, 1971, **7**, 72.
[93] N. A. Pangarov, *Zashch. Metal*, 1969, **5**, 467.
[94] E. A. Mamontov, *Trudy Kishinev. Sel'skokhoz Inst.*, 1970, **59**, 163.

The Solid Metal Electrode in Aqueous Solution 57

structural defects on the rate of electrochemical reactions on the basis of delayed discharge; however, Zubov et al.[95] have concluded that the true rate of electrodeposition of metals under the conditions of 'self adaptation' of their growing surfaces to the polarizing current strength is limited by the diffusion of foreign particles. Braunsburger[96] has described the electrocrystallization of a single metal ion electrode under a.c. near the equilibrium potential. The dependence of the effective valence (n_i) on the time for various mechanisms of electrode reactions was examined by Malodov et al.,[97] and it was shown that to establish the mechanism both n_i itself as well as the dependence of n_i on t are of great significance. The conditions for obtaining undistorted polarization curves, during analysis of which multistage criteria can be used, were discussed. The results of radiochemical experiments, using copper and indium amalgams, for the dependence of n_i on t during the anodic dissolution reaction showed that the dependence confirmed a multistep mechanism for the dissolution process. It was concluded that generally, for systems with low concentrations of intermediate species, polarization data for steady- and non-steady-state experiments should coincide, whereas for systems with high concentrations of intermediates this coincidence is not to be expected. A paper by Davydov et al.[98] describes peculiarities of the anodic activation–passivation phenomena, explaining the connection between passivation and the diffusion conditions at the electrode surface, and they further discuss the possibility of convective passivation. The electrocrystallization process for both films and metals has been discussed in connection with recently proposed models.[99,100] The processes are classified as (i) those with slow step nucleation and growth, (ii) those involving surface diffusion, and (iii) those involving diffusion–precipitation.

Results concerning special cases wherein the centres interact solely with themselves or with the boundaries and those which interact with bulk diffusion were presented. Theoretical models for these processes have been critically examined. Armstrong[100] argues that the most useful technique for distinguishing between the various mechanisms is the observation of the dependence on rotation speed of the steady current–voltage curve. For irreversible dissolution the curve will be independent of rotation speed for a solid-state reaction, whilst for a dissolution–precipitation mechanism a significant dependence on rotation speed will be found since the surface concentration of the metal cations will be reduced. For the case of a reversible dissolution the curves are dependent on rotation speed irrespective of the mechanism; however, in this case a decision of mechanism may be made on the basis of the effect of rotation speed on the active–passive transition, which is dependent

[95] M. S. Zubov, D. V. Fedoseev, L. A. Uvarov, and A. T. Vagramyan, *Elektrokhimiya*, 1970, **6**, 1690.
[96] S. Braunsburger, *Z. phys. Chem.* (*Frankfurt*), 1970, **69**, 221.
[97] A. I. Malodov, V. I. Barmashenko, and V. V. Losev, *Elektrokhimiya*, 1971, **7**, 18.
[98] A. D. Davydov, V. D. Koshcheev, and B. N. Kabanov, *Elektrokhimiya*, 1970, **6**, 1760.
[99] R. D. Armstrong, J. A. Harrison, and H. R. Thirsk, *Trans. SAEST*, 1970, **5**, 65.
[100] R. D. Armstrong, *Corrosion Sci.*, 1971, **11**, 693.

on rotation speed in the case of dissolution–precipitation and not in the other case.

Stress in Electrodeposits.—Stress in electrodeposits is important industrially. Sykes and Rothwell[101] showed that even in null-deflection methods for stress measurement there was some stress relaxation arising from the bending of the test strip. Similarly there would be some relaxation because constraint was not applied across the strip (*i.e.* the strip is not clamped), but for thin deposits this will only lead to a small decrease in measured stress. It was concluded after a theoretical argument that the only method that properly represents true plating conditions is that in which the specimen is clamped completely flat during plating and the curvature measured upon subsequent release. For continuous measurement some stress relaxation in the deposit cannot be avoided. Dvořák and Vrobel[102] have described a new method for measurement of internal stress in electrodeposits which involves plating both sides of a thin steel strip, accurate measurements of changes in its length being simultaneously made by means of a precision dial gauge. Internal stress was calculated from measured changes in length. A feature of electroforming is the low bond strength between the base metal and the deposit. Consequently, when subjected to bending deformations, the deposit may exhibit shear (creep) relative to the base metal. Gofman and Moldaver[103] have considered that the degree of creep of the deposit for similar (bending) deformations is a measure of the deposit to base-metal bond strength. Consequently there exists a quantitative relationship between the deposit creep and the strength of this bond. No previous work has been published concerned with this relation and the authors have studied this quantitative relation between deposit creep and the strength of the bond between deposit and base metal. Silicon strain-gauge measurements provided a sensitive means of studying these problems, and encouraging results are reported to have been obtained.

4 Reactions of Individual Metal Systems

Aluminium.—Papers on aluminium have mainly been concerned with the formation of films on the metal. An exception is the work of DiBari and Read[104] on the electrochemical behaviour of high-purity aluminium in chloridic solutions. Here the corrosion diagram was consistent with the observed relation between the corrosion potential and pH. The rate of dissolution of high-purity aluminium was found to increase slightly between pH 0 and 4, to decrease between pH 4 and 8, and to increase from pH 8—14. The minimum at pH 8 in the rate of dissolution may correspond to the point at which the formation of $Al(OH)_3$ on the electrode surface equals the rate of formation of the aluminate ion.

[101] J. M. Sykes and G. P. Rothwell, *J. Electrochem. Soc.*, 1971, **118**, 91.
[102] A. Dvořák and L. Vrobel, *Trans. Inst. Metal Finishing*, 1971, **49**, 153.
[103] Ya. A. Gofman and T. I. Moldaver, *Elektrokhimiya*, 1970, **6**, 860.
[104] G. A. DiBari and H. J. Read, *Corrosion*, 1971, **27**, 483.

The Solid Metal Electrode in Aqueous Solution

Alloys.—The electrochemical behaviour of alloys has been the subject of a large number of papers published during the review period. Wood and Khoo[105] have surveyed the various mechanisms by which metals in general oxidize to give barrier-type anodic films. The alloy constituents enter the resulting film virtually in their alloy proportions, and a further interesting conclusion was that there appeared to be substantial anionic movement during film growth. The anodic behaviour of brasses in ammoniacal solutions has been considered by Hoar and co-workers.[106, 107] The results presented show that Cu_2O can be formed on the alloy surface by both anodic and cathodic reactions at various pH values. For the yield-assisted dissolution of brass,[107] the greatest effect occurs at pH 7.3, since broken protective surface films cannot readily re-form. The corresponding studies of activation kinetic analyses suggest that static metal dissolves from few sites with low activation energy but that this mechanism is overshadowed on metal that is yielding under stress by dissolution from many sites with higher activation energy. Flatt and Brook[108] have studied $i-E$ data and corresponding films on copper, zinc, and brass in chloridic solutions. The surface finishes were found to be related to the anode potential, and the results showed that copper anodes were covered with a film of Cu_2Cl_2 at all potentials just above the rest potential but that on zinc the only film is metallic zinc. On brasses, duplex films of zinc and Cu_2Cl_2 are found over a copper-rich alloy. The results support the theory that copper and zinc are oxidized simultaneously at all potentials more positive than the rest potential. In a further paper[109] the concentration profiles in solution at dissolving brass surfaces are examined by freezing techniques.

Nickel-containing alloys have been investigated.[110—113] Aleikina et al.[110] have made anodic potentiodynamic experiments on Ni–Al alloys in both acid and alkali. The preferential removal of aluminium from alloys leads to the formation of a defect lattice in the surface layer, which gradually becomes enriched with nickel. The influence of the phase composition of Ni–Al alloys on the electrochemical properties of nickel-based electrode catalysts has been considered by Dousek and Jansta.[111] The electrodeposition of nickel alloys has been investigated.[112, 113] Experiments with a.c. superimposed on d.c. indicated that it was possible, by the proper choice of conditions, to deposit Ni–Fe alloys with a uniform composition throughout the deposit thickness, and also that the incorporation of hydroxides into the deposit may be avoided. From the same electrolyte solution the composition of an alloy can be varied

[105] G. C. Wood and S. W. Khoo, *J. Appl. Electrochem.*, 1971, **1**, 189.
[106] T. P. Hoar, J. J. Podesta, and G. P. Rothwell, *Corrosion Sci.*, 1971, **11**, 231.
[107] J. J. Podesta, G. P. Rothwell, and T. P. Hoar, *Corrosion Sci.*, 1971, **11**, 241.
[108] R. K. Flatt and P. A. Brook, *Corrosion Sci.*, 1971, **11**, 185.
[109] R. K. Flatt, R. W. Wood, and P. A. Brook, *J. Appl. Electrochem.*, 1971, **1**, 35.
[110] S. M. Aleikina, I. K. Marshakov, A. B. Fasman, and I. V. Vavresyuk, *Elektrokhimiya*, 1970, **6**, 1648.
[111] F. P. Dousek and J. Jansta, *Coll. Czech. Chem. Comm.*, 1971, **36**, 2115.
[112] Z. Kovac, *J. Electrochem. Soc.*, 1971, **118**, 51.
[113] R. S. Vakhidov, V. I. Popov, and A. A. Starchenko, *Elektrokhimiya*, 1970, **6**, 1720.

in a rather wide range (8—60% Fe) merely by varying the frequency. Vakhidov et al.[113] have described an electrolyte for the electrodeposition of Ni–P alloys. In connection with experiments on Ni–Fe alloy electrodeposition, Fedoseeva et al.[114] have proposed a satisfactory method for the approximate calculation of the percentage composition of an alloy, taking into account the effect of the electrode surface condition and the rates of the linked electrochemical reactions. It was shown from an analysis of the equation proposed that the proportion of the more electropositive elemental component of the alloy should increase with increase in temperature. The experimental results showed that at low temperature deposits contained mostly iron, whereas with increasing temperature the amount of nickel in the deposit increased. At temperatures above 100 °C the deposition of nickel predominated. A detailed sequence of experiments showed that the percentage composition of an alloy under given experimental conditions could be predicted.

Lead alloys are of interest in connection with their applications in energy storage cells; Burbank[115] has shown that apparently the discharge of antimonial lead coatings appears to be limited by the crystal growth rate. Azim and El-Sobki[116] have reported the electrochemical behaviour of Pb–Cd alloys in H_2SO_4 solutions. The effects of alloy composition, acid concentration, stirring, and anealing on the dissolution rates in the active and passive state were investigated. A linear relationship was reported between the critical current for passivation and the alloy composition. At compositions 1.8—17.3% cadmium (eutectic) the electrode behaviour is dominated by lead, and at higher cadmium content by cadmium. Johnson et al.[117] have studied Pb–Tl alloys. In H_2SO_4 (0.05 mol l^{-1}) solutions the nature of the film on the surface (oxide or sulphate) depended on the potential. In the PbO_2 (1.52—1.97 V) region the relative amounts of PbO_2 on the surface decreased with increasing amounts of thallium in the alloy, and PbO (tetragonal) developed on the surface. It was concluded that the presence of Tl^{III} in the electrolyte indicated that the PbO_2 may have been reduced by the Tl^I–Tl^{III} redox couple.

Other alloys have been mentioned, and Miyashita and Kurihara[118] have described solutions for the deposition of a Sn–Co alloy. Woods[119] discusses the surface composition of Pt–Au alloys and Tourky et al.[120] present data for the anodic and cathodic behaviour of Cu–Ag alloys.

The formation of intermetallic compounds of sodium and cadmium by cathodic penetration has been studied by Kabanov and co-workers[121] by investigating the decomposition of cathodically formed intermediates when

---

[114] T. A. Fedoseeva, L. A. Uvarov, D. V. Fedoseev, and A. T. Vagramyan, *Elektrokhimiya*, 1970, **6**, 1841.
[115] J. Burbank, *J. Electrochem. Soc.*, 1971, **118**, 525.
[116] A. A. A. Azim and K. M. El-Sobki, *Corrosion Sci.*, 1971, **11**, 821.
[117] J. W. Johnson, P. J. Aragon, and W. J. James, *Corrosion*, 1971, **27**, 107.
[118] H. Miyashita and Sh. Kurihara, *J. Metal Finishing Soc. Japan*, 1970, **21**, 79.
[119] R. Woods, *Electrochim. Acta*, 1971, **16**, 655.
[120] A. R. Tourky, L. A. Shalaby, and M. Elsaid, *J. Chem. U.A.R.*, 1969, **12**, 41.
[121] I. G. Kiseleva, B. N. Kabanov, and D. N. Machavariani, *Elektrokhimiya*, 1970, **6**, 905.

only the more negative component – the alkali metal – ionizes. It was concluded that the intermetallic compound was formed by the penetration of alkali-metal atoms into the cathode interior along intercrystalline boundaries, at which points plane crystals of the intermetallic compound form. Crystals grow to such a size that intersection of a new boundary occurs when growth becomes retarded. The penetration conforms to the parabolic growth law $a^2 = Pt$, where P = penetration coefficient and a = thickness of growing layer. The sodium penetration coefficient was found to be 5×10^{-10} cm^2 s^{-1}. The self-dissolution of intermetallic compounds of antimony is of importance in the semiconductor field. Polarization curves at low current densities showed[122] that for InSb, CdSb, and ZnSb the anodic dissolution was controlled by antimony dissolution, and the curves were in the same region as those for antimony. As the potential was made more positive, dissolution became more uniform. The dissolution of Cd$_4$Sb$_3$ and Zn$_4$Sb$_3$ was rather more complicated.

Bismuth.—Ammar and Khalil[123] have investigated the anodic oxidation of bismuth. In H$_2$SO$_4$ solutions (0.005—3 mol l^{-1}) anodic charging curves were measured at 30 °C, using a variety of experimental procedures. The anodic behaviour depended markedly on the acid concentration. In the higher concentration range the dissolution followed a Tafel relation. In dilute solution, growth of oxide film occurred and the potential rose rapidly with time, reaching over 100 V at sufficiently high current densities. Sparking and oxide breakdown occurred at ~150 V. Oxide growth followed the high-field ionic conduction; the reciprocal capacitance was linearly related to log (current density). The effect of halide ions indicated that oxide films could be grown on bismuth to an appreciable thickness, the film thickness depending upon the halide, the concentration of sulphuric acid, and the magnitude of the current density. In solutions (0.05 mol l^{-1}) of Na$_2$SO$_4$, Na$_2$CO$_3$, and NaOH the anodic behaviour was dominated by anodic film growth; however, it was noted that at higher temperatures this film growth is retarded due to the increased solubility of the oxide. A paper by Weber[124] deals with the electrolytic deposition of bismuth from iodide solutions. Bismuth was deposited from stagnant complex iodide solutions as a powder; however, with vigorous stirring this was prevented and good coatings were obtained. Bismuth was present in the electrolyte as Bi^{II} and Bi^{III}, and about 30% I$^-$ was present in the deposit. An analysis of cathodic efficiencies showed that bismuth was apparently deposited mainly as (Bi I)$^+$. The behaviour of bismuth as a valve metal where solid-phase film growth is the main process has been considered.[125] The electrosorption of bismuth on palladium in 1 M-HClO$_4$

[122] I. K. Marshakov, Ya. A. Ugai, S. M. Aleikina, T. A. Bessonova, and I. V. Vavresyuk, *Elektrokhimiya*, 1970, **6**, 1865.
[123] I. A. Ammar and M. W. Khalil, *Electrochim. Acta*, 1971, **16**, 1379, 1601.
[124] J. Weber, *Prace Inst. Mechan. Precyzyjnej*, 1967, **15**, 10.
[125] I. A. Ammar and M. W. Khalil, *J. Electroanalyt. Chem. Interfacial Electrochem.*, 1971 **32**, 373.

has been studied in detail by Mikuni and Takamura.[126] Potential-sweep voltammetry was used in addition to constant potential–current curves, which showed several peaks ascribable to Bi^{3+}. When the potential of the electrode was more negative than 0.22 V, faradaic adsorption of bismuth proceeded *via* the reaction $Bi^{3+} + 3e^- = Bi_{ad}$. The electrode, when kept at a sufficiently negative potential, gave several anodic current peaks during successive anodic potential sweeps. The height of these anodic peaks depended on the bismuth ion concentration and the other experimental variables. X-Ray analysis showed that anodic peaks more positive than 0.8 V were due to the dissolution of Bi–Pd alloy. The main cathodic peak was attributed to the reduction of the oxides of palladium, but when bismuth was adsorbed on palladium the magnitude of the current peaks increased and the position shifted in the negative direction. This change was explained by the increase of the surface area and by the co-deposition of bismuth and palladium.

Cadmium.—Hampson and Latham[127] have examined the exchange reaction at a polycrystalline cadmium electrode in electrolyte solutions based on the perchlorate ion at pH ~3. The faradaic impedance method was used to estimate exchange currents, and the apparent transfer coefficient (from a change of i_0 with cupric ion concentration) was estimated as 0.35. The apparent enthalpy of activation for the charge-transfer process was ~29 kJ mol^{-1}. The process of exchange was controlled by the release (and inclusion) of atoms from the lattice and their diffusion to sites on the electrode surface at which charge transfer occurs with transport across the double layer. For the exchange in alkaline solution it was reported,[128] from the results of faradaic impedance experiments, that the reaction was complicated by the formation of a surface film which progressively retarded the electrode reaction. The presence of a film at the cadmium electrode in alkali is emphasized by Armstrong and West,[129] who studied ring and ring-disc electrodes in 1 and 10 M-KOH. A region of active dissolution was observed prior to film formation, which occurred through a solid-state mechanism. Film thickening followed an approximately parabolic law at short reaction times; however, at longer times a steady-state condition was reported when the dissolution of the film at the outer surface was balanced by the rate of formation. Because of the reversibility of film dissolution the steady-state current and film thickness were functions of the mass-transport conditions at the electrode surface. The impedance of a cadmium electrode covered with a relatively thick film of $Cd(OH)_2$ (passive) in 0.1—5 M-KOH has been studied by Gravchev and L'Vova,[130] and it was concluded that the

---

[126] F. Mikuni and T. Takamura, *Denki Kagaku*, 1971, **38**, 237.
[127] N. A. Hampson and R. J. Latham, *J. Electroanalyt. Chem. Interfacial Electrochem.*, 1971, **32**, 175.
[128] N. A. Hampson and R. J. Latham, *J. Electroanalyt. Chem. Interfacial Electrochem.*, 1971, **32**, 337.
[129] R. D. Armstrong and G. D. West, *J. Electroanalyt. Chem. Interfacial Electrochem.*, 1971, **30**, 385.
[130] D. K. Gravchev and L. A. L'Vova, *Elektrokhimiya*, 1971, **7**, 230.

observed frequency plots arose only from a dispersion of the impedance components, so that the faradaic impedance was not a satisfactory method to apply in studying the passage of current through passivating layers. Kang,[131] for the case of cadmium in 8 M-KOH solution, has shown that at low temperature the passivation time (t_p)–current density data could be correlated in the form $i\,t_p^{\frac{1}{2}} = a+bi$ (a and b were constants). The relationship was taken as support for a dissolution–precipitation model, and if this is really the case the passivation time is controlled by the rate of precipitation and/or the rate of mass transfer of dissolved cadmium species.

Porous cadmium electrodes have been the subject of papers by Bro and Kang[132] and Vijayavalli et al.[133] Bro and Kang[132] have discharged porous cadmium electrodes in KOH (30%) at 25 °C under galvanostatic conditions. Electrodes were made by sintering loose powder compacts in 3 M-HCl. Using conventional electrochemical techniques, current densities between 2 mA cm^{-2} and 200 mA cm^{-2} were used, and discharge profiles were obtained by the chemical analysis of thin sections sliced from the electrode. The discharge profiles could be described by a hyperbolic cosine function:

$$q/q_0 = A\,[\cosh B\,(1-x)]/\cosh B$$

where q = discharge density/mAh cm^{-3}, q_0 = stoicheiometric charge density of 'fresh' anode/mAh cm^{-3}, A,B are constants, and x = fractional depth in anode from front surface. The product formed during discharge remained inside the electrode and eventually choked the porous structure. The discharge of the cadmium anodes was limited by both faradaic and transport factors associated with the discharge. Vijayavalli et al.[133] studied the efficiency of charging of porous cadmium electrodes in the presence of an imposed a.c. Vinet et al.[134] have reported the results of a study of the anodic redissolution of deposits of cadmium on a gold vibrating electrode. The results may provide a basis for the determination of Cd^{2+} in the 10^{-4} mol l^{-1} range.

Cobalt.—The anodic oxidation of cobalt in KOH electrolytes has been examined by Behl and Toni.[135] Cyclic voltammetry and electrometric experiments at rotating disc and rotating ring-disc electrodes were employed in 0.2—8.0 M-KOH. The equilibrium potentials of several associated electrode reactions were calculated from thermodynamic data and used to support an overall mechanism for the electro-oxidation process. It was shown that this occurred only in the solid phase, to produce a $Co(OH)_2$ film which was subsequently oxidized to Co_3O_4 and $CoOOH$. Experiments with rotating electrodes indicated that the species predominantly produced in the KOH solutions were Co^{2+} species arising from dissolution of $Co(OH)_2$. The

---

[131] H. Y. Kang, *J. Electrochem. Soc.*, 1971, **118**, 462.
[132] P. Bro and H. Y. Kang, *J. Electrochem. Soc.*, 1971, **118**, 519.
[133] R. Vijayavalli, P. V. Vasudeva Rao, and H. V. K. Udupa, *Electrochim. Acta*, 1971, **16**, 1197.
[134] D. Vinet, C. Ducauze, and D. Jacquot, *Bull. Soc. chim. France*, 1971, **1**, 340.
[135] W. K. Behl and J. E. Toni, *J. Electrochem. Soc.*, 1971, **31**, 63.

ultimate passivation of the cobalt electrode was attributed to the formation of a thin layer of CoO next to the metal surface. A reactivation of the cobalt electrode was observed to occur in 5.0 and 8 M-KOH solutions in stationary conditions as well as at rotating electrodes at all potentiodynamic scanning rates studied, whereas in 1.0 and 2.5 molar solutions reactivation was observed only at very low scan rates (5—10 mV s^{-1}). A reaction scheme was proposed for the process of electro-oxidation:

$$Co + 2OH^- \rightarrow Co(OH)_2 + 2e^- \rightarrow Co(H_2O)_x^{2+} \text{ in solution}$$
$$Co + 2OH^- \rightarrow CoO + H_2O + 2e^-$$
$$3\,Co(OH)_2 + 2OH^- \rightarrow Co_3O_4 + 4H_2O + 2e^-$$
$$Co(OH)_2 + OH^- \rightarrow CoOOH + H_2O + e^-$$
$$CoOOH + OH^- \rightarrow CoO_2 + H_2O + e^-$$

Experimental results indicated that although the oxidation of Co(OH)$_2$ to CoOOH occurred readily, the reduction of CoOOH to Co(OH)$_2$ did not occur at any appreciable rate. The charge used in the oxidation of cobalt was not completely recovered and the electrode was always covered either with a brown film (CoOOH) or blue film [Co(OH)$_2$], depending on the experimental potential-scanning conditions. Sukhatin and Pyzhkov[136] have made a study of the anodic behaviour of cobalt in trichloroacetic acid solutions. It was found that a change of pH has a more pronounced effect than that of the total acid concentration.

Copper.—By far the largest group of papers published during the review period have concerned copper. Most of these have been concerned with aspects of electrodeposition; however, a few more fundamental papers have appeared and these will be dealt with first. Hampson and Latham[137] reported the results of an investigation of the Cu^{II}–Cu exchange in sulphate electrolytes. There was evidence at the equilibrium potential for the participation of Cu^{I} species, and it was argued that the reaction clearly involved adspecies and was also complicated by the adsorption of the anion. The enthalpies of activation for the three processes of charge transfer, adatom diffusion, and adsorption were ~31, ~51, and ~27 kJ mol^{-1}, and these values indicate that the adatom incorporation step is energetically the most difficult. The interpretation of the impedance of a copper electrode in CuSO$_4$ solution has also been studied by Grachev.[138] Eichkorn, Fischer, and Mache[139] have analysed the overpotential–time curves corresponding to the galvanostatic depositions of copper. Using a current-step technique, the crystallization overpotential was separated from the total overpotential. Kinetic equations were developed which included both the discharge of Cu^{2+} ions in two steps and

---

[136] A. M. Sukhatin and E. M. Pyzhkov, *Zashch. Metal*, 1969, **5**, 559.
[137] N. A. Hampson and R. J. Latham, *Trans. Faraday Soc.*, 1971, **67**, 1440.
[138] D. K. Grachev, *Elektrokhimiya*, 1971, **9**, 527.
[139] G. Eichkorn, H. Fischer, and H-R. Mache, *Ber. Bunsengesellschaft phys. Chem.*, 1971, **75**, 482.

The Solid Metal Electrode in Aqueous Solution 65

the permanent change in the surface morphology during the deposition. By examination of transient curves from experiments in which fine crystalline copper deposits were formed and comparison with those in which relatively coarse copper deposits were formed, a characteristic increase in η_k was observed which was tentatively ascribed to an increase in the surface excess of adatoms. The anodic dissolution of copper in acid sulphate solution has been considered[140] and the results are in the main complementary to those of Nigretto and Jozefowicz[141] on the reduction of Cu^{II} at rotating copper electrodes. It was considered that the primary step was $Cu^{2+} + e^- \rightleftharpoons Cu^+$ in the electrochemical reduction of Cu^{II} salts (in the presence of alkali-metal sulphates) at a rotating copper electrode on which had been predeposited a layer of copper. The results indicated that in certain circumstances the primary process could be masked by other processes such as Cu^I disproportionation.

An interesting paper by Koryushin[142] dealt with the electrodeposition of copper from acid $CuSO_4$ solution under non-isothermal conditions. The technique used was to examine the effects of the direction and intensity of the temperature field and of deposit thickness on the condition of the surface of the metal deposit. The way in which a temperature field affects the metal electrodeposition is not understood. However, it was possible to show that in the case of the electrolysis with a heat flux directed from the cathode surface into the electrolyte a smoothed deposit of copper was obtained, in contrast to that obtained under isothermal conditions. Haruyama et al.[143] have investigated the effect of plastic deformation on the electrochemical behaviour of the Cu/Cu^{2+} system. The changes in reversible potential on plastic deformation could be explained in terms of the anodic dissolution of active sites formed by the deformation. The exchange current density on an annealed specimen decreased with increasing stress, whilst that on a rolled specimen was apparently unchanged by increasing stress, a result difficult to explain by existing theory. A coupled morphological and impedance study with parallel rotating-disc measurements has been recorded by Gorbunova and Tkachik[144] on the (111) plane of a copper single crystal. At high overpotentials the deposition process was considered to be controlled by the rate of the one-electron transfer step $Cu^{2+} \rightarrow Cu^+$. Close to the equilibrium potential there was evidence of mass-transfer control, i.e. control by hindered transport of Cu^+ ions from the bulk to the electrode surface and by hindered diffusion of copper adatoms on the surface to the sites of their incorporation into the lattice. The low-defect-density electrodeposits of copper on single crystals have been investigated by Bertocci and Bertocci.[145] The orientation of the substrate has been found to affect the perfection of the deposits, and it was concluded that the mechanism of step generation (where orientation is very

[140] L. Kiss and J. Farkas, *Acta Chim. Acad. Sci. Hung.*, 1970, **66**, 395.
[141] J. M. Nigretto and M. Jozefowicz, *Electrochim. Acta*, 1971, **16**, 297.
[142] A. P. Koryushin, *Elektrokhimiya*, 1970, **6**, 1844.
[143] S. Haruyama, S. Asawa, and K. Nagasaki, *Denki Kagaku*, 1971, **39**, 564.
[144] K. M. Gorbunova and Z. A. Tkachik, *Electrochim. Acta*, 1971, **16**, 191.
[145] U. Bertocci and C. Bertocci, *J. Electrochem. Soc.*, 1971, **118**, 1287.

close to {111}) was random nucleation. The morphology of copper electrodeposits obtained under potentiostatic control was studied by Budewskii et al.[146] The influence of anions and additives on the nature (strength, morphology, stress, etc.) of copper deposits has been the subject of a number of papers in the review period. Gurevich and Pomosov[147] have reported on the influence of halides on the formation of friable copper deposits, and it was shown that the effect was governed both by the adsorption of the anions themselves at the cathode and by ultra-small particles of the dispersed phase (CuCl and CuI) that formed in the solution. The effect of organic additives containing —SO_2OH, —NO_2, —CO_2H, and amino-groups on the cathodic polarization curves of copper in $CuSO_4$ electrolytes have been discussed by Russian workers;[148] all the compounds tested except those containing —SO_2OH groups shifted the curves in the negative direction. Shapnik et al.[149] made a potentiometric study of the copper deposition from baths containing ethylenediamine, $CuSO_4$, Na_2SO_4, and $Na_2C_4H_4O_3$ at pH 4.4—11.9. The formation of complexes between ethylenediamine and other compounds as a function of concentration was discussed. Loutfy and Sukava[150] examined the effect of monocarboxylic acids on the electrodeposition of copper. The effects of monocarboxylic acids on electrosorption and cathodic overpotential were studied at current densities up to 20 mA cm^{-2}. By the application of Traubes' rule, the separate adsorption free-energy contributions of the carbonyl and methylene groups in the additive molecule were calculated. These were obtained from dipole–dipole interaction free-energy, which results in a coverage-lateral interaction free-energy between adjacent carbonyl groups in the adsorbed phase. The carboxy-group contribution was found to be -1570 cal mol^{-1} at zero coverage whilst the —CH_2— contribution was -704 cal mol^{-1} and independent of coverage. Walker and Benn[151] have noted that copper deposits produced with ultrasonic vibrations from benzonitrile-containing acid $CuSO_4$ solutions contained more benzonitrile than from a simple stirred bath. Because vibrations also raise the limiting current density and give smoother and finer surfaces, and reduce the internal stress and brittleness, the workers have recommended that ultrasonics should be used to produce smooth, hard, copper deposits. The effect of the addition agents thiourea and benzonitrile in electrolyte solutions (0.7 M-$CuSO_4$+1.5 M-H_2SO_4) in conditions of turbulent flow was studied by Gabe and Robinson.[152] The incidence of powdery or dendritic deposits has been related to a transition

---

[146] E. Budewskii, V. Bostanov, and P. Rhotenbacher, Z. Metallk., 1970, 60, 840.
[147] L. I. Gurevich and A. V. Pomosov, Elektrokhimiya, 1971, 7, 158.
[148] L. Valentukeviciute-Sliesaraviciene, J. Matulis, and J. Bubelis, Liet. T.S.R. Mokslu Akad. Darb. Ser. B, 1968, 3, 65.
[149] M. S. Shapnik, N. V. Gudin, E. M. Gamburg, A. V. Il'Yasov, and N. N. Satrikova, Zashch. Metal., 1969, 5, 434.
[150] R. O. Loutfy and A. J. Sukava, J. Electrochem. Soc., 1971, 118, 216.
[151] R. Walker and R. C. Benn, Electrochim. Acta, 1971, 16, 1081.
[152] D. R. Gabe and D. J. Robinson, Trans. Inst. Metal Finishing, 1971, 49, 17.

from charge-transfer to mass-transfer control which is thought to occur at ~ 0.41 I_L (I_L = limiting current density). In practice and with thin deposits the transition appears more likely at ~ 0.7 I_L. Whilst additions of thiourea and benzotriazole in acid $CuSO_4$ both affect the deposit structure, they do so by different mechanisms and have little effect on the onset of dendritic or powdery growths. Galinker et al.[153] have examined the effect of some organic additives on the electrodeposition of copper from a tripolyphosphate electrolyte solution. Thiourea and diphenylguanidine were shown to increase dispersibility and grain size and to improve the microhardness of the copper deposit. It was reported that in the presence of thiourea the activation enthalpy for the discharge of Cu^{2+} ions is higher than in the presence of diphenylguanidine. The micro-relief of electrolytic deposits of copper was investigated by Kovarskii and Golubev,[154] who found that in the absence of micro-ripple of the plating surface the micro-relief observed with copper systems was similar to those obtained with tin and zinc systems. The electrodeposition of copper onto steel has been studied by Inui et al.,[155] using the ethylenediamine bath. The formation of $[Cu(edta)]^{2+}$ and $[Cu(edta)]_2^{2+}$ in the electrolyte solution was detected. The results showed that bright and well-adhering deposits are obtained from $[Cu(edta)]_2^{2+}$ solutions and that the best electrolysis conditions obtained at pH 6.5—8, at temperatures within the range 10—50 °C, and at current densities of 1.5—2.5 A dm^{-2}, all of which gave a current efficiency of 95%. Šinkovic and Leskovšek[156] have determined the mean limiting current density and current distribution on 500 mm high vertical copper electrodes with natural convection in 0.1 M-$CuSO_4$+0.5 M-H_2SO_4 under potentiostatic conditions. Twenty-segment electrodes were used to determine the vertical current density distribution by weighing the copper deposited on each segment. The laminar and non-laminar flow regions along the cathode in natural convection were characterized by dimensionless numbers.

A rotating ring-disc study of the underpotential corresponding to copper deposition on platinum from 0.5 M-HCl solutions has been made by Cadle and Bruckenstein,[157] who reported that for platinum with a roughness factor of 1.3 a monolayer of copper corresponded to 540 µC cm^{-2}. Monolayer coverage of the electrode occurred at 0.2 V, which is the potential for the Cu^{II}–Cu^{I} reduction. Maximum coverage was observed at ~ 0.0 V (n.h.e.), i.e. at the limiting current for the convective diffusion-controlled production of Cu^I. The quantity of copper so deposited was found to be independent of cupric ion concentration in the range 2×10^{-6}—2×10^{-4} mol l^{-1}. Byallozor,[158] however, claimed for the deposition of copper on to platinum from perchloric

[153] V. S. Galinker, V. P. Milovzorov, and P. V. Savenko, *Visn. Kiiv. Politekhn. Inst., Ser. Khim. Mashinobuduv. Tekhnol.*, 1968, **5**, 134.
[154] N. Ya. Kovarskii and V. N. Golubev, *Elektrokhimiya*, 1970, **6**, 762.
[155] T. Inui, K. Hosono, and K. Ozaki, *Kinzoku. Hyomen Gijutsu*, 1969, **20**, 248.
[156] J. Šimkovic and D. Leskovšek, *Electrochim. Acta*, 1971, **16**, 2125.
[157] S. H. Cadle and S. Bruckenstein, *Analyt. Chem.*, 1971, **43**, 932.
[158] S. G. Byallozor, *Elektrokhimiya*, 1971, **7**, 103.

acid solutions that, in the immediate vicinity of the electrode surface, ions of the type $[Cu_2(OH)_2]^{2+}$ dissociate such that two Cu^{2+} ions are discharged simultaneously at the electrode. Schmidt et al.[159] have also encountered specific copper adsorbates in the undervoltage region of the Cu^{II}–Cu equilibrium in association with the course of electrodeposition of copper on gold and silver electrodes from $Cu(ClO_4)_2$ and $CuSO_4$ solutions. A voltammetric technique was used with a thin layer configuration to show that the charge stoicheiometry of adsorption corresponded approximately to the apparent discharge of adsorbed Cu^{2+}. Copper monolayers were thus shown to be formed on gold. Twin electrode techniques[159] were used to separate the charge due to adsorption from that due to the faradaic formation of Cu^+. Gold electrodes were used by Jacquot[160] as vibrating bases for the cathodic deposition and anodic dissolution of copper, studied by stripping voltammetry. In the concentration range 10^{-5}—10^{-3} M-Cu^{2+} the anodic peak was proportional to the cupric ion concentration. Electron diffraction studies showed that the cathodic deposit consisted of copper and/or Cu_2O.

For the dissolution of copper, Landolt et al.[161] have considered crystallographic factors and anode potentials in connection with the processes occurring at high rates of metal removal. Surface textures were studied which had resulted from the electrodissolution of polycrystalline and single-crystal copper at 50 A cm^{-2} in 2M-KNO_3. Electrolyte flow velocities of 2500 and 400 cm s^{-1} have been employed for dissolution in the active and transpassive modes. The resulting surface topography depended on the crystal orientation. Flow streaks were observed with transpassive dissolution, and transpassive pitting was observed only in the case of polycrystalline samples. In the active dissolution region submicroscopic facets led to ridging, and resulted in a differentiation between grains in polycrystalline material, in agreement with a dissolution mechanism based on the motion of atomic ledges on tightly packed lattice planes. Experimental work on the dissolution of copper in phosphoric acid has been presented by Valeev et al.[162] and Varenko et al.[163] The first of these papers describes measurements of refractive index during potentiostatic polarization, and it was concluded that the concentration gradient at the electrode is four times smaller in the case of 4 molar solutions than in 11.5 molar electrolyte. This was considered to be the cause of the absence of smoothing of subrelief in the dilute solution. In the second paper, experiments at rotating electrodes were described, and the importance of diffusion as the rate-controlling mechanism for oxidation of copper in H_3PO_4 solutions at concentrations greater than 2.5 mol l^{-1} was emphasized. Tvarusko and Watkins,[87] using Mach–Zehnder interferometry to study the

---

[159] E. Schmidt, P. Beutler, and W. J. Lorenz., *Ber. Bunsengesellschaft phys. Chem.*, 1971, **75**, 71.
[160] D. Jacquot, *Ann. Chim. (France)*, 1970, **5**, 99.
[161] D. Landolt, R. H. Muller, and C. W. Tobias, *J. Electrochem. Soc.*, 1971, **118**, 36, 40.
[162] A. Sh. Valeev, L. V. Khlopotina, and L. V. Chugunova, *Elektrokhimiya*, 1970, **6**, 985.
[163] E. S. Varenko, V. P. Galushko, V. N. Kouton, P. M. Fedash, and Yu. M. Loshkarev, *Zashch. Metal*, 1970, **6**, 103.

deposition of copper in neutral sulphate electrolytes, showed that the diffusion-layer thickness increased linearly with $t^{\frac{1}{2}}$ up to the potential of the h.e.r. Interferometric holography has also been applied to the study of natural convection at a copper electrode during deposition of copper from $CuSO_4$ by Duron and Makenc,[164] who showed that it was possible to detect concentration profiles of diffusion layers at both anode and cathode. Fouad et al.[165] described a series of experiments in which limiting current was measured for the anodic polishing of vertical copper electrodes in H_3PO_4. The effect of initial roughness on the rate of mass transfer and limiting current was also studied, and it was found that, within the limits studied, surface roughness has no substantial effect.

The behaviour of copper in alkaline solution has been studied.[166, 167] Kim and Nobe[166] have studied the Tafel region over a considerable range of pH. The rate of anodic dissolution was increased by increasing hydroxide ion concentration and the reaction order was reported to be 0.5. Linear-sweep voltammetry[167] showed two major reaction peaks corresponding to the formation of Cu^I and Cu^{II} oxides. The Cu^{II} peak may be resolved at high sweep rates into two components corresponding to the formation of soluble and insoluble Cu^{II} species. Evidence was presented that the transference of ionic species in the solid phase was the rate-controlling process in the formation of Cu_2O.

Chromium.—The mechanism and kinetics of chromium electrodeposition from aqueous chromic acid has been investigated by Usachev.[168] The dependence of the partial formation rates on the concentration of the solution components has been examined, making the assumption that the reactions occur concurrently. The production of decorative chromium coatings from solutions of Cr^{III} compounds was investigated by Ward and Christie.[169] A solution based on aqueous $CrCl_3$, containing 40% v/v dimethylformamide, was reported to possess excellent electrodeposition characteristics. Work on aqueous sulphate electrolytes showed that pH maintenance was the most important factor governing the production of decorative deposits. Solutions were difficult to stabilize using conventional buffers, but dimethylamido buffers were found to provide adequate pH control of chromic chloride solutions.

The dissolution of chromium from an anode in the transpassive potential region has been studied in both acid and alkaline solutions by Armstrong and Henderson.[170] The a.c. impedance and potentiostatic pulse methods were

[164] C. Durou and J. Makenc, *Compt. rend.*, 1971, **272**, C, 2035.
[165] M. G. Fouad, F. N. Zein, and M. I. Ismail, *Electrochim. Acta*, 1971, **16**, 1477.
[166] C. P. Kim and K. Nobe, *Corrosion*, 1971, **27**, 382.
[167] N. A. Hampson, J. B. Lee, and K. I. MacDonald, *J. Electroanalyt. Chem. Interfacial Electrochem.*, 1971, **32**, 165.
[168] D. N. Usachev, *Elektrokhimiya*, 1971, **7**, 38.
[169] J. J. B. Ward and I. R. A. Christie, *Trans. Inst. Metal Finishing*, 1971, **49**, 148.
[170] R. D. Armstrong and M. Henderson, *J. Electroanalyt. Chem. Interfacial Electrochem.*, 1971, **32**, 1.

used with stationary and rotating ring-disc techniques. It was shown that in the transpassive potential region chromium had a thin layer of oxide covering the surface, the mean valency of this oxide layer depending on the electrode potential. Genshaw and Sirohi[171] have used *in situ* ellipsometry to investigate the mechanism of passivation of chromium in H_2SO_4 solution. They found that at -410 mV a region of active dissolution at an apparently oxide-free surface was observed, whereas when the electrode potential was stepped to -360 mV, passivation occurred with less than 1 Å of oxide film. The film was reported to thicken at more anodic potentials and was considered to be Cr_2O_3. The mechanism of passivation was apparently that of simultaneous oxide formation and metal dissolution, with passivation resulting from the formation of a monolayer of passivating film.

Carbon.—The use of graphite anodes in electrochemical science has been of some interest. Eberil' and Elina[172] have considered the behaviour of graphite anodes in the electrolysis of solutions of NaCl in connection with the production of chlorates. The disintegration of anodes remains a principal problem, and from this investigation it is clear that lowering the temperature of electrolysis, especially at high current densities, is disadvantageous from the standpoint of electrode wear and may even be hazardous. An interesting application of porous carbon electrodes has been reported by Johnson and Newman,[173] in which ionic adsorption occurs on porous carbon. Preferential adsorption of divalent ions was observed and it was concluded that the potential-dependent adsorption of ions on carbon surfaces gave them an ion-exchange function. Thus a negatively charged carbon surface in NaCl solution absorbs Na^+. These ions exchange with Ca^{2+} ions from a $CaCl_2$ solution without a flow of current. A positively charged carbon surface should behave as an anion-exchanger. Preliminary work by Jenkins and Weedon[174] on the examination of electrodes which had had prolonged contact with carbon-purified electrolytes would seem to warrant further study.

Gallium.—Armstrong, Race, and Thirsk[175] have investigated the anodic behaviour of gallium in alkaline solution using galvanostatic, linear-sweep voltammetry, potential pulse, and impedance measurements. For the active dissolution a Tafel slope (30 ± 2 mV per decade) suggested that the mechanism was:

$$Ga \rightarrow Ga^I + e^-$$
$$Ga^I \rightarrow Ga^{II} + e^-$$
$$Ga^{II} \rightarrow Ga^{III} + e^-$$

[171] M. A. Genshaw and R. S. Sirohi, *J. Electrochem. Soc.*, 1971, **118**, 1558.
[172] V. I. Eberil' and L. M. Elina, *Elektrokhimiya*, 1970, **6**, 782, 1010.
[173] A. M. Johnson and J. Newman, *J. Electrochem. Soc.*, 1971, **118**, 511.
[174] D. A. Jenkins and C. J. Weedon, *J. Electroanalyt. Chem. Interfacial Electrochem.*, 1971, **118**, app. 13.
[175] R. D. Armstrong, W. R. Race, and H. R. Thirsk, *J. Electroanalyt. Chem. Interfacial Electrochem.*, 1971, **31**, 405.

The final anodic product of the active dissolution was $Ga(OH)_6^{3-}$. High values of double-layer capacitance in the active dissolution region were interpreted as indicating the adsorption of OH^-. The formation of a two-dimensional anodic film marked the onset of the passive condition. This film, it was recorded, drastically reduced the rate of the dissolution reaction. Passivating films on gallium have been considered by Popova and co-workers.[176, 177] The differences in the nature of the passivating oxide films formed in alkali saturated with gallate have been noted. It is claimed that for the transpassive dissolution of gallium, two parallel reactions occur: (i) an increase in the content of super-stoicheiometric oxygen in the oxide film as a result of OH^- discharge, and (ii) the direct transport of gallate ion into solution. Capacitance measurements in alkaline and neutral gallium-containing solutions established that the true surface area of highly passivated gallium is approximately that of the liquid metal. Selekhova and Lyubimova[178] investigated the cathodic deposition of gallium from gallate solutions using polarization and impedance-measurement techniques. The process was found to be limited by the discharge step. Bagotskaya and Khalturina,[179] in investigating the h.e.r. on gallium, found that for $HClO_4$ solutions in the presence of halide ion the adsorption of the halide ion was practically reversible. In H_2SO_4 solutions, when the chloride ion concentration was decreased, there was evidently a partial replacement of previously adsorbed ions in the surface layer. The results confirmed the reported specific adsorption of sulphate ions on solid gallium.

Germanium.—The cathodic polarization of p-type germanium has been studied in solutions of tetraethylammonium bromide, tetraethylammonium iodide, and HBr.[180] The Tafel relation for the h.e.r. was found not to be valid over a wide range of current densities. The parameters b, α, and v support the mechanism:

$$Ge_s + H_3O^+ + e^- = Ge_sH + H_2O \text{ (slow)}$$
$$Ge_sH + Ge_sH = 2\,Ge_s + H_2 \text{ (fast)}$$

where Ge_s denotes a surface atom. The hydrogen overpotential was found to decrease with solution pH.

Gold.—Gedansky and Hepler[181] have reviewed the thermodynamic and electrochemical data for a number of gold derivatives and aqueous ions.
In a detailed examination of some of the electrochemical and structural aspects of the electrodeposition of gold, Cheh and Sard[182] employed rotating

---

[176] T. I. Popova, N. A. Simonova, Z. I. Moiseeva, and N. G. Bardina, *Elektrokhimiya*, 1970, **6**, 1125.
[177] T. I. Popova and N. A. Simonova, *Elektrokhimiya*, 1970, **6**, 1378.
[178] N. P. Selekhova and N. A. Lyubimova, *Elektrokhimiya*, 1970, **6**, 1199.
[179] I. A. Bagotskaya and T. I. Khalturina, *Elektrokhimiya*, 1970, **6**, 1013.
[180] D. Singh and A. N. Dwivedi, *Z. phys. Chem. (Frankfurt)*, 1970, **72**, 259.
[181] L. M. Gedansky and L. G. Hepler, *Engelhard Ind. Tech. Bull.*, 1969, **10**, 5.
[182] H. Y. Cheh and R. Sard, *J. Electrochem. Soc.*, 1971, **118**, 1737.

disc electrodes for which i–E curves were determined at 60 °C, using 0.005 M-Au$^+$ solutions. The morphology of deposits (1 μm thick) was determined mainly by scanning electron microscopy; however, transmission electron micrography was used to study the early stages of film growth. The cyanide system ($i_0 = 0.82$ mA cm^{-2}, $\alpha = 0.7$) showed the greatest tendency for the formation of spiked deposits. An increase in current density and a reduction in rotation speed favoured spikes, which occurred over a range of potential ($150 < |\eta| < 300$ mV) and not at a particular concentration of gold at the interphase. Spiky deposits typically ~1 μm high (population $> 10^8$ cm^{-2}) were observed in deposits from the citrate system but to a much smaller extent than from the phosphate system. At low overvoltages a smooth layer type of growth was formed. The morphology of gold deposits may also be improved by pulsed current,[183] and in this connection it was concluded that the effect was not to increase the limiting over-all deposition rate but to raise the limiting current density.

It has been noted[184] that adsorbed sulphur favours the dissolution of gold in HCl, and in the case of oxide layers chemisorbed O^{2-} ions are in equilibrium with the electrolyte: $H_2O(aq.) \rightleftharpoons O^{2-}_{ad} + 2H^+(aq.)$. The replacement of gold atoms by oxide ions in the surface layer, $O^{2-}_{ad} + Au^{3+} \rightleftharpoons O^{2-}_{ox} + Au^{3+}_{ad}$, is rate-determining.[185] As the result of a potentiostatic investigation of the reduction of an oxide film on gold, Ogura et al.[186] concluded that the potential-determining reaction was $Au_2O_3 + 6H^+ + 6e^- \rightarrow 2Au + 3H_2O$ (in strong acid) and $Au(OH)_3 + 3H^+ + 3e^- \rightarrow Au + 3H_2O$ (in weak acid). Electrode films on gold have also been studied by ellipsometry.[187]

In an interesting paper Breiter and Luborsky[188] described the identification of gold islands on copper-plated wire surfaces. Periodic potentiostatic i–E curves were used to show the existence of gold islands with an average thickness of from 0—1400 Å.

Iron.—The majority of papers in connection with iron deal with aspects of corrosion and passivation. These papers are discussed in Section 5 (p. 83).

An aspect of iron that is rapidly assuming great technological importance is that of electrochemical machining. Chin[189] has investigated the anodic mechanism of electrochemical machining by studying current transients on a rotating steel electrode under potentiostatic control in NaCl and NaClO$_3$ solutions. In the presence of chloride ions, transient experiments revealed the formation of a porous, non-protective film from the back-precipitation of the anode product. In the presence of chlorate a much less porous film, about 1000 Å thick, was formed. The form of the transients revealed that the film

---

[183] H. Y. Cheh, *J. Electrochem. Soc.*, 1971, **118**, 551.
[184] M. Kostelitz and N. Barbouth, *Compt. rend.*, 1971, **272**, C, 1619.
[185] J. W. Schultze and K. J. Vetter, *Ber. Bunsengesellschaft phys. Chem.*, 1971, **75**, 470.
[186] K. Ogura, S. Haruyama, and K. Nagasaki, *J. Electrochem. Soc.*, 1971, **118**, 531.
[187] F. Chao, M. Costa, and A. Tadjeddine, *Bull. Soc. chim. France*, 1971, 2465.
[188] M. W. Breiter and F. E. Luborsky, *J. Electrochem. Soc.*, 1971, **118**, 867.
[189] D. T. Chin, *J. Electrochem. Soc.*, 1971, **118**, 174.

formed by anodic oxidation and the rate of film growth increased with the increase in the rate of dissolution of iron. The importance of electrolyte type, of concentration, and of operating voltage has been investigated by LaBoda, Hoare, and Beacom,[190] using an electrolytic grinding rig employing steel tube cathodes in solutions of NaCl, NaClO$_3$, and Na$_2$Cr$_2$O$_7$. Solutions of NaClO$_3$ gave superior results with respect to surface finish and dimensional control at high metal-removal rates. It was concluded that the ECM behaviour observed in the case of steels in the three electrolytes was accounted for in terms of the build-up and breakdown of protective films on the steel surface. (N.B. Breakdown may well be linked with anion penetration of the oxide film.) Hoare et al.[191] considered that NaClO$_3$ electrolytes were good for electrochemical machining but underwent some degradation. NaClO$_4$ was found to be a longer-lived electrolyte, but surface finishes from NaClO$_3$ were more reflecting than those from NaClO$_4$. Apparently, the film formed on steel in NaClO$_4$ was not uniform, and this permitted localized attack of the metal surface. Whereas an oxide film must be present in order to provide good dimensional control and machining geometry, that obtained in the case of NaClO$_4$ was evidently unsatisfactory. Chikamori and Ito[192] have claimed that good electrochemical machining characteristics could be obtained by the addition to the NaNO$_3$ electrolyte of other oxygen-containing anions such as SO$_4^{2-}$, BrO$_3^-$, and ClO$_3^-$. On the other hand, chloride ions were found to be detrimental. The throwing power of electrolytes used in electrochemical machining could readily be obtained by the measurement of the time-variation of the space gap under machining conditions, and thus the tool feed-rate could be optimized in order to maintain a constant gap.[193] Stray currents were found to promote inferior surface finishes and poor dimensional tolerances in machining. Boden and Evans[194] have made a study of the problem using a segmented anode to measure weight losses in the stray-current region as a function of chloride ion concentration. It was observed that at intermediate concentrations the attack can be decreased by the formation of a passive film, although the effect was marred by deep pitting. This pitting may be minimized by the addition of anions that are capable of forming insoluble compounds within the pits, and it was concluded that a reduction in stray-current attack on nickel during electro machining may be obtained by the addition of CO$_3^{2-}$, PO$_4^{3-}$, or Fe(CN)$_6^{3-}$ to the chloride electrolyte.

Iridium.—A paper[195] of some interest in connection with iridium is an a.c. study of the I$_2$/I$^-$ reaction at an iridium electrode, where the results were

---

[190] M. A. LaBoda, J. P. Hoare, and S. E. Beacom, *Coll. Czech. Chem. Comm.*, 1971, **36**, 680.
[191] J. P. Hoare, K-W. Mao, and A. J. Wallace, *Corrosion*, 1971, **27**, 211.
[192] K. Chikamori and S. Ito, *Denki Kagaku*, 1971, **39**, 493.
[193] D. T. Chin and A. J. Wallace, *J. Electrochem. Soc.*, 1971, **118**, 831.
[194] P. J. Boden and J. M. Evans, *Electrochim. Acta*, 1971, **16**, 1071.
[195] A. M. Trukhan ,Yu. M. Povarov, and P. D. Lukovstev, *Elektrokhimiya*, 1970, **6**, 1734.

found to be explained by the inhomogeneity of the surface of the electrode.

Lead.—The galvanostatic polarization of lead in $HClO_4$ has been studied[196] and the results have been compared with theoretically obtained values for the transition (passivation) times. The theoretical model was extended to the case of interrupted polarization, i.e. to an open-circuit period sandwiched between two galvanostatic polarizations at different rates. Good agreement was obtained between the theoretically and experimentally predicted transition times. For the case of a lead electrode in alkaline solution, Carr and Hampson[197] have studied the anodic oxidation of the metal by linear-sweep voltammetry. It was observed that the process was rate-controlled by the diffusion of OH^- in the solution phase. The major interest in connection with lead continues to be the lead–acid cell, and papers have appeared during the review period on linear-sweep voltammetry applied to polycrystalline lead in H_2SO_4,[198] on the formation of $PbSO_4$ crystallites on lead in the lead–acid cell,[199] and on the measurement of the effective resistance of the liquid phase applied to the study of porous electrodes.[200] These latter three papers are only of minor importance to the present review since they mainly deal with the development of the sulphate phase.

Manganese.—Qazi and Leja[201] have described an interesting process for the production of 98% manganese with a current efficiency of 76%. $MnCl_2$ dissolved in an organic amide, e.g. formamide, forms a catholyte which was separated by a diaphragm cell from an anolyte of aqueous NH_4Cl. The cell was simple to operate and the product had satisfactory physical characteristics. It was claimed that, by comparison with the 'total' aqueous ($MnCl_2$) process, the formamide process gave higher conductivity and higher current efficiency. Moreover, catholytes up to pH ~ 8 could be used without precipitating $Mn(OH)_2$.

Molybdenum.—In a preliminary note Hull[202] described linear-sweep voltammetry measurements which apparently show that the surface of a molybdenum electrode undergoing anodic dissolution in OH^- (0.1—10.0 mol l^{-1}) is covered with an oxide film only for electrode potential more positive than $\sim +0.4$ V (vs. Hg|HgO). Two distinct films were identified on the electrode at potentials more positive than 0.4 V and these were presumed to be MoO_2 and MoO_3.

Nickel.—A large number of investigations of the electrochemical behaviour

---

[196] C. J. Bushrod and N. A. Hampson, *Brit. Corrosion J.*, 1971, **6**, 87.
[197] J. P. Carr and N. A. Hampson, *J. Electrochem. Soc.*, 1971, **118**, 1262.
[198] J. P. Carr, N. A. Hampson, and R. Taylor, *J. Electroanalyt. Chem. Interfacial Electrochem.*, 1971, **33**, 109.
[199] T. Chiku and K. Nakajima, *J. Electrochem. Soc.*, 1971, **118**, 1395.
[200] I. L. Romanova and I. A. Selitskii, *Elektrokhimiya*, 1970, **6**, 1776.
[201] M. A. Qazi and J. Leja, *J. Electrochem. Soc.*, 1971, **118**, 548.
[202] M. N. Hull, *J. Electroanalyt. Chem. Interfacial Electrochem.*, 1971, **30**, app. 1.

of nickel have appeared during the review period. The electrodeposition of nickel has been the subject of a cautionary note by Brownsword and Farr,[203] who emphasize that slow solidification, as occurs in zone refining or crystal growing, may result in local segregation of impurity, which makes imperative careful metallographic examination of substrate surfaces, particularly in studies of nucleation and growth. Epelboin and Wiart[204] have studied the electrodeposition of nickel in acidic solution by an analysis of the polarization characteristics and of cathodic impedance. Results from acid solutions support the idea that the cathodic deposition occurs in several steps probably involving $(NiOH)_{ads}$. The results could be accounted for by a mechanism implying two successive electron-transfer steps in which $(NiOH)_{ads}$ acts as an intermediate compound. The presence in the electrolyte solution of either sodium benzenesulphonate or 2-butyne-1,4-diol (additives that are commonly used to improve nickel electrodeposits) specifically modified the kinetic parameters of the transfer reactions and diminished the double-layer capacitance. Another view of the electrodeposition of nickel is put forward by Le Gorrec and Guitton,[205] who used the rotating-disc method. It was reported that at constant pH the rate of reaction was independent of the concentration of Ni^{2+}. It was proposed that the electro-deposition proceeded *via* the chemical reduction of Ni^{2+} by H_2, probably at the interphase. In a later contribution[206] it was shown that H_2 adsorbed at the electrode participated in the reduction of Ni^{2+}, and that H^+ and Ni^{2+} were the only ionic species arriving by diffusion at the interphase. The preferential deposition of nickel has also been noted.[207] The behaviour of added toluene-*p*-sulphonamide (*p*tsa) during the deposition of nickel from $NiSO_4$ electrolyte was studied by Kruglikov and Volkov,[208] who determined the removal of the additive spectrophotometrically ($\lambda = 2230$ Å). It was found that all sulphur atoms from *p*tsa removed cathodically were incorporated into the nickel deposit, probably as the sulphide. An equilibrium was assumed to exist between adsorbed and solution concentrations, and the derived adsorption isotherms conformed to the Frumkin equation at low and medium surface coverages. The Temkin equation for a uniform–nonuniform surface described the experimental curves. However, the calculated constants depend on the hydrodynamic conditions. It was concluded that the assumption that the rate of sulphur occlusion is proportional to the degree of coverage was not justified, and that the processes occurring on the cathode are more complex in nature, especially in relation to the changes in the nature of particles being adsorbed.

Stress in both chemical[209] deposits and electrochemical deposits[102] of nickel has been considered.

[203] R. Brownsword and J. P. G. Farr, *Electrochim. Acta*, 1971, **16**, 845.
[204] I. Epelboin and R. Wiart, *J. Electrochem. Soc.*, 1971, **118**, 1577.
[205] B. Le Gorrec and J. Guitton, *Compt. rend.*, 1971, **272**, C, 1784.
[206] B. Le Gorrec and J. Guitton, *Compt. rend.*, 1971, **272**, C, 2031.
[207] J. Amblard, M. Fromont, and G. Maurin, *Compt. rend.*, 1971, **272**, C, 995.
[208] S. S. Kruglikov and V. A. Volkov, *Elektrokhimiya*, 1970, **6**, 1033.
[209] Y. Shibasaki and Y. Inaba, *Denki Kagaku*, 1971, **5**, 413.

The anodic behaviour of nickel has been studied. Vagramyan et al.[210] have measured the potentials of a nickel electrode and kinetic parameters of dissolution (and deposition) in aqueous chloride solutions at 25—275 °C. A comparison of the results from H_2SO_4 solutions confirmed the view that the cause of the unique electrochemical behaviour of nickel in aqueous solution lay in the inhibiting effect of foreign substances adsorbed on the electrode. The acceleration of the rate of the electrochemical reaction at higher temperatures could be explained by an increase in the activity of the electroactive ion and a lower inhibiting effect due to foreign particles. Kesten and Feller[211] observed that nickel electrodes undergoing linear-sweep polarization in 0.5 M-H_2SO_4 showed different sulphate coverages depending upon the history of the H_2SO_4 solution. The influence of foreign substances was again apparent. The presence of sulphide ion and NiS on the electrode surface was demonstrated by chemical analyses and by electron diffraction. It was also shown that the HSO_4^- and HS^- ions are adsorbed differently on the crystal planes. In all, nickel appeared to be an extremely complicated system when studied in acidic electrolytes.

The nickel electrode is also of interest in alkaline electrolytes, particularly in relation to battery applications. Armstrong[212] has pointed out that the passivation of nickel reported as:

$$NiOH^+ \rightleftharpoons Ni^{2+} + OH^-$$

$$NiOH^+ + OH^- \rightarrow Ni(OH)_2 \text{ (surface film)}$$

apparently caused by the conversion of this film into an electronic conductor, may be incorrect. An ellipsometric study was referred to and some doubt was cast on the equations used. Oshe and Lobachev[213] have investigated the anodic oxidation of nickel in the regions of potential preceding the basic passivation of nickel. This region was characterized by an increase in current with time occurring in the region of 'active' dissolution. If the rate of oxidation of nickel is determined by the concentration of three-dimensional holes in the reaction zone, then an increase in the latter, during the formation of a solid solution of Ni_2O_3 in NiO, was the cause of the acceleration of oxidation. The kinetics of a nickel hydroxide electrode in alkaline solution are affected by the addition to the electrolyte of Li^+ and rare-earth compounds,[214] which when incorporated into the nickel hydroxide have the effect of producing an n-type semiconductor during the charge period and a p-type semiconductor during the discharge period. The addition of rare-earth compounds causes an increase in the population of active centres on the electrode surface.

Palladium.—The galvanostatic study of kinetic currents during the electro-

---

[210] A. T. Vagramyan, M. A. Zhamagortsyan, L. A. Uvarov, and A. A. Yavich, *Elektrokhimiya*, 1970, **6**, 755.
[211] M. Kesten and H. G. Feller, *Electrochim. Acta*, 1971, **16**, 763.
[212] R. D. Armstrong, *J. Electroanalyt. Chem. Interfacial Electrochem.*, 1970, **28**, 221.
[213] A. I. Oshe and V. A. Lobachev, *Elektrokhimiya*, 1970, **6**, 1419.
[214] Z. Takehara, M. Kato, and S. Yoshizawa, *Electrochim. Acta*, 1971, **16**, 833.

reduction of chloride complexes of palladium by Zelenskii and Kravstov[215] showed that $i\tau^{\frac{1}{2}}$ (where τ was the transition time) was linear with i. The dependence of $\partial(i\tau^{\frac{1}{2}})/\partial i$ on chloride ion concentration showed that the deposition of palladium was preceded by the slow chemical reaction $PdCl_4^{2-} \rightleftharpoons PdCl_3^- + Cl^-$, where Cl^- is in large excess. For the reaction $PdCl_3^- \rightleftharpoons PdCl_2 + Cl^-$, $k^- \sim 5 \times 10^2$ s^{-1}. The electrochemical behaviour of the metal is, however, very much influenced by adsorption. In the case of chemisorbed oxygen atoms, the coverage may reach 20 atoms per surface metal atom, an actual oxide phase developing only under severe conditions of anodic polarization.[216] For ionic adsorption[217] it is well known that surface-active ions have a substantial effect upon electrode reactivity, although for ions such as SO_4^{2-} the bonding is of the weak electrostatic type.

Platinum.—Of the many papers dealing with platinum the major preoccupation has been with adsorption at the electrode and the activation of the surface in connection with electro-catalysis. Shibata and Sumino[218] have shown that the activation of platinum resulted from the formation of a superficial active monolayer of platinum atoms which arose through being forced to rearrange to new positions by oxygen adsorbed during oxidation cycles or exposure to air. Chao et al.[219] have shown, however, that repeated oxidation cycles may completely degrade the surface. Khazova et al.[220] considered that the differences in catalytic action at different potentials are due to a change in the surface concentration of reacting materials. The surface of platinum electrodes was also considered by Bagotskii.[221] The anodic oxidation of platinum in 0.5 M-H_2SO_4 has been studied[222] by voltage step and decay in order to construct Tafel curves, and the film development was followed using an ellipsometer.

Adsorption on platinum has been studied.[223—227] An interesting paper[224] deals with the effect of Cd^{2+} on the adsorption on platinum of anions of the type HSO_4^- and Cl^-. It was reported that the adsorption of Cd^{2+} was accompanied by that of anions. Other papers deal with reduction[228] and oxidation[229]

[215] M. I. Zelenskii and V. I. Kravstov, *Elektrokhimiya*, 1970, **6**, 793.
[216] D. A. J. Rand and R. Woods, *J. Electroanalyt. Chem. Interfacial Electrochem.*, 1971, **31**, 29.
[217] M. I. Kulezneva and N. A. Balashova, *Elektrokhimiya*, 1970, **6**, 1057.
[218] S. Shibata and M. P. Sumino, *Electrochim. Acta*, 1971, **16**, 1511.
[219] F. Chao, M. Costa, and A. Tadjeddine, *Compt. rend.*, 1971, **272**, C, 821.
[220] O. A. Khazova, Yu. B. Vasil'ev, and V. S. Bagotskii, *Elektrokhimiya*, 1971, **7**, 1367.
[221] V. S. Bagotskii, Yu. B. Vasil'ev, and I. I. Pyshnogreeva, *Elektrokhimiya*, 1971, **7**, 2141.
[222] J. L. Ord and F. C. Ho, *J. Electrochem. Soc.*, 1971, **118**, 46.
[223] G. Horányi, J. Solt, and F. Nagy, *J. Electroanalyt. Chem. Interfacial Electrochem.*, 1971, **31**, 87, 95.
[224] G. Horányi, J. Solt, and G. Vértes, *J. Electroanalyt. Chem. Interfacial Electrochem.*, 1971, **32**, 271.
[225] F. Mikuni and T. Takamura, *Denki Kagaku*, 1971, **39**, 579.
[226] M. Bonnemay, G. Bronoël, and G. Peslerbe, *Compt. rend.*, 1971, **272**, C, 724.
[227] C. D. Zakumbaeva, F. M. Toktabaeva, and D. V. Sokol'skii, *Elektrokhimiya*, 1970, **6**, 777.
[228] A. J. Arvia, J. S. W. Carrozza, and H. A. Garrera, *Electrochim. Acta*, 1971, **16**, 79.
[229] W. Smit and J. G. Hoogland, *Electrochim. Acta*, 1971, **16**, 1.

of sulphuric derivatives at platinum. For the platinum electrode in alkaline solution the state of the electrode surface has been examined in various potenttial regions.[230] Although some of the papers mentioned are not really pertinent to this Report, practically all emphasize the importance of the surface structure and the need properly to characterize the surface.

Feltham and Spiro[231] have reviewed the literature on the platinized platinum electrodes. In a separate paper dealing with this type of electrode, Feltham and Spiro[232] have shown that lead (from the platinizing solution) was incorporated into the spongy deposit. This lead was leached from the electrode in solution and the effect of the leached Pb^{2+} was to cause the electrodes to behave in an irreversible manner.

Ruthenium.—Greenspan[233] has reviewed the composition of solutions used for the deposition of ruthenium from the 1930s to 1969. The most useful solutions seemed to be based upon ruthenium nitrosyl chloride and sulphonic acid. Efficiencies appear to be at best ~30%. One major drawback appeared to be the inability for deposits over 2.5 μm thick to be crack-free.

Silver.—The deposition of sub- and mono-layer amounts of silver has been studied using the rotating platinum ring-disc electrode on 0.2 M-H_2SO_4 containing 2×10^{-5} M-Ag^+, and the well-known phenomenon of underpotential was demonstrated.[234] Budewski[235] has studied the electrolytic growth of silver on the faces of a single crystal of silver. The nucleation of silver on platinum has been considered by Klapka,[236] using galvanostatic techniques, who has shown that the dependence of crystal growth on overpotential, after the development of a galvanostatic pulse, was explained in terms of a two-dimensional nucleation at an overpotential exceeding 8 mV. The supersaturation during nucleation was 12—18 and the activational energy for the formation of nuclei 4.1×10^{-14} erg. The behaviour of the system indicated that the electrode material participated in the electrode-nucleation process, rather than behaving, as expected, simply as an inert substrate.

In alkaline solution the effect of dissolved oxygen on the oxidation of silver was to change markedly the electrochemical properties and conductance of the silver oxide layer formed in the oxidation process.[237] An examination by Wales[238] of the microstructure of silver electrodes prepared by sintering silver powder showed that almost all silver particles were completely coated with Ag_2O when the electrode was half charged. The effect of temperature on

[230] A. A. Yakovleva and V. I. Veselovskii, *Elektrokhimiya*, 1970, **6**, 967.
[231] A. M. Feltham and M. Spiro, *Chem. Rev.*, 1971, **71**, 117.
[232] A. M. Feltham and M. Spiro, *J. Electroanalyt. Chem. Interfacial Electrochem.*, 1970, **28**, 151.
[233] L. Greenspan, *Engelhard Ind. Tech. Bull.*, 1970, **11**, 76.
[234] G. W. Tindall and S. Bruckenstein, *Electrochim. Acta*, 1971, **16**, 245.
[235] E. Budewski, *Trans. SAEST*, 1970, **5**, 55.
[236] V. Klapka, *Coll. Czech. Chem. Comm.*, 1971, **36**, 1181.
[237] N. Yu. Osipchuk, T. I. Dunaeva, and M. F. Skalozubov, *Elektrokhimiya*, 1971, **7**, 106.
[238] C. P. Wales, *J. Electrochem. Soc.*, 1971, **118**, 7.

The Solid Metal Electrode in Aqueous Solution 79

the Ag|Ag$_2$O electrode has been studied,[239, 240] and some thermodynamic data have been reported.

Sodium.—Chernomorskii and Kabanov[241] reported that sodium may be introduced into lead at potentials more negative than -2.0 V (n.h.e.). Chronopotentiometric analyses of lead for incorporated alkali metal were made at 2.5×10^{-4} A cm^{-2}. An increase of an order of magnitude in the quantity of incorporated alkali metal at the lead electrode corresponds to an increase in potential of ~ 0.45 V for 10 M-NaOH. The rate of introduction of alkali-metal ions, i_{in}, was found to be expressed by the relationship:

$$i_{in} = k\, c_{Na^+} \exp(-\alpha FE/RT)$$

where k is the specific rate constant for introduction. The rate was determined by the activation energy for the discharge of the ion with simultaneous introduction into the solid cathode.

Tantalum.—Work reported on tantalum has been confined to the formation of films.[242]

Thallium.—The kinetics of the anodic dissolution of thallium in perchloric solutions have been reported by Bailey and Wright.[243] Using a rotating disc electrode and a potentiostatic sweeping technique, they determined the orders of reaction. It was observed that the reverse reaction due to the deposition of Tl$^+$ made a significant contribution to the observed current, and it was found to be impossible to measure the anodic process in isolation. Faradaic currents were independent of pH in the range 0—12.3 and it was concluded that the mechanism could not involve Tl$_2$O or TlOH. The current was also apparently independent of added surfactants such as thiourea or H$_4$ edta at pH 0; dissolved oxygen, on the other hand, showed a marked effect.

The mechanism proposed was a simple one-electron transfer which is rate-determining:

$$\text{Tl} \rightarrow \text{Tl}^+(\text{aq}) + e^-$$

so that the anodic current density could be expressed by:

$$i = Fk \exp[(1-\alpha) FE/RT]$$

and from the experimental data α could be reasonably assumed to be 0.5.

Tin.—The anodic and cathodic behaviour of tin in alkali have been investigated in order to optimize the stripping of tinplate.[244] Anodic current can be

---

[239] E. G. Gagnon and L. G. Austin, *J. Electrochem. Soc.*, 1971, **118**, 497.
[240] M. I. Gillibrand, L. Langrish, and G. R. Lomax, *J. Appl. Electrochem.*, 1971, **1**, 9.
[241] A. I. Chernomorskii and B. N. Kabanov, *Elektrokhimiya*, 1970, **6**, 1224.
[242] J. Zahavi and J. Yahalom, *Electrochim. Acta*, 1971, **16**, 89.
[243] P. C. A. Bailey and G. A. Wright, *Electrochim. Acta*, 1971, **16**, 865.
[244] H. Barbré, C. Bagger, and E. Maahn, *Electrochim. Acta*, 1971, **16**, 559.

accounted for on the basis that Sn^{II} as well as Sn^{IV} is produced. It was found possible to maximize the cathodic current efficiency at 100% from stannate and stannite solutions. However, in stannite solutions it was possible to run at much higher current densities. It was confirmed that iron and the Fe–Sn alloy layer of the tinplate did not dissolve at potentials where the anodic dissolution of the tin occurred.

Nechaev et al.[245] reported the results of experiments designed to investigate the influence of organic compounds on the electrodeposition of tin from sulphate solutions. Using the expression:

$$E_n = \alpha + k\,\beta$$

where E_n is the energy of n electrons in the organic compounds and α and β are the coulombic and resonance integrals for atoms of carbon in aromatic compounds, it was shown that substances having electron energies equal to 0.5— 0.6 β may be adsorbed on the surface of tin in acid-tinning solutions, independent of their structure and the nature of their substituted groups.

Evidence for the photoelectric effect in the case of a plane electrode in the $Sn^{2+}|Sn$ system has been presented by Jeanne and Laforge-Kantzer.[246] A study of the current–potential–time relationship showed that the kinetics of establishment of potential are different in the presence and absence of light. The E–i data in the high overpotential region ($\eta > 10$ mV) were plotted as E vs. $\log[i/(1 - i/i_\infty)]$ and the 'illuminated electrode' data formed a series of points lying above those corresponding to the 'dark reaction'.

Titanium.—Work reported[247] was confined to a study of oxide film growth.

Tungsten.—In the first of two papers Heumann and Stolica[248] described experiments on the anodic dissolution of tungsten in solutions of pH 0—12. Potential–current density curves consisted of a Tafel region and a plateau, showing no discontinuity characteristic of oxygen evolution. The plateau current was proportional to $(C_{OH^-})^{0.3}$ for pH 2.2—8 and the open-circuit potential decreased by 43 mV for each pH unit. A second paper[249] showed that at pH 13—14 the open circuit became more positive as time elapsed up to 18 h, becoming constant thereafter. The Tafel slope indicated an αz value of 0.74 and was independent of the electrode pretreatment. The order of reaction with respect to OH^- was unity and the valency of the dissolved tungsten was 6.0. The rate-determining step for the dissolution of tungsten was considered to be:

$$WO_2 + OH^- \rightarrow WO_3 + H^+ + 2e^-$$

[245] E. A. Nechaev, N. T. Kudryavtsev, and G. I. Medvedev, *Elektrokhimiya*, 1971, **7**, 383.
[246] A. Jeanne and D. Laforge-Kantzer, *Compt. rend.*, 1970, **271**, C, 1502.
[247] I. A. Ammar and I. Kamal, *Electrochim. Acta*, 1971, **16**, 1539, 1555.
[248] T. Heumann and N. Stolica, *Electrochim. Acta*, 1971, **16**, 643.
[249] T. Heumann and N. Stolica, *Electrochim. Acta*, 1971, **16**, 1635.

($z = 2$, $\alpha = 0.37$, $n = 1$). The WO_3 so formed dissolved according to:

$$WO_3 + OH^- \rightarrow HWO_4^- \rightleftharpoons WO_4^{2-} + H^+ \text{ for pH} > 8$$

The authors considered other possible reactions to be less probable.

In acid and alkaline solutions Johnson and Wu[250] have made studies of the faradaic efficiency for the dissolution of tungsten, and the results also showed that tungsten was oxidized to the W^{6+} state. In acidic solutions it was reported that a thick film of WO_3 was formed. An anodic dissolution mechanism was proposed which involved a sequence of reactions in which a surface film W_2O_5 was further oxidized to WO_3, which then dissolved upon hydrolysation.

$$2W_{(s)} + 5H_2O(aq) \rightarrow W_2O_{5(s)} + 10H^+(aq) + 10e^-$$
$$W_2O_{5(s)} + H_2O(aq) = W_2O_5OH_{(s)} + H^+(aq) + e^-$$
and/or $W_2O_{5(s)} + OH^-(aq) = W_2O_5OH_{(s)} + e^-$
$$W_2O_5OH_{(s)} + W_2O_{5(s)} \rightarrow W_2O_5\text{—OH—}W_2O_{5(s)} \text{(rate-determining step)}$$
$$W_2O_5\text{—OH—}W_2O_{5(s)} \rightarrow W_2O_3\text{—O—}W_2O_{5(s)} + H^+(aq) + e^-$$
$$W_2O_5\text{—O—}W_2O_{5(s)} \rightarrow 2WO_{5(s)} + W_2O_{5(}$$
$$WO_{3(s)} + H_2O(aq) \rightarrow H_2WO_{4(s}$$
$$H_2WO_{4(s)} \rightarrow H^+(aq) + HWO_4^-(aq)$$
and/or $H_2WO_{4(s)} + 2OH^-(aq) \rightarrow 2H_2O(aq) + WO_4^{2-}(aq)$

The extent of the attack at constant current upon a tungsten surface by aqueous solutions of $HClO_4$, HNO_3, 1 M-HCl, 0.5 M-H_2SO_4, and 0.33 M-H_3PO_4 was examined.[251] Under the experimental conditions Ammar and Salim concluded that the corrosion of the electrode accounted for a relatively small amount of dissolution.

Uranium.—Jouve et al.[252] have published details of a study of the electrochemical behaviour of uranium in aqueous solutions of simple salts. It was reported that the potential of dissolution of -1.2 V (s.c.e.) is a mixed potential determined by the two equilibria:

$$U \rightarrow U^{3+} + 3e^-; E = -1.798 + 0.0197 \log[U^{3+}]$$

and

$$U^{3+} + 2H_2O \rightleftharpoons UO_2 + 4H^+ + e^-; E = -0.38 - 0.2364(pH) - 0.059 \log[U^{3+}]$$

The polarization (i–η) curves obtained depended upon the previous history of the electrode, for example whether it was mechanically polished or electropolished. The reactions were irreversible, as shown by curves which shifted with dE/dt. The constitution of the film produced at pH ≤ 7 was UO_2; however, according to the results of electron diffraction studies, at pH 13 the

---

[250] J. W. Johnson and C. L. Wu, *J. Electrochem. Soc.*, 1971, **118**, 1909.
[251] I. A. Ammar and R. Salim, *Corrosion Sci.*, 1971, **11**, 591.
[252] G. Jouve, M. Aucouturier, and P. Lacombe, *Compt. rend.*, 1971, **272**, C, 598.

oxide was $UO_3, 2H_2O$. The behaviour of the system was very complex and was further complicated by the thinness and friable nature of the electrode layers.

Zinc.—Electropolishing of zinc in aqueous $ZnCl_2$ solutions has been studied using spectral measurements. Dmitriev and Gubskaya[253] have established that an oxide-type film was formed on the surface of the dissolving metal and that the presence of this film was necessary to achieve polishing. Photoelectrochemical studies established that in the presence of the (white) film a photoeffect was observed which pointed to semiconductor properties of the film.

The anodic dissolution of zinc in aqueous salt solutions frequently results in anodic efficiencies greater than 100%, and often an 'apparent' valency is used to express the deviation from Faraday's laws. Johnson, Sun, and James[254] have studied the anodic dissolution in aqueous solutions containing Cl^-, Br^-, I^-, Ac^-, SO_4^{2-}, and NO_3^-. Significant deviations from Faraday's law were found only in the case of NO_3^-; these were functions of concentration, current density, i, and temperature. The experimentally determined apparent valency Z_i was correlated by the equation:

$$Z_i = 2 - 58.8 \, i^{0.52} (c_{NO_3^-})^{0.33} \exp(-1960/T)$$

The mechanism of the dissolution process was that local corrosion occurred simultaneously with anodic dissolution, and the combined action of these undermined and detached areas of the surface, causing surface disintegration and an apparent valency < 2. A zinc dissolution scheme was proposed:

$$Zn_{(s)} + H_2O = ZnOH_{(s)} + H^+(aq) + e^-$$
$$ZnOH_{(s)} = ZnO + H^+(aq) + e^-$$
$$ZnO_{(s)} + H_2O(aq) = Zn^{2+}(aq) + 2OH^-(aq)$$

The corresponding cathodic reaction associated in the case of 'local cell' reaction would be:

$$NO_3^- + H_2O + 2e^- = NO_2^- + 2OH^-$$

It was agreed by the authors that this proposed reaction sequence provided the reasons for the dependence of Z_i on i and $c_{NO_3^-}$.

With zinc the importance of adsorbed species present at the interphase has repeatedly been emphasized. In a paper dealing with the behaviour of zinc in alkaline solutions, Hull and Toni[255] have made measurements using linear-sweep voltammetry at rotating disc and rotating ring-disc electrodes. The results indicated that the formation of films occurs by an 'adsorption' rather than a 'dissolution–precipitation' model. The presence of two different types of film at unamalgamated electrodes was confirmed; only one type of film occurred at amalgamated electrodes.

[253] V. A. Dmitriev and V. P. Gubskaya, *Elektrokhimiya*, 1971, **6**, 1350.
[254] J. W. Johnson, Y. C. Sun, and W. J. James, *Corrosion Sci.*, 1971, **11**, 153.
[255] M. N. Hull and J. E. Toni, *Trans. Faraday Soc.*, 1971, **67**, 1128.

The control of dendritic growth of zinc from alkaline zincate electrolyte solutions still attracts attention. This problem, in a number of cases, can be controlled to some extent by the use of periodically changing currents. The morphology of zinc (and copper) deposits obtained by mass-transport-controlled electrodeposition has been examined by Despić and Popov.[256] It was found that by imposing a pulsed cathodic current of varying frequency on electrodes in the potential region where the resulting current densities were large enough to produce complete concentration polarization at the electrode, the morphology of the deposit could be affected. At low frequency (~ 10 Hz), deposits exhibited a surface roughness which increased with time in a manner similar to that observed with galvanostatic deposition; on the other hand, at higher frequency ($\sim 10\,000$ Hz) no increase in roughness was noted. It was considered that the results could be explained in terms of reaction-layer thickness. At very high frequency this layer is thin and follows the microprofile of the surface so closely that the diffusion flux and the resulting deposit are even. No amplification of surface irregularities therefore takes place. For the dissolution of zinc in alkaline solution, Kano et al.[257] produced evidence, from sweeping potential measurements at different concentrations of hydroxide ion, that the process was a simultaneous dissolution in the solution phase and formation of $Zn(OH)_2$ adsorbed on the zinc surface. A number of papers concerning the passivation of zinc electrodes in alkali have appeared during the review period.[258-260] The effect of CO_3^{2-} on the anodic behaviour of horizontal zinc anodes in ~ 8M-KOH, investigated by observing the galvanostatic passivation time, was to reduce the active dissolution region.[260] The phenomenon was explained as a result of the increase in viscosity of the solution. Passivation experiments have been extended[261] to a study of porous electrodes, and an investigation of electrodes prepared from pressed powdered zinc metal showed that the ohmic resistance was a major factor in the polarization process. Thermodynamic data concerning the 'Ag|Zn' cell have been calculated.[240]

5 Corrosion Processes

Theoretical.—The teaching of corrosion sciences using potential–pH diagrams was the subject of a paper by Bockris.[262] The burden of the argument was that corrosion technologists should rather be introduced to the concepts of electrochemical kinetics with emphasis upon connections with solid-state physics and quantum mechanics than to traditional thermodynamics. The principles of

---

[256] A. R. Despić and K. I. Popov, *J. Appl. Electrochem.*, 1971, **1**, 275.
[257] G. Kano, K. Horita, M. Okazaki, and Y. Matsubara, *Kinzoku Hyomen Gijutsu (J. Metal Finish. Soc. Japan)*, 1970, **21**, 60.
[258] T. P. Dirkse and N. A. Hampson, *Electrochim. Acta*, 1971, **16**, 2049.
[259] T. P. Dirkse and D. J. Kroon, *J. Appl. Electrochem.*, 1971, **1**, 293.
[260] Y. Sato, H. Niki, and T. Takamura, *J. Electrochem. Soc.*, 1971, **118**, 1269.
[261] R. N. Elsdale, N. A. Hampson, P. C. Jones, and A. N. Strachan, *J. Appl. Electrochem.*, 1971, **1**, 213.
[262] J. O'M. Bockris, *Corrosion Sci.*, 1971, **11**, 889.

electrode kinetics have been used to provide a better understanding of the fundamentals of the theory and practice of cathodic protection.[263] It was shown that the kinetics of anodic dissolution of the corroding metal determine the cathodic potential required for adequate protection. The great usefulness of this approach was demonstrated by the discussion of sacrificial anodes, resistance interferences, and constant-potential cathodic protection. Unz[264] has pointed out that the interpretation of polarization measurements obtained in galvanic corrosion studies may be misleading when care is not taken to consider circulating currents which may be developed during the tests. A method was proposed for determining the true values of the 'self potential' of the specimen and the induced 'polarization potential'. It was apparent that, if circulating currents are present, errors should be computed. However, in a uniform environment, surface-potential readings can be directly evaluated by means of general theory of potential, although when the conditions are complex an interrupted (step-function) test can be applied. Mansfeld[265] has considered area relationships in galvanic corrosion, and in a further paper with Oldham[266] deals with the linear polarization method for measurement of corrosion rates. It was pointed out that only in exceptional circumstances will the polarization curve be linear at the corrosion potential (it would be at the equilibrium potential). In fact the presence of linearity would be suspect since it probably indicates the presence of significant ohmic control. A further related paper[267] considered the Stern–Geary equation:

$$S = I_k\{(1/b'_{1a}) + (1/b'_{2c})\}$$

where S is the slope of the i–E (polarization) curve in the vicinity of the corrosion potential, I_k is the corrosion current density, and the factors b' are the Tafel slopes of the respective reactions:

$$M = M^{n_1+} + n_1 e^- \ (1a); \text{ and } \tfrac{1}{2}O_2 + H_2O + 2e^- = 2OH^- \ (2c)$$

It was shown that this relation applies if and only if the corrosion potential lies far from the reversible potentials for the two reactions.

The relatively recent (1960—1970) literature on pitting corrosion has been reviewed[268] and it was concluded that none of the existing theories explains all the known phenomena of pitting. Vermilyea[269] defined the critical pitting potential as 'the least positive potential at which pits can be grown'. He considered that this may be the potential at which the metal salt of the aggressive ion in solution is in equilibrium with the metal oxide. Such an equilibrium may be established when the activity of the anion in the pits has been increased because of the potential difference between the inside and outside of the pit.

---

[263] D. A. Jones, *Corrosion Sci.*, 1971, **11**, 439.
[264] M. Unz. *Corrosion Sci.*, 1971, **11**, 169.
[265] F. Mansfeld, *Corrosion*, 1971, **27**, 436.
[266] K. B. Oldham and F. Mansfeld, *Corrosion*, 1971, **27**, 434.
[267] F. Mansfeld and K. B. Oldham, *Corrosion Sci.*, 1971, **11**, 787.
[268] Z. Szklarska-Smialowska, *Corrosion*, 1971, **27**, 223.
[269] D. A. Vermilyea, *J. Electrochem. Soc.*, 1971, **118**, 529.

Vermilyea shows that the theory predicts the critical pitting potentials for Al, Mg, Fe, and Ni but fails for Zn, Ti, and Ta.

The dependence of the electrode potential of a metal on mechanical stress determines the importance of stressed areas to the overall pattern of corrosion. Karpenko et al.[270] have concluded that the dependence of potential on stress can be used to develop sensitive and relatively simple methods for determining the stress concentration in microscopic volumes of the metal. This, it was suggested, could be used for studying various types of loading in active media, and for studying the kinetics of crack formation during corrosion cracking, corrosion fatigue, and so on. The stress corrosion of metals has been considered from the viewpoint of linear elastic fracture mechanisms.[271] In his lecture, Nielsen[272] speculated on stress-corrosion mechanisms and emphasized that although the incidence of failure could be controlled by judicious design, construction, operation, and maintenance of equipment, complete understanding of stress-corrosion-cracking mechanisms remained to be achieved. The broad features of the mechanism of corrosion cracking involve anodic dissolution of a ferrous alloy whereby nickel or another 'noble' metal species is locally enriched, becoming sufficiently cathodic for evolution of hydrogen to occur and so embrittle the de-alloyed structure. The role of tensile stress in the system is to give continuality and directionality to the corrosion process. Most phenomenological features observed in alloys which have failed by stress-corrosion cracking can be matched to this model. The conclusion has been drawn from the growing body of data that the role of hydrogen as an embrittling agent may provide a possible unifying mechanism in stress corrosion, and is worthy of reconsideration. The influence of chloride ions on the anodic behaviour of metals has been considered in the case of some selected metals.[273] High metal–metal bond energies and high lattice energies of metal oxides were ascribed as reasons for the small tendency of Ti, Ta, Mo, W, and Zr to leave the lattice of the corresponding metal or metal oxide, respectively. Furthermore, even when their atoms do pass into solution they are irreversibly stabilized by the formation of stable complexes of the 'inner orbital' type. Opposite considerations apply for the case of easily corroding metals such as Al, Fe, Cr, and Fe–Cr alloys. The exceptional ease with which aluminium corrodes on anodization in the presence of a variety of anions was discussed in terms of the foregoing and other related concepts.

Corrosion inhibition is of considerable technological importance. The general physico-chemical properties of corrosion inhibitors were considered by Vosta and Eliásek,[274] who claimed that such quantities as the dipole moment or electron density played an important rôle. From these and other assumptions an attempt was made to apply quantum-chemical considerations

[270] G. V. Karpenko, I. E. Zamastyanik, Yu. I. Babei, and V. I. Pakhmurskii, *Elektrokhimiya*, 1971, **7**, 220.
[271] C. Tyzack, *Brit. Corrosion J.*, 1971, **6**, 219.
[272] N. A. Nielsen, *Corrosion*, 1971, **27**, 173.
[273] A. K. Vijh, *Corrosion Sci.*, 1971, **11**, 161.
[274] J. Vosta and J. Eliásek, *Corrosion Sci.*, 1971, **11**, 223.

to the selection and evaluation of corrosion inhibitors, e.g. of substituted pyridine derivatives. It was verified that the most reactive site in the compounds studied is the O in the N—O and S—O groups. The influence of γ-substituents upon the N-oxide inhibition centre was examined and the dimethylamino-group appeared to be the most effective compared with such groups as —SH, —C≡N, —CH$_3$, and —NO$_2$. Evans [275] has reviewed the fields of inhibition and passivity.

Within the review period few new instrumental procedures have emerged. However, an electronic zero-resistance ammeter with instantaneous characteristics has been described,[276] in which an operational amplifier was used as a voltage controller to replace the manual balance part of a conventional instrument. A continuous readout of corrosion current was therefore obtained under short-circuit coupling.

Corrosion of Individual Metal Systems.—*Aluminium.* Johnson[277] has emphasized the practical importance of pitting, and proposed that pits start from flaws in the surface oxide film. The effect of the aggressive ions Cl$^-$ and SO$_4^{2-}$ and the presence of metallic copper was also stressed. Intergranular corrosion has been shown to be a function of lattice disorientation.[278]

Antimony–Lead Alloy. Mao and Rao[279] have discussed the mechanism by which silver additions (up to ~0.2%) inhibit the anodic corrosion of 4.5% Sb–Pb in H$_2$SO$_4$ solutions. (This is a technologically important system in the energy-storage field.) The corrosion resistance of these alloys was shown to be concentration-dependent and to increase with increasing silver concentration. Microstructures indicated that in the case of 4.5% Sb alloys corrosion takes place by selective dissolution of the antimony-rich phase which forms a continuous network in cast alloys and a discontinuous network in homogenized quenched alloys. Whereas in Ag (4.5%) Sb–Pb alloys corrosion took place mostly by the oxidation of the lead-rich phase, no preferential attack of the antimony-rich phase occurred in homogenized quenched alloys. The improvement in the structure of the metal by the distribution and location of antimony and silver in ternary alloy structures was therefore the principal mechanism by which silver inhibits anodic attack in these alloys.

Copper and Related Alloys. Taylor[280] has studied the corrosion behaviour of copper and naval brass in 0.5 M-NaCl solutions at room temperatures. It was shown that pure copper corroded only *via* Cu$^+$ over the range of corrosion current ~2.5 to ~900 × 10^{-6} A cm^{-2}. No evidence was presented for the presence of Cu^{2+} or for film-forming reactions. Naval brass (60.54% Cu,

---

[275] U. R. Evans *Electrochim. Acta* 1971, **16**, 1825.
[276] W. D. Henry and B. E. Wilde, *Corrosion*, 1971, **27**, 479.
[277] K. W. Johnson, *Brit. Corrosion J.*, 1971, **6**, 200.
[278] M. Froment and C. Vignaud, *Compt. rend.*, 1971, **272**, C, 165.
[279] G. W. Mao and P. Rao, *Brit. Corrosion J.*, 1971, **6**, 122.
[280] A. H. Taylor, *J. Electrochem. Soc.*, 1971, **118**, 854.

38.7% Zn, 0.76% Sn) corroded slowly via Cu^+ and Zn^{2+} at rates up to $\sim 450 \times 10^{-6}$ A cm^{-2}, and again the absence of Cu^{2+} was reported. The selective dissolution of zinc was not observed. The results suggested that a homogeneous solution reaction occurred in which soluble Cu^+ species underwent subsequent hydrolysis, and no evidence apparently could be found for the direct formation of CuOH and Cu_2O on the corroding surface. Benzotriazole, an effective corrosion inhibitor for copper, has been examined by Mansfeld et al.[281] The mechanism of inhibition was shown to be via its chemisorption at the interphase, preventing the adsorption of oxygen and the formation of a prenucleation layer, which is the forerunner of oxide formation. The rate of formation of tarnish film has been shown to be an important factor in the stress-corrosion cracking of brass.[282] Irradiation which retarded film formation on copper, but not the corrosion of copper, prevented stress cracking of copper in cupric acetate solution. In the absence of stress, oxide formed all over the metal surface, whereas the effect of stress was to concentrate oxide formation at the grain boundaries. This type of experiment pointed to a brittle-film rupture mechanism of stress cracking.

Copper–nickel alloys are important technologically. Uhlig et al.[283] have alloyed manganese with Cu–Ni alloys and studied the passivity of the resulting ternary alloys. Potentiostatic polarization curves and potential decay techniques were used for alloys in H_2SO_4 solutions. It was established that manganese increases the passive properties of the alloys, but less so than does nickel. The results were interpreted as arising from contributed d-electron vacancies which favoured adsorption of oxygen and the formation of a passive film. Uhlig[284] returned to this aspect of Cu–Ni alloys in a later paper. Chromium has been shown to have a harmful effect on the rate of corrosion of copper.[285]

Iron, Steel, and Associated Alloys. The majority of the published papers connected with corrosion fall into this group. However, the papers selected for mention in this Report have been chosen in view of their electrochemical content rather than the engineering or technological value, and this section is extremely selective.

In phosphate solutions the anodic behaviour of iron has been examined.[286] Anodic potential data showed that the passivation process of iron is preceded by the formation of a layer of iron phosphate, although the process was apparently more complex than a simple coating of the electrode by a precipitated salt. Akiyama and Saji[287] studied zone-refined iron in electrolytes based on $1M-H_3PO_4$. The corrosion potentials and currents were linearly related to

---

[281] F. Mansfeld, T. Smith, and E. P. Parry, *Corrosion*, 1971, **27**, 289.
[282] E. Escalante and J. Kruger, *J. Electrochem. Soc.*, 1971, **118**, 1062.
[283] H. H. Uhlig, F. Mansfeld, and M. Kesten, *Z. phys. Chem. (Frankfurt)*, 1971, **74**, 216.
[284] H. H. Uhlig, *Electrochim. Acta*, 1971, **16**, 1939.
[285] L. Giuliani, A. Tamba, and C. Modena, *Corrosion Sci.*, 1971, **11**, 485.
[286] F. Zucchi and G. Trabenelli, *Corrosion Sci.*, 1971, **11**, 141.
[287] A. Akiyama and T. Saji, *Denki Kagaku*, 1971, **39**, 210.

the pH of the solution. The steady-state anodic polarization curves were composed of two regions (40 mV per decade at low current density and 60 mV per decade at high current density), and were observed to be similar to those corresponding to H_2SO_4. The results of the cathodic polarization study in H_3PO_4 were different from those of the cathodic study in H_2SO_4 owing to the development of an insulating film of (apparently) ferrous phosphate. Dissolved oxygen did not affect the dissolution mechanism of iron in H_3PO_4. Nord and Bech-Nielsen[288] have reported that in several electrolytic systems i–E curves on iron exhibited two anodic maxima. In the case of acetate solutions at rather low pH both maxima were analysed. It was shown that each of the maxima could be accounted for by assuming adsorption of a layer on the metal surface. The degree of coverage at equilibrium was related to potential, pH, and anion concentration. It was shown that a Langmuir-type adsorption, combined with some increase in Tafel slope for the uninhibited process in the range of adsorption, probably gives the best explanation of the deviations from ideal shape that are observed with the directly calculated isotherms. At pH > 3—4 the anodic behaviour of pulse-polarized iron[289] conformed to the formation of a surface oxide or hydroxide, in addition to the dissolution process. Under these conditions, arrest potentials were reported in the anodic pulse-polarization curves. These potentials corresponded to the Flade potential of passive iron. The dissolution of iron in sulphuric acid solutions has been considered[290] in terms of observed Tafel relationships, and for the simple corrosion process some doubt has been cast on the validity of the steady-state data for mechanistic interpretations.[291]

In unbuffered neutral chloride solutions[292] iron shows anodic Tafel slopes of 40 mV per decade for fast polarization and 80 mV per decade for slow or steady-state polarization. Steady-state anodic characteristics, studied by superimposing short cathodic potential and current pulses, showed a Tafel relation, with a Tafel slope of ~ 120 mV per decade for both potentiostatic and galvanostatic pulses. The differences in anodic slopes have been interpreted as indicating the formation of a hydroxo-chloro-iron complex, $[Fe(OH)_m(Cl)_n]^{2-m-n}$. The cathodic pulse-polarization results indicate that $m = 1$. Asakura and Nobe[293] also considered alkaline solutions, and using potential-step and linear-sweep voltammetry they explained their results by assuming the existence of complex ions $Fe(OH)_6^{4-}$ and $Fe(OH)_6^{3-}$ in the oxide film, the electrometric results being governed by composition changes in the mixed oxide layer due to the diffusion of Fe^{II}.

The anodic polarization of stainless steels has been studied.[294—296] Lizlovs

[288] H. Nord and G. Bech-Nielsen, *Electrochim. Acta*, 1971, **16**, 849.
[289] Y. Torigoe, *Denki Kagaku*, 1971, **39**, 572.
[290] W. E. Reid, *Corrosion*, 1971, **27**, 407.
[291] S. Barnartt, *Corrosion*, 1971, **27**, 467.
[292] S. Asakura and K. Nobe, *J. Electrochem. Soc.*, 1971, **118**, 13.
[293] S. Asakura and K. Nobe, *J. Electrochem. Soc.*, 1971, **118**, 536.
[294] L. A. Lizlovs and A. P. Bond, *J. Electrochem. Soc.*, 1971, **118**, 22.
[295] P. Forchhammer and H.-J. Engell, *Corrosion Sci.*, 1971, **11**, 49.
[296] G. Bombara, A. Tamba, and N. Azzerri, *J. Electrochem. Soc.*, 1971, **118**, 676.

The Solid Metal Electrode in Aqueous Solution 89

and Bond[294] have made potentiodynamic polarization experiments in 0.5 M-H_2SO_4 and 1 M-HCl, using 25% Cr ferrite stainless steels containing 0—5% molybdenum and 0—4% of nickel. High-purity alloys and some containing C, N, Si, and Mn were investigated. The critical current density for both acids was decreased by the addition of molybdenum or molybdenum and nickel. All the high-purity 25% chromium steels containing 3.5 or 5% molybdenum were immune to pitting corrosion and highly resistant to crevice corrosion in aggressive solutions (*e.g.* 1 M-HCl, 0.33 M-$FeCl_3$); with alloys containing impurities, however, resistance to pitting corrosion was also improved by adding molybdenum. Forchhammer and Engell[295] have made linear-sweep voltammetry measurements between potentials at which the h.e.r. and the o.e.r. become important. The experiments did not show conclusively any preferred adsorption of chloride ions on 18/8 Cr–Ni or Cr–Ni–Mo steels, although there was evidence that chloride ions increase the dissolution current of steel in the passive region, and that SO_4^{2-} and NO_3^- lower the dissolution current. The potentiostatic anodic pickling of stainless steels has also been considered.[296]

A large number of the investigations reported deal with passivation and the formation of films on iron,[297—304] and are on the whole outside the present Reporter's brief. Some of these papers deal with studies in the rapidly developing ellipsometric technique,[297, 298, 300, 303] and this type of approach seems very promising in the elucidation of reactions at iron surfaces. Another important aspect which is of marginal interest in the present Report, although of considerable technological importance, is the study of crevice and pitting corrosion of iron and steels. Typical papers[305—310] contain such speculative theory as is inevitable in the face of very complex observations.

Asakura and Nobe[310] have studied the anodic behaviour of iron in alkaline solutions of chloride ion, using potential-sweep and galvanostatic methods. In the active dissolution region Tafel slopes of 65 and 66 mV per decade were obtained from the galvanostatic and potential-sweep experiments, respectively. The peak current i_p from potentiodynamic curves was proportional to the square root of the sweep rate, and the peak potential varied linearly with the

---

[297] J. L. Ord and D. G. De Smet, *J. Electrochem. Soc.*, 1971, **118**, 206.
[298] J. O'M. Bockris, M. A. Genshaw, V. Brusic, and H. Wroblowa, *Electrochim. Acta*, 1971, **16**, 1859.
[299] Y. Torigoe, *Denki Kagaku*, 1970, **38**, 772.
[300] H. Wroblowa, B. Brusic, and J. O'M. Bockris, *J. Phys. Chem.*, 1971, **75**, 2823.
[301] R. P. Frankenthall, *Electrochim. Acta*, 1971, **16**, 1845.
[302] G. Kreysa, U. Ebersbach, and K. Schwabe, *Electrochim. Acta*, 1971, **16**, 1489.
[303] N. Sato and K. Kudo, *Electrochim. Acta*, 1971, **16**, 447.
[304] M. Okuyama, T. Tsuru, S. Haruyama, and K. Nagasaki, *Denki Kagaku*, 1971, **39**, 8.
[305] G. Karlberg, *Coll. Czech. Chem. Comm.*, 1971, **36**, 377.
[306] H. H. Strehblow, K. J. Vetter, and A. Willgallis, *Ber Bunsengesellschaft phys. Chem.*, 1971, **75**, 822.
[307] M. Jamk-Czachor, *Brit. Corrosion J.*, 1971, **6**, 57.
[308] G. Karlberg and G. Wranglen, *Corrosion Sci.*, 1971, **11**, 499.
[309] F. G. Wilson, *Brit. Corrosion J.*, 1971, **6**, 100.
[310] S. Asakura and K. Nobe, *J. Electrochem. Soc.*, 1971, **118**, 19.

log of the sweep rate, with a slope of 30 mV per decade. The results could be interpreted by assuming that the electrode became passive when the anodically produced iron chloride complex attained the solubility limit.

The inhibition of corrosion brought about by the presence of added substances is generally explained by the adsorption of ions or molecules at the electrode and their interfering with aggressive attack. Ammar and Khalil[311] present evidence for the adsorption of Br^- and I^- onto iron from H_2SO_4 solutions and have reported the effect of temperature on the cathodic polarization of the electrode. Both Br^- and I^- decrease the rate of the h.e.r., not through changes in the apparent heat of activation as much as through causing a decrease in the pre-exponential factors. It was concluded that the effect of the halide was due to blocking of the electrode surface, and a self-consistent calculation of coverage based on the Langmuir isotherm as a function of temperature enabled an estimate of the enthalpy and entropy of adsorption to be made. Tsai and Hackerman[312] have studied the surface coverage by amides on iron, and their results revealed that (since a coverage maximum was always observed in the E_{crit}–E_p region) the adsorption and desorption processes were potential-dependent. In the active region, the adsorption of amide was predominant and reversible. However, beyond the critical potentials the adsorption of amide is predominant and is irreversible in the passive region. In connection with adsorption properties of iron it has been shown that the adsorption properties of iron in acid solution are influenced by the mechanical and thermal treatment.[313] Thus it was shown that the texture and concentration of defects at the surface exerted a strong effect on the adsorption properties of the metal, and it was noted that the steps of the dislocation appearing on the deformation of the crystallites may also be potential centres of high adsorption energy. Annealing of deformed iron at 750—800 °C led to a rapid decrease in the adsorption properties although the microrelief of the surface changed to a smaller degree. The mechanism of the corrosion inhibition of iron by sodium benzoate has been of interest for some years. The most recent papers [314, 315] deal with the effect of air on the inhibitive properties, and it was reported that in air-saturated solutions in which sodium-benzoate-inhibited iron is corroding, increasing the concentration of sodium benzoate affects the rate of oxygen diffusion to the electrode surface. Experiments in de-aerated and air-saturated solutions showed clearly that sodium benzoate is unable to inhibit in de-aerated solutions. Allyl halides act as inhibitors of the corrosion of iron in dilute hydrochloric acid.[316] The order of relative bonding abilities of π-electrons in the double bond with metal atoms at the surface seemed to be $I > Br > Cl$, while that of the halogen atoms was $Cl > Br > I$. The inhibitive efficiency of allyl halides on the corrosion of

---

[311] I. A. Ammar and M. W. Khalil, *Electrochim. Acta*, 1971, **16**, 939.
[312] K-C. Tsai and N. Hackerman, *J. Electrochem. Soc.*, 1971, **118**, 28.
[313] V. V. Batrakov, Yu. A. Nikiforova, and Z. A. Iofa, *Elektrokhimiya*, 1970, **6**, 1017.
[314] D. E. Davies and Q. J. M. Slaiman, *Corrosion Sci.*, 1971, **11**, 671.
[315] Q. J. M. Slaiman and D. E. Davies, *Corrosion Sci.*, 1971, **11**, 683.
[316] S. Nakagawa and G. Hashizume, *Denki Kagaku*, 1970, **38**, 831.

iron was in the order I > Br > Cl; consequently it was argued that the allyl halides caused inhibition not by the function of the halogen but by the π-electrons. In a similar vein, the inhibiting efficiency of furan derivatives on the anodic dissolution of iron was found to be explicable on the basis of structure-dependent electron-donor properties.[317] It is interesting that this class of inhibitor brought about marked changes in the Tafel parameters of the dissolution reactions of iron. Cathodic and anodic polarization experiments have been made for the electrochemical behaviour of iron in 2M-HCl containing oximes and quinuclidine. It was concluded[318] that the inhibition observed, due to the presence of these compounds, of both the anodic dissolution of iron and of the cathodic h.e.r. was caused by the adsorption of organic molecules on the metal surface. Similarly, Chin and Nobe[319] have shown that the effect of benzotriazole in Fe–H$_2$SO$_4$ systems was to reduce the rate of the h.e.r., possibly by the effect of adsorption on the slow discharge step. Surface coverage determined from capacitance fitted a Langmuir adsorption isotherm, and corrosion rate data were consistent with the extent of surface coverage.

The influence of Cl$^-$ and SO$_4^{2-}$ on the corrosion of iron in H$_2$SO$_4$ (pH 3) has been investigated by Traubenberg and Foley,[320] and it was concluded that both ions could be aggressive or not, depending on conditions. Thus Cl$^-$ accelerates corrosion considerably when the anodic reaction contributes significantly to the over-all kinetics. On the other hand, SO$_4^{2-}$ accelerates corrosion when the reaction is under cathodic control. The protective action of the (so-called) anodic inhibitors (K$_2$Cr$_2$O$_7$ and NH$_2$·CH$_2$·CH$_2$OH) is decreased by Cl$^-$; the protective action of anodic inhibitors (thiourea) is decreased by SO$_4^{2-}$. The corrosion potential was found to be independent of chloride ion concentration but was shifted in the negative direction when SO$_4^{2-}$ was added. These observations have been interpreted in terms of the different properties of the complexes Fe–Cl$^-$ and Fe–SO$_4^{2-}$. Corrosion inhibited by adsorbed hydrogen, H$_{ads}$, in H$_2$SO$_4$ has been studied at a rotating disc electrode.[321] Hysteresis when the corrosion rate was low was due to the partial coverage of the surface with H$_{ads}$, in agreement with the role of (Fe H$_{ads}$) as a corrosion inhibitor. The effect of inhibitors in connection with oxide film formation has also been investigated using ellipsometric and potentiostatic techniques.[322]

Iridium. ^{192}Ir and ^{194}Ir isotopes with half lives of 74.4 days and 19 h, respectively, may be obtained by the irradiation of Ir with thermal neutrons. Of these, the ^{192}Ir isotope is suitable for radiochemical measurement of corrosion, and corresponding data have been presented by Llopis *et al.*,[323] who claimed that the sensitivity of the method allowed studies of the corrosion rates of

---

[317] H. Vaidyanathan and N. Hackerman, *Corrosion Sci.*, 1971, **11**, 737.
[318] E. Laengle and N. Hackerman, *J. Electrochem. Soc.*, 1971, **118**, 1273.
[319] R. J. Chin and K. Nobe, *J. Electrochem. Soc.*, 1971, **118**, 545.
[320] S. E. Traubenberg and R. T. Foley, *J. Electrochem. Soc.*, 1971, **118**, 106.
[321] I. Epelboin, P. Morel, and H. Takenouti, *J. Electrochem. Soc.*, 1971, **118**, 1282.
[322] B. J. Bornong and P. Martin, *Corrosion*, 1971, **27**, 315.
[323] J. Llopis, J. M. Gamboa, and V. Menéndez, *Coll. Czech. Chem. Comm.*, 1971, **36**, 528

iridium (said to be the most corrosion-resistant member of the Platinum Group) to be made even under conditions of passivation. When corrosion was more pronounced, spectrographic analyses of solutions showed that at concentrations of 6 mol l^{-1} or higher of HCl, [IrCl$_6$]$^{3-}$ and [IrCl$_6$]$^{2-}$ complexes were formed. However, it was reported that for concentrations lower than 1 mol l^{-1} the formation of aquo-chloro-complexes of Ir^{III} and Ir^{IV} takes place; a higher proportion of water molecules was present in the co-ordination shell at lower concentrations of HCl.

Molybdenum. Molybdenum remains inert in a dry environment and corrodes in wet media. In addition, molybdenum forms a non-protective porous corrosion film and dissolves in deionized water at 85 °C. Lee and Doo[324] have shown that the simultaneous presence of water and oxygen in these environments is the controlling factor for the corrosion of molybdenum films. An effective passivation of thin films could be provided by putting either a single layer of SiO$_2$ or a composite layer of SiN over SiO$_2$ on to the metal film.

Nickel. Tokuda and Ives[325] proposed a model for the aggressive chloride ion corrosion of nickel in which atoms are postulated to be removed according to a 'one by one' process that is governed by accessibility. It was proposed that the state of close-packing of a crystal surface was not the only factor to be considered in determining the relative dissociation rates of crystallographic surfaces when aggressive ions are present.

Experiments on the plastic deformation of nickel show that the process shifts passivation potentials to more negative values and increases the current density required for passivation.[326] However, recrystallization of a deformed specimen may effect a partial return of the passivation potential to its original (undeformed) value.

The pitting corrosion of nickel is of technological interest, and it was shown[327] that the active sites for the pitting of nickel are identical with the active sites for dissolution in the active region. Szklarska-Smialowska[328] has studied the kinetics of the pitting process of nickel in solutions containing SO$_4^{2-}$ and Cl$^-$, and it was reported that, depending on the Cl$^-$/SO$_4^{2-}$ concentration ratios, the exponent b in the equation $i = at^b$ had values from 0.15 to 2.5. At low Cl$^-$/SO$_4^{2-}$ ratios the exponent is >1.0, while at ratios >0.25 it is <1.0. Pits were formed near the breakdown potential at low Cl$^-$/SO$_4^{2-}$ ratios. The major application of nickel is as an alloy component with other metals in connection with corrosion-resistant materials, and it has been demonstrated[329] that alloys with the highest nickel content generally exhibit the best corrosion resistance to certain corrosive environments.

[324] L. H. Lee and V. Y. Doo, *J. Electrochem. Soc.*, 1971, **118**, 443.
[325] T. Tokuda and M. B. Ives, *J. Electrochem. Soc.*, 1971, **118**, 1404.
[326] I. Garz and U. Häfke, *Corrosion Sci.*, 1971, **11**, 329.
[327] T. Tokuda and M. B. Ives, *Corrosion Sci.*, 1971, **11**, 297.
[328] Z. Szklarska-Smialowska, *Corrosion Sci.*, 1971, **11**, 209.
[329] A. C. Hart, *Brit. Corrosion J.*, 1971, **6**, 205.

Zinc. Kawai et al.[330] have shown that quinoline and derivatives are effective inhibitors in the corrosion of zinc. This effect has been traced to the formation of insoluble zinc compounds on the metal surface. Zinc alloy behaviour has been considered by Williams et al.[331]

[330] S. Kawai, H. Kato, and Y. Hayakawa, *Denki Kagaku*, 1971, **39**, 288.
[331] P. N. Williams, B. G. Koehl, W. E. Berry, and E. S. Bartlett, *J. Electrochem. Soc.*, 1971, **118**, 1684.

4
Ionic Double Layers and Adsorption

BY D. J. SCHIFFRIN

1 Introduction

This Report covers the same topics as were presented in Volumes 1[1] and 2[2] of this series, and reference to previous work has been kept to a minimum.

In the period under review, the first fully automatic capillary electrometer has been successfully put into operation and this, coupled with on-line computational facilities, will very likely change both the amount and the quality of the interfacial information presently available. Something similar can be said about the application of phase-detection techniques to electrode impedance determinations, although very little systematic work seems to have been done up to the present time.

There has been considerable interest in the adsorption behaviour on platinum and on the platinum-group metals. The general picture which emerges is still rather sketchy, in particular in relation to the nature of the adsorbed state of atoms and molecules.

The English translation of a book by Damaskin, Petrii, and Batrakov[3] on the adsorption of organic compounds on mercury and the applications of interfacial thermodynamics to adsorption on solids has been published. Also, another review on organic adsorption by Damaskin and Frumkin has appeared.[4]

2 Ionic Adsorption and Double-layer Structure at a Liquid Metal–Solution Interface

General Characteristics.—The geometry of the measuring cell is often neglected in double-layer studies. Armstrong and Race[5] have shown that dispersion effects are to be expected when working with dilute electrolyte solu-

---

[1] D. J. Schiffrin, 'Electrochemistry', ed. G. J. Hills (Specialist Periodical Reports), The Chemical Society, London, 1970, Vol. 1, chap. 6.
[2] D. J. Schiffrin, 'Electrochemistry', ed. G. J. Hills (Specialist Periodical Reports), The Chemical Society, London, 1971, Vol. 2, chap. 4.
[3] B. B. Damaskin, O. A. Petrii, and V. V. Batrakov, 'Adsorption of Organic Compounds on Electrodes', ed. R. Parsons, Plenum Press, New York, 1971.
[4] B. B. Damaskin and A. N. Frumkin, 'Reaction of Molecules at Electrodes', ed. N. S. Hush, Interscience, London, 1971.
[5] R. D. Armstrong and W. P. Race, *J. Electroanalyt. Chem. Interfacial Electrochem.*, 1971, **33**, 285.

tions. In this case, the series combination of the interfacial capacitance and solution resistance is shunted by the geometrical capacitance of the cell. When the interfacial impedance is measured as a series combination, this results in an apparent relaxation at a concentration-dependent frequency, and a considerable dispersion is noted above 1 kHz. Measurements were performed in dilute NH_4F solutions (0.01—0.0002M), and the geometric capacitance calculated from these results agreed with that calculated from the electrode dimensions and the geometry of the cell. The need to take into account this experimental complication when studying relaxation effects and in studies involving dilute solutions was stressed.

It is perhaps unfortunate that glass capillaries are such an important chattel in the electrochemist's trade; their behaviour in electrocapillary studies tends to be erratic and irreproducible, and there are almost as many different ways of preparing them as workers in the field. It is for this reason that the less empirical approach by Trassati[6] to the origin of the errors commonly observed in interfacial tension measurements with the Lippmann electrometer in dilute solutions is very welcome. Trassati proposes a qualitative explanation of the discrepancies based on the theory of heterocoagulation between dissimilar particles, *i.e.* glass and mercury. A close parallel could be established between the strength of adhesion of mercury to glass at anodic potentials and the difference in the interfacial tension values measured with the capillary electrometer and those obtained by double integration of the capacitance data. No difference between the latter quantities could be found for concentrations greater than 0.15M, in agreement with detachment experiments, and this gives the lower concentration limit for the use of a conventional electrometer. Both capillary-size and anion-nature effects were found to be in accordance with the DLVO theory predictions. The above-mentioned discrepancies had also been attributed in the past to the a.c. signal present across the interface in double-layer capacitance measurements. Lawrence and Mohilner[7] have checked this point by measuring the electrocapillary curves of mercury in contact with 0.01M-NaF, 1.085M-NaCl, and 0.5M-KI solutions by the maximum bubble-pressure method, both with and without an a.c. signal superimposed on the d.c. polarizing potential. These authors found no a.c. effect and no difference between the electrocapillary curves obtained in this way and those calculated from capacitance measurements, giving further evidence that the discrepancies are only due to experimental inadequacies.

The usefulness of automatic drop-time measurements to obtain electrocapillary data has been exemplified by Verdier and Vanel[8] in a surface excess study of aqueous LiF and CsF. Since these measurements were performed at 25 °C, no anionic adsorption was detected (see ref. 7). The results indicated some specific adsorption of Cs^+ ion at cathodic potentials, as is already well documented.[2] An empirical relationship between the interfacial tension, γ,

[6] S. Trasatti, *J. Electroanalyt. Chem. Interfacial Electrochem.*, 1971, **31**, 17.
[7] J. Lawrence and D. M. Mohilner, *J. Electrochem. Soc.*, 1971, **118**, 259.
[8] E. Verdier and P. Vanel, *J. Chim. phys.*, 1971, **68**, 1437.

and the potential has been proposed and tested by Grand and Privat.[9] Their equation gave a good representation of the electrocapillary curve, but its second derivative did not reproduce accurately the experimental values of the capacitance. Nevertheless, its use might be envisaged in computer-fitting methods for electrocapillary curves, rather than the use of moving least-squares error polynomial fits.

Hills and Reeves[10] have attempted to rationalize the difference in adsorption behaviour of anions on mercury when measurements are made at constant ionic strength or on single-salt solutions. The specific adsorption of PF_6^- ion from 0.4M constant ionic strength KPF_6–KF mixtures was studied. An interesting feature of this system is that the *apparent* amount adsorbed, q'_{app}, as calculated from the Parsons–Hurwitz treatment, passes through a maximum value at a surface charge density on the metal, q^m, of ~ 8 μC cm^{-2}, and then decreases at values of q^m anodic to this, becoming negative at sufficiently anodic potentials. This effect was attributed to a significant specific adsorption of the fluoride ion in this potential region (see below). It was stressed that, in this treatment, what is really measured is the *relative* adsorption of one ion with respect to the other, that is,

$$q'_{app} = q'_{PF_6^-} - \frac{x}{M-x} q'_{F^-} \tag{1}$$

where x is the KPF_6 concentration and M the total molarity of the salt mixture. Thus the potential dependence of q'_{app} should be influenced by the values of q'_{F^-}. This dependence manifests itself more strongly for weakly adsorbed ions, as is the case for PF_6^-, and with some simplifying assumptions, allows an estimation of the values of q'_{F^-} to be made. As has been noticed for other mixtures, the value of the inner-layer capacitance at constant charge, $_{q'}C_i$, assumes negative values at anodic potentials. This result would imply, from a simple dielectric model of the inner region, that the inner Helmholtz plane (IHP) lies outside the inner region, or that the solvent desorption contribution to the potential drop there, ϕ^{m-2}, due to ionic adsorption, makes a significant contribution. This effect was attributed both to a combination of errors in the values of potential and adsorbed charge and to the resulting errors in the calculated diffuse-layer correction.

The influence of the form of the adsorption isotherm of two adsorbing anions on the calculated surface excess and double-layer parameters for salt mixtures at constant ionic strength has been analysed by Damaskin.[11] Modified virial isotherms of the form

$$\ln (\beta_1 IM) = \ln \Gamma_1 + 2B_{11}\Gamma_1 + 2B_{12}\Gamma_2 \tag{2}$$
$$\ln (\beta_2 IM) = \ln \Gamma_2 + 2B_{22}\Gamma_2 + 2B_{12}\Gamma_1 \tag{3}$$

were employed. Here, β is the adsorption equilibrium constant, I the ionic

---

[9] R. Grand and M. Privat, *J. Chim. phys.*, 1971, **68**, 1543.
[10] G. J. Hills and R. M. Reeves, *J. Electroanalyt. Chem. Interfacial Electrochem.*, 1971, **31**, 269.
[11] B. B. Damaskin, *Soviet Electrochem.*, 1971, **7**, 776.

strength, Γ_1 and Γ_2 the anionic surface excesses of species 1 and 2, and B_{11}, B_{12}, and B_{22} the virial coefficients reflecting the ionic interactions in the adsorbed layer. From equations (2) and (3), the errors introduced into the calculation of one ionic excess when the other is assumed negligible were evaluated. Care should be exercised, however, in the choice of the form of the compound adsorption isotherms. If the adsorbed layer is regarded as a surface phase, the Gibbs–Duhem relationship for its components must be consistent with the chosen isotherms. Three components must be considered, anions 1 and 2 and the solvent. At constant temperature and pressure,*

$$X_1^\sigma \mathrm{d}\ln a_1^\sigma + X_2^\sigma \mathrm{d}\ln a_2^\sigma + X_3^\sigma \mathrm{d}\ln a_3^\sigma = 0 \quad (4)$$

where X^σ represents the mole fraction of a component in the surface phase and a^σ its activity. Since the salts used for the mixtures are usually chosen with identical activity coefficients, $\mathrm{d}\ln a_3^\sigma = 0$, and

$$\left(\frac{\partial \ln a_1^\sigma}{\partial \ln a_2^\sigma}\right)_{T,P} = -\frac{1 - X_1^\sigma - X_2^\sigma}{X_1^\sigma} \quad (5)$$

For the virial equations (2) and (3), equation (5) is only valid for the trivial situation when $B_{11} = B_{22}$, i.e. when the two salts are the same. It would appear that the reason for this discouraging result is the incorrect way of expressing the interactions 1–2 through a term of the form $2B_{ij}\Gamma_j$, and perhaps a better approach would be to introduce these interactions through a single virial coefficient depending on the relative surface concentration of anions 1 and 2.

The specific adsorption of azide ion has been studied by D'Alkaine, Gonzalez, and Parsons[13] by measurements of electrocapillary curves and from capacitance data. N_3^- has a behaviour intermediate between that of Cl^- and Br^- ions. An interesting feature of this system is the tendency of the adsorption isotherms at constant charge in dilute solutions to reach a limiting value which is close to q^m. Other ions, such as Cl^- and NO_3^-, show a similar behaviour, and this was attributed to the influence of the outer Helmholtz plane (OHP) potential on the micropotential. A simple modified ideal-law type isotherm was used. This isotherm had already been employed (see ref. 23 in this work), and in it were considered the contributions to the micropotential due to both q^m and q'. When these and the OHP potential were taken into consideration, a virial-type isotherm was found, which reproduced reasonably well the experimental results. The importance of this approach is twofold: it introduces explicitly the OHP potential in the isotherm and gives an insight

---

[12] J. E. B. Randles and B. Behar, *J. Electroanalyt. Chem. Interfacial Electrochem.*, 1972, **35**, 389.
[13] C. V. D'Alkaine, E. R. Gonzalez, and R. Parsons, *J. Electroanalyt. Chem. Interfacial Electrochem.*, 1971, **32**, 57.

* Relations (4) and (5) are approximate since the surface tension is also a function of the surface composition, but, for the purpose of this discussion, this variation can be regarded as a second-order effect. For a more rigorous treatment see ref. 12.

into the meaning of the virial coefficient in terms of discreteness-of-charge theory. The parameter g, expressing the self-atmosphere potential in Levine's theory, was found to be $\sim 30\%$ lower than calculated, suggesting an overestimation in the theory. The question of the correct description of ionic adsorption has also been discussed by Schiffrin[14] in a study of the specific adsorption of fluoride ions on mercury at 0 and 15 °C. A significant amount of adsorption was found at anodic potentials and a Freundlich-type isotherm was found to describe correctly this system (see also ref. 10). This was interpreted in terms of the ionic equilibrium prevailing across a charged interface, leading to a concentration variable depending on the diffuse-layer charge of the form $\tau_+ \ln a_\pm^2$, where τ_+ has been termed the transport number across the double layer.[15] The potential drop across the inner layer was found to be uncharacteristic of ionic adsorption in that it passed through a maximum value at a concentration of ~ 0.3M. The negative values of $_{q'}C_i$ thus obtained in dilute solutions could be caused by the lack of coincidence of the OHP with the surface equipotentials corresponding to ϕ^{2-s} due to the discrete nature of the adsorbed charges. From the decrease of $_{q^m}C_i$ with q', values of the inner-layer capacitance in the absence of adsorption were estimated by extrapolation. In spite of the uncertainties in such procedure, it can be seen that the charge dependence of this quantity still shows a residual hump. In contrast with the inner-layer capacitance calculated on the assumption of the absence of specific adsorption C_i^*, the maximum of the hump appeared at potentials close to the p.z.c., and its temperature dependence did not show the anomalous behaviour of C_i^* at anodic potentials. It was suggested that this residual hump is a true dielectric one due to the solvent. This idea would appear to be in better accordance with the suggested preferred orientation of the water dipoles at the p.z.c. (see ref. 19). Some capacitance measurements of mercury in contact with SiF_6^{2-} solutions were also made and showed conclusively that the origin of the anodic rise in the capacitance at anodic potentials cannot be caused by the adsorption of this anion, since its adsorption leads to an actual decrease in the double-layer capacitance compared with results obtained using KF solutions.

Continuing with some previous work, Polyanovskaya, Frumkin, and Damaskin[16] have reported electrocapillary curves of mercury in contact with thallium salt solutions. The mixtures studied were $TINO_3 + KNO_3$, $TIF + NaF$, and $TINO_3 + KNO_3 + C_6H_5OH$. This non-equilibrium study yielded electrocapillary curves presenting two maxima in accordance with previous considerations.[2] There was some evidence for anion-induced cation adsorption and the effect of phenol was to reduce the adsorption of Tl^+ ions. Evidence for anion-induced cation adsorption of thallium ions on mercury was also found by Kivalo, Lindström, and Sundholm,[17] using a chronocoulometric

[14] D. J. Schiffrin, *Trans. Faraday Soc.*, 1971, **67**, 3318.
[15] D. C. Grahame, *J. Chem. Phys.*, 1948, **16**, 1117.
[16] N. S. Polyanovskaya, A. N. Frumkin, and B. B. Damaskin, *Soviet Electrochem.*, 1971, **7**, 558.
[17] P. Kivalo, M. Lindström, and G. Sundholm, *Suomen Kem.*, 1971, **44**, 151.

technique. The coadsorption of Tl^+ and I^- ions was studied and it was shown that the adsorption of the cation requires the initial adsorption of the anion. The relationships between work function, electronegativity, and potentials of zero charge have been extensively reviewed by Trassati,[18] and some interesting correlations have been brought to light. It is concluded that rather than a single linear relationship between work function (Φ) and the p.z.c. potential ($E_{p.z.c}$) for all metals, as has been suggested in the past, there is a relationship of the form

$$E_{p.z.c} = \Phi - \text{constant} \qquad (6)$$

the value of the constant depending on the electronegativity (χ_M) of the metal in question. The relationship proposed, based on a selection of literature values,[19] was

$$E_{p.z.c} = \Phi - 4.61 - 0.666(2.10 - \chi_M) \qquad (7)$$

From a combination of electrochemical and physical data, it was concluded that the orientation of water molecules at the mercury–aqueous solution interface is with the oxygen atom pointing towards the metal. Furthermore, the contribution of this oriented layer to the inner potential difference between the two phases was estimated as 80 mV, in agreement with values obtained from the adsorption of organic molecules. In relation to this, it is interesting to note that the value of the inner potential difference $\Delta\Phi_{Hg-soln.}$ is of the order of -0.3 V, for a non-adsorbed electrolyte at the point of zero charge.[2] This value has been estimated assuming a reasonable value (0.1 V) for the surface potential of water. We can thus estimate a contribution by the metal to $\Delta\Phi_{Hg-soln.}$ of the order of ~ -0.2 V, the negative sign implying that the centres of charge of the Hg^+ ions in the metal are closer to the solution side than their corresponding electron clouds. A way of evaluating the degree of orientation of adsorbed water was also proposed. The degree of orientation was defined as

$$\alpha - 4.0 - 2\chi_M \qquad (8)$$

and gives for Hg a value of 0.08, in accordance with the proposed weak orientation of water in this interface, a result which has also received some confirmation from work function variations with adsorption.[1] However, this relationship does not appear to give good results in the case of lead; Kalish and Burshtein[20] have recently measured the change in work function of Pb when water is adsorbed, and only observed a very small change in the value of the work function, whereas equation (8) would predict a value of α of 0.40.

Ionic Adsorption and Double-layer Theory.—The influence of discreteness-of-charge effects on the adsorption isotherm of ions has been discussed by

---
[18] S. Trasatti, *J. Electroanalyt. Chem. Interfacial Electrochem.*, 1971, **33**, 351.
[19] S. Trasatti, *Chimica e Industria*, 1971, **53**, 559.
[20] T. V. Kalish and R. Kh. Burshtein, *Soviet Electrochem.*, 1971, **7**, 108.

Levine.[21] The modifications introduced in the treatment presented, as compared with previous work by the same author, consist in treating the inner-layer dielectric constant as a function of the distance to the metal and considering that the adsorbed ions can be located not only at the IHP, but can also be distributed between the IHP and the OHP. A self-atmosphere potential contribution due to the ions in the diffuse double layer was introduced to take account of imperfect imaging conditions in the solution. The derived adsorption isotherms were found to be in good agreement with experimental results, with a value of the screening constant equal to unity (however, see ref. 13).

Lorenz[22] has formulated the idea of partial charge transfer on adsorption, in terms of the thermodynamics of irreversible processes near equilibrium. The electrochemical potential of an adsorbed ion ($\bar{\mu}_{ads}$) was considered as being determined by both a chemical and an electrical contribution, as has been done in the past in the formulation of isotherms for ionic adsorption.[1] The difference in this presentation is, however, in the physical meaning given to the potential dependence of the electrical contributions to $\bar{\mu}_{ads}$: in the Lorenz treatment this latter quantity involves only the part of the charge remaining on the ion (see also ref. 17 in ref. 2). In this connection, Barclay and Caja[23] have advanced some ideas regarding the nature of the adsorption process at a metal–solution interface. These authors propose that the adsorption of anions should be treated as a co-ordination reaction. It is suggested that the strongest co-ordination bond should be formed when the energies of the donor and acceptor orbitals on the anion and the metal are equal, and a bond could only be formed when these orbitals have the same symmetry. Cyanide ion is given as an example of this: this ion is more strongly adsorbed than I⁻ on platinum, whereas the reverse is the case on mercury. Cyanide ion is a good π-electron acceptor and some degree of interaction between unfilled levels having this symmetry and the t_{2g} levels of platinum is feasible, whereas such a possibility is not present in the case of mercury.

The origin of the hump in the capacitance curves on mercury has been discussed by Buhl.[24a,b] The double-layer capacitance of mercury in contact with aqueous KCl solutions was measured at 0, 25, and 50 °C. The height of the anodic hump was found to decrease with an increase in temperature. The Watts–Tobin model for the inner region (involving adsorbed water molecules in two different orientations) was modified to account for the effect of ionic adsorption. The main idea here was to consider that the inclusion of an anion in the inner region must result in a more intense electric field in the region between the IHP and the metal surface. The calculated overall dielectric polarization of the inner region differed from that evaluated from the Watts–Tobin model in that the position of the maximum in polarizability is closer to

[21] S. Levine, *J. Colloid Interface Sci.*, 1971, **37**, 619.
[22] W. Lorenz, *Z. phys. Chem.* (*Leipzig*), 1971, **248**, 161.
[23] D. J. Barclay and J. Caja, *Croat. Chem. Acta*, 1971, **43**, 221.
[24] H. Buhl, (*a*) *Electrochim. Acta*, 1971, **16**, 1495; (*b*) *Coll. Czech. Chem. Comm.*, 1971, **36**, 781.

the p.z.c. potential. The derivation of the proposed equation was not presented in these papers, but it is noticeable that q' does not enter explicitly into the proposed relationship, a situation which appears to be in contradiction with the original assumptions.

This type of formulation suffers from another basic difficulty; if the hump is only due to dielectric effects, the orientation of the solvent at the p.z.c. should be with the hydrogen atoms pointing towards the metal. All the evidence available at present indicates, however, the opposite natural orientation.[1, 2, 18] In spite of this, it should be noted that the apparent inner-layer capacitance in the absence of adsorption[14] for fluorine solutions still shows a hump close to the p.z.c., perhaps at a potential slightly anodic to the null-charge potential. It must be concluded then, that, although orientational effects must have some influence, there are other factors, such as inner-layer thickness values, which must be taken into consideration to understand the origin of the hump. It should be stressed that some anions lead to capacitance curves with pronounced humps; this is the case for NO_3^- and ClO_4^-, and especially for the PF_6^- ion.[10] A clear understanding of the dependence of $_qmC_i$ on variations in q' is essential in this connection. The differences found for this dependence between studies performed in salt mixtures and single-salt solutions is surprising. For NO_3^-, $_qmC_i$ does not vary with concentration in studies in single-salt solutions, whereas a strong decrease is noted for the mixtures.[1] In contrast with this, F^- adsorption appears to decrease strongly the values of $_qmC_i$. A rationalization of these differences is required to understand the structure of the inner layer.

Theory of the Diffuse Double Layer.—Very little work has been published in 1971 on the theory of the diffuse double layer. Krylov and Kir'yanov[25] suggested that p.z.c. potential shifts can be produced by non-absorbed electrolytes under two circumstances, *i.e.* when there is a difference in the values of either (*a*) the distance of the plane of closest approach to the metal or (*b*) the charge of the anion and cation. For the first case, $E_{p.z.c.}$ shifts were of the order of a few μV up to a solution concentration of 2×10^{-3} M, when the difference in the distance of closest approach was of the order of 0.1 nm. These values were calculated using the statistical theory of the diffuse layer previously developed. A somewhat bigger effect was predicted for the second case. Calculations at higher concentrations were not reported. It should be mentioned here that the physical origin of these effects is the existence of attractive image forces at the metal–solution interface, but that so far there has been no experimental verification of this type of interaction. To do this would require very accurate measurements of either $E_{p.z.c.}$[25] or $\gamma_{p.z.c.}$[1]

The extension of the statistical theory to diffuse-layer calculations for asymmetric electrolytes has been considered by Kir'yanov and Krylov.[26]

[25] V. S. Krylov and V. A. Kir'yanov, *Soviet Electrochem.*, 1971, **7**, 262.
[26] V. A. Kir'yanov and V. S. Krylov, *Soviet Electrochem.*, 1971, **7**, 999.

Gallium.—There has been very little work reported on the interfacial properties of this metal in the period under review. Grigor'ev, Fateev, and Bagotskaya[27] studied the changes in differential capacitance occurring during the liquid–solid transition. Ga metal of 99.998% purity was employed and measurements were performed both in acidified (0.01M acid concentration) and neutral solutions. The differential capacitance for the solid was lower than that for the liquid, and the degree of adsorption of ions, as determined by the onset of the anodic rise, was in the same order as in mercury, $ClO_4^- < Cl^- < I^-$.

Continuing with some previous work on thiourea adsorption, the same authors[28] have investigated the effect of the adsorbed molecules on the surface activity of several anions. KCl and KI solutions at different concentrations were studied. A diffuse-layer minimum was noticed at a potential of ~ -1.2 V vs. S.C.E. for the salts mentioned, and also for $LiClO_4$ and Na_2SO_4 solutions, indicating that the adsorption of thiourea decreases ionic adsorption substantially, and probably eliminates it at the lowest concentration studied.

Non-aqueous Solvents.—The adsorption of chloride ion at the mercury–solution interface in LiCl–methanol and HCl–ethanol systems was investigated by Jastrzębska.[29] It was found that the specific adsorption of Cl^- is stronger in methanol than in ethanol solutions. When the results presented here are compared with those in aqueous solutions at the same value of the mean ionic activity,[30] it is found that the values of q' are higher for aqueous solutions than those reported for the alcoholic solutions. This result reflects not only the differences in real potential of the anion in these different media, but also the difference in the work of removing adsorbed solvent molecules from the interface. It should be noted here that the use of the thermodynamic coefficient $(\partial E_+/\partial \mu)_{q^m,T,p}$ to compare the differences in degree of adsorption is incorrect and only gives an idea of the charge dependence of the free energy of adsorption (ΔG_{ads}°).

The real free energy of hydration of ions in non-aqueous solvents is a quantity of great interest, since it can yield information on orientational effects at interfaces, and can serve as a basis for comparison of the adsorption behaviour of ions from different solvents. Zagórska and Minc[31] have measured the outer potential difference ($\Delta\Psi_{Hg/Cl^-}$) between a calomel electrode and a chloride solution in methanol. Two experimental techniques were employed, Kenrick's and the vibrating-plate method. The results obtained with these techniques differed widely, and from a comparison with literature data it was concluded that the vibrating-plate method gave the correct results. The failure of the jet method was attributed to solvent adsorption on the flowing mercury jet. A value of $\Delta\Psi_{Hg/Cl^-}$ of -0.24 ± 0.025 V was obtained.

[27] N. B. Grigor'ev, S. A. Fateev, and I. A. Bagotskaya, *Soviet Electrochem.*, 1971, **7**, 1788.
[28] N. B. Grigor'ev, S. A. Fateev, and I. A. Bagotskaya, *Soviet Electrochem.*, 1971, **7**, 206.
[29] J. Jastrzębska, *Electrochim. Acta*, 1971, **16**, 1693.
[30] D. C. Grahame and R. Parsons, *J. Amer. Chem. Soc.*, 1961, **83**, 1291.
[31] I. Zagórska and S. Minc, *Electrochim. Acta*, 1971, **16**, 609.

The specific adsorption of SCN⁻ ions on mercury from water–acetone mixtures has been studied by Behr, Dojlido, and Stroka.[32] The purpose of this work was to investigate the effects brought about by gradual changes in the solvent composition on the adsorption behaviour of anions. It was found that the amount of specifically adsorbed SCN⁻ ion decreased with increasing acetone content, but it increased for solutions having more than 10% by volume of acetone. The proposed explanation of this was the interplay of the effects of adsorption of the organic solvent at the interface and the changes in $\Delta G^\circ_{hydr.}$ of SCN⁻. Some partial confirmation of these ideas can be found in the significantly lower values of the real potential of anions in acetone compared with those in aqueous solutions,[33] even taking into account the difference in surface potentials between these two solvents (for instance, for the Cl⁻ ion, $\Delta\alpha$(water–acetone) = −96.0 kJ mol⁻¹, whereas $-F\Delta\chi$(water–acetone) ≈ −50 kJ mol⁻¹).

Damaskin, Ivanov, Balashov, and Khonina[34] have studied the electrocapillary and capacitance curves for mercury in contact with LiNO₃ solutions in pyridine. From the general considerations advanced by Payne,[35] the adsorption of the solvent should be significantly higher than that of water (interfacial tension values of 36.34 and 42.54 μJ cm⁻² respectively). Thus, it would not be expected that ionic adsorption would be higher in this solvent than in water. In spite of the low value of the dielectric constant of pyridine, this appeared to be the case for NO₃⁻ adsorption. Damaskin et al. also found a marked diffuse-layer minimum in the capacitance close to the p.z.c., which reflected the poor dissociation of LiNO₃ in this solvent.

It has been observed that the order of solvation energies of anion in dimethylformamide (DMF) decreases in the sequence Cl⁻ > Br⁻ > I⁻ > SCN⁻, which is the reverse of what is known to occur in aqueous solutions. This order of solvation was inferred from the relative surface activity of these ions, and, in order to further confirm this, Ganshina, Damaskin. Kaganovich, and Ivanova[36] measured the surface potential and surface tension variations caused by these ions at the DMF–air interface. The adsorption of all these anions leads to negative values of $\Delta\chi$, whereas the reverse is true for aqueous solutions. Ion–dipole repulsion forces were invoked to explain this reversal of sign. The orientation of a DMF molecule at the surface of its aqueous solution was found to be with the positive end pointing to the gas phase. However, there appears to be little foundation to the claim that this is the actual orientation in pure DMF, since orientational interactions are bound to occur in the former case. It also seems difficult to understand the way in which a weak preferential orientation of dipoles at a surface can be the over-

---

[32] B. Behr, J. Dojlido, and J. Stroka, *Coll. Czech. Chem. Comm.*, 1971, **36**, 1317.
[33] I. Zagórska and Z. Koczorowski, *Roczniki Chem.*, 1970, **44**, 1559.
[34] B. B. Damaskin, V. F. Ivanov, V. F. Balashov, and V. F. Khonina, *Soviet Electrochem.*, 1971, **7**, 121.
[35] R. Payne, 'Advances in Electrochemistry and Electrochemical Engineering' ed. P. Delahay, Wiley–Interscience, New York and London, 1970, Vol. 7, p. 1.
[36] I. M. Ganshina, B. B. Damaskin, R. I. Kaganovich, and R. V. Ivanova, *Soviet Electrochem.*, 1971, **7**, 342.

riding factor in determining an ionic distribution which is also assumed to be generated by strong ion–solvent interactions. The reason for the negative values of $\Delta\chi$ therefore remains obscure.

Dzhaparidze, Dzhaparidze, and Damaskin[37] have measured the specific adsorption of I^- from mixtures at constant ionic strength in ethylene glycol. A 'square root' isotherm previously used for adsorption studies from DMF was found to fit the experimental results. $\Delta G°_{ads}$ was linearly dependent on the electrode charge, and the interaction coefficient representing interionic repulsions in the adsorbed layer was found to be charge dependent. The value of $\Delta G°_{ads}$ calculated was -108.7 kJ mol^{-1}, which is lower than the values found in water and DMF (-155.1 and -121.2 kJ mol^{-1} respectively). An interesting suggestion is made in this paper regarding the dielectric properties of the inner layer in different solvents. On the assumption that the distance of the IHP to the metal depends only on the nature of the anion, values of the ratios of dielectric constants and thickness with respect to water can be evaluated. This presupposes that $\Delta\phi^{m-2}$ can be divided into separate contributions due to q^m and q', an assumption usually made in double-layer studies. It was found that the ratios of dielectric constants of the inner region was independent of the solvent, whereas the thickness ratio changed in the order (for instance) ethylene glycol (1.3) < methanol (1.4) < DMF (1.9). It was concluded that the inner-layer capacitance changes observed from one solvent to another are then only due to changes in the inner-layer thickness. This idea had already been proposed by Payne[35] in a discussion of the interfacial properties of the amides, where no correlation could be found between the inner-layer capacitance values and the bulk dielectric constants.

The specific adsorption of Cs^+ ion from the same solvent was studied by the same authors.[38] A 'square root' isotherm was again found to describe the results, with a value of $\Delta G°_{ads}$ of -69.8 kJ mol^{-1}, which is smaller than $\Delta G°_{ads}$ for I^- by 38.9 kJ mol^{-1}. The adsorption of Cs^+ ion from this solvent is very similar to that already noted from DMF solutions and, surprisingly enough, it is higher than that observed from aqueous solutions.[39] This appears to depart markedly from the known behaviour of the anions, and the origin of this difference is not clear.

3 The Adsorption of Organic Molecules on to Mercury

Aliphatic.—Continuing previous work on the adsorption of ethers and alcohols, Gerovich, Damaskin, and Kaganovich[40] investigated the behaviour of diethylene glycol (DEG) and dipropylene glycol (DPG), the structures of

---

[37] D. I. Dzhaparidze, Sh. S. Dzhaparidze, and B. B. Damaskin, *Soviet Electrochem.*, 1971, **7**, 1259.
[38] Sh. S. Dzhaparidze, D. I. Dzhaparidze, and B. B. Damaskin, *Soviet Electrochem.*, 1971, **7**, 1486.
[39] R. Parsons and A. Stockton, *J. Electroanalyt. Chem. Interfacial Electrochem.*, 1970, **27**, App. 1.
[40] V. M. Gerovich, B. B. Damaskin, and R. I. Kaganovich, *Soviet Electrochem.*, 1971, **7**, 1301.

which have similarities to both ethers and alcohols. As has been noted for a wide variety of organic compounds, the adsorption at the p.z.c. was substantially higher at the mercury–solution interface than at the air–solution interface. The difference in adsorption energies was of the order of 3.8 kJ mol^{-1} for both DEG and DPG. This value was higher than that for the corresponding ethylene and propylene glycols, revealing that the effect of the additional oxygen atom in the molecule is to increase the interaction of the adsorbed species with the metal. The concentration dependence of the p.z.c. potential gave additional evidence for the influence of the C—O—C group interaction with the metal. The orientation at the air–solution boundary was the same, as shown by the surface-potential results. The analysis of the capacitance curves was based on a nine-parameter generalized model of the surface layer, which yielded reasonable values for the constants.

The adsorption of polyethylene glycols with a molecular weight of ~300 was studied by Dobren'kov and Guseva.[41] It was expected that these compounds of general formula OH—(CH$_2$—CH$_2$—O)$_r$H with $r = 6$ or 7, would adopt a planar orientation at the interface, thus displacing a large number of water molecules on adsorption and allowing then the test of isotherms based on the Flory–Huggins statistics.

In the study of the adsorption of organic compounds on to a mercury electrode, there exists very often the possibility of competition between the adsorbate and specifically adsorbed ions from the base electrolyte, as well as with the adsorbed solvent molecules; this is particularly important at anodic potentials. Shallal, Bauer, and Britz[42] made an extensive study in this respect, and examined the adsorptive behaviour of several alcohols (butan-1-ol, pentan-1-ol, and cyclohexanol), dissolved in aqueous solutions of KF, KCl, NaClO$_4$, or Na$_2$SO$_4$ at several concentrations. The surface coverage was evaluated assuming the validity of a two-parallel-condenser model for the interface. The Frumkin isotherm was modified to account for changes in the activity coefficient of the dissolved alcohols with changes in salt concentration, and a reasonably good fit was found in most instances. The adsorption coefficients were calculated from

$$BcS_0/S = [\theta/(1-\theta)]\exp(-2a\theta) \qquad (9)$$

where B is the adsorption coefficient, c the concentration, S_0 and S the solubilities of the alcohols in water and the salt solution respectively, θ the surface coverage, and a the interaction parameter. The term S_0/S represents an activity coefficient that allows for salting effects in the bulk phase. The use of a Flory–Huggins isotherm did not yield better fits, and, what is more important, the value of the isotherm parameter r was found to be close to unity. In all cases, the B values decreased with increasing supporting electrolyte concentration and this effect was the greater the larger the degree of ionic adsorption. This work should be compared with similar studies by

[41] G. A. Dobren'kov and L. T. Guseva, *Soviet Electrochem.*, 1971, **7**, 544.
[42] A. K. Shallal, H. H. Bauer, and D. Britz, *Coll. Czech. Chem. Comm.*, 1971, **36**, 767.

Trasatti,[2] in which negligible ionic-adsorption effects were detected in the ΔG_{ads}° values for glycol (see also ref. 54).

Shallal and Bauer[43] also studied the desorption potential of several alkyl alcohols for a variety of base electrolyte solutions. As previously reported, the specific adsorption of the anions shifted the positions of the anodic desorption peaks in the order expected from their adsorptabilities, i.e. $Cl^- > SO_4^{2-} > ClO_4^- > F^-$.

A very thorough experimental study of the adsorption of 5-chloropentan-1-ol was conducted by Doblhofer and Mohilner.[44] NaF was employed as a base electrolyte in order to avoid ionic co-adsorption complications; electrocapillary and capacitance curves for 13 and 22 different alcohol concentrations respectively were measured. The existence of a value of q^m (-1.00 μC cm^{-2}) at a given value of the potential, common to all the concentrations studied, indicated that the surface coverage passed through a maximum value at this point at all the concentrations studied. 'S' shaped Esin and Markov plots were observed in this case, and the adsorption isotherm was found not to be congruent with respect to either the charge or the potential.

Borovaya and Damaskin[45] continued with previous work on the adsorption behaviour of long-chain amino-acids. ω-Aminoenanthic acid, $NH_2(CH_2)_7$-CO_2H, and its salts were studied. Its interfacial properties appeared to be influenced by the possibility of lactam formation. It is interesting to notice that in this case changes in the value of the pH of the solution have a comparatively small effect on the surface tension, showing here that the length of the paraffinic chain determines this molecule's interfacial properties, rather than the presence of ionizable groups. The surface potentials of solutions of the acidic form were found to be greater than those of the neutral or alkaline ones. This was in marked contrast to the behaviour of the corresponding C_7 compound and was attributed to orientational effects of the adsorbed molecules on increasing surface crowding. This explanation is somewhat difficult to accept in the case of the more concentrated solutions studied, where the area per molecule is of the order of 2 nm^2 molecule^{-1}, unless very strong lateral interactions in the adsorbed state are operative. The adsorption on to mercury is further discussed and some capacitance results are presented. Joshi, Bapat, Dhawle, and Rajagopalan[46] have studied the adsorption of propionic acid from KCl and KBr base electrolyte solutions. The anionic and cationic surface excess of the base electrolyte was measured both with and without adsorbed organic material, and some degree of anion desorption was noted, which was due to the influence of the surface-active acid (for instance, for 0.1M-KBr at the p.z.c. it amounted to 2.2 μC cm^{-2}). The standard free energy of adsorption was calculated on the assumption that the adsorption process is equivalent to a chemical reaction involving the displacement of

[43] A. K. Shallal and H. H. Bauer, *Analyt. Letters*, 1971, **4**, 371.
[44] K. Dolofer and D. M. Mohilner, *J. Phys. Chem.*, 1971, **75**, 1698.
[45] N. A. Borovaya and B. B. Damaskin, *Soviet Electrochem.*, 1971, **7**, 552.
[46] K. M. Joshi, M. R. Bapat, S. W. Dhawle, and S. Rajagopalan, *J. Indian Chem. Soc.*, 1971, **48**, 549.

adsorbed water molecules from the surface layer. Maximum adsorption was noted at -2 μC cm^{-2} and a linear dependence of $\Delta G°_{ads}$ with $\theta^{3/2}$ found. Considering the accuracy of these results, it is very difficult to reach any conclusion from this dependence concerning the type of dipolar interactions in operations in the adsorbed state.

The simultaneous adsorption of two organic compounds was considered by Jehring;[47] a.c. polarography was used to record apparent capacitance curves for several systems. The desorption peak of a mixture of t- and s-butyl alcohols occurred at potentials cathodic to those observed for either of the pure substances. Although these two alcohols have desorption peaks separated by ~0.25 V, only one desorption wave is observed for the mixture, and from this point of view it behaves as a single chemical compound. It is worth mentioning here that something similar happens in relation to the anodic hump from aqueous solutions of salt mixtures. For mixtures of iso-octylsulphonate and dodecylpyridinium ions, a strong decrease of the capacitance at anodic potentials was observed. This was considered to be caused by strong attractive interactions in the absorbed layer. The behaviour of other mixtures containing polyethylene glycols was also described.

Some theoretical aspects of the adsorption of dipolar molecules have been considered by Krylov and Grigor'ev.[48] This was an extension of a previous formulation by the same authors, where it was considered that the adsorbed solvent molecules behave as a dielectric continuum. The Esin–Markov coefficient was predicted for the adsorption of thiourea and water on gallium and the experimental results were in good agreeement with the predictions.

Aromatic and Heterocyclic.—Bezuglyi, Titova, and Korshikov[49] found a marked difference in surface activity for the reduction products of the isomers *o*- and *p*-nitroaniline. At potentials anodic to the half-wave potential, the reduction products of the former did not appear to be adsorbed, whereas for the latter the adsorption of the products was evident. There was, however, no such difference in the reduction products, *viz*. *o*- and *p*-phenylenediamine, when studied by themselves. This raises the question of the real significance of an interfacial tension measurement in the presence of an electrode reaction, especially, as is pointed out in this paper, when the reduction mechanism can involve the formation of surface complexes.

The adsorption of *o*-cresol from neutral and alkaline solutions on to mercury was studied by Manohar and Sathyanarayana.[50] Capacitance and electrocapillary curves were measured, but only the results at the p.z.c. were analysed. As has been observed for other aromatic alcohols, well-defined cathodic desorption capacitance peaks were observed. At the p.z.c. a limiting coverage corresponding to a co-area of 0.38 nm^2 was reached at very low concentrations

[47] H. Jehring, *Z. phys. Chem. (Leipzig)*, 1971, **246**, 1.
[48] V. S. Krylov and N. B. Grigor'ev, *Soviet Electrochem.*, 1971, **7**, 487.
[49] V. D. Bezuglyi, V. B. Titova, and L. A. Korshikov, *Soviet Electrochem.*, 1971, **7**, 1698.
[50] G. Manohar and S. Sathyanarayana, *J. Electroanalyt. Chem. Interfacial Electrochem.*, 1971, **30**, 301.

in neutral solutions. This value of the co-area corresponds to a flat orientation of the *o*-cresol molecule, which is conditioned by the π-interaction between the ring and the metal. Some evidence for multilayer formation was presented and it was suggested that the complete analysis of the capacitance curves would require an equivalent model of the interface of six capacitors connected either in series or in parallel. These suggestions may lead the reader to wonder as to the physical reality of such models, particularly on account of the very large number of adjustable parameters involved. Undoubtedly, given enough parameters, any capacitance curve can be accurately fitted, even to a polynomial!

The comparison of the surface activity of organic compounds at different interfaces often yields a great wealth of information regarding orientational and structural effects. This point is exemplified in a paper by Kaganovich, Damaskin, and Suranova[51] on the adsorption of the 2- and 4-propyl derivatives of pyridine at the air- and mercury–solution interfaces. These derivatives were found to be more strongly surface active than pyridine, but the activity of the corresponding cations was nearly the same. However, the change in adsorption energy for the transition from the free solution surface to the interface with mercury was greatest for pyridine (3.72 *vs.* 2.81 and 2.35 kJ mol^{-1} respectively), indicating that the π-electron interaction with the metal surface is weakened by the presence of the alkyl groups in the molecule. The existence of strong π-interactions gives rise to a preferential orientation of these derivatives, *viz.* flat on the surface, causing a lower inner potential difference across the Hg–solution interface than that observed for in the air–solution interface.

The adsorption of 2,1,3-benzoxadiazole (benzofuran), 2,1,3-benzothiadiazole (piazothiol), and 2,1,3-benzoselenadiazole (piaselenol) on mercury from aqueous solutions was studied by Tsveniashvili, Zhdanov, and Todres.[52] The structure of these molecules is shown in the formula, where X = O

$X = O, S,$ or Se

(benzofurazan), S (piazothiol), or Se (piaselenol). From the molecular structure of these compounds, it might be expected that they would be strongly surface active owing both to the presence of a π-electron system and to the interactions of the heteroatoms with the mercury surface. This was indeed the case and the values of $\Delta G°_{ads}$ found at the p.z.c. were -27.2, -33.4, and -36.0 kJ mol^{-1} respectively. These compounds appear to undergo reorientation in the adsorbed state at anodic potentials, in common with many other

---

[51] R. I. Kaganovich, B. B. Damaskin, and M. A. Suranova, *Soviet Electrochem.*, 1971, **7**, 1109.

[52] V. Sh. Tsveniashvili, S. I. Zhdanov, and Z. V. Todres, *Soviet Electrochem.*, 1971, **7**, 24.

aromatic compounds. The origin of the cathodic peaks is probably related to the electroreduction of these compounds. (The half-wave potentials at pH 7 are -1.05, -1.01, and -0.65 V *vs.* S.C.E. for benzofuran, piazothiol, and piaselenol respectively,[53] the last value corresponding to the first reduction wave.)

The electrosorption on mercury of a series of furan derivatives and cyclopentanone was studied by Barradas and Sedlak.[54] The compounds studied were tetrahydrofuran (THF), tetrahydrofurfurylamine (TFFA), tetrahydrofurfuryl alcohol (THFA), cyclopentanone (CP), furfurylamine (FFA), and furfuryl alcohol (FA). A striking difference was observed in the behaviour of THFA in 1M-Na$_2$SO$_4$ and 1M-KCl solutions: whereas in the former case a limiting value of the coverage was observed at each value of q^m at sufficiently high concentrations, in the latter case that limiting behaviour was not reached even at the highest concentrations studied (1.5M). This difference was attributed to the interaction of the hydrated sulphate anion present in the diffuse double layer with the polar end of the adsorbed THFA molecules. In the case of chloride-ion solutions, it was argued that this effect is absent owing to specific adsorption of the ion. Actually, sulphate ions are fairly strongly adsorbed on mercury, and, for instance, in a 1M-MgSO$_4$ solution, $q' \sim -13$ µC cm^{-2} for $q^m = 6$ µC cm^{-2}.[55] If it is considered that MgSO$_4$ is a poorly dissociated electrolyte, an even stronger adsorption should be expected from Na$_2$SO$_4$ solutions, and the origin of the above-mentioned differences is not very clear. The values of ΔG°_{ads} for these compounds are shown in the Table.

Table *Free energies of adsorption of furan derivatives*

Compound	ΔG°_{ads}/kJ mol^{-1}	Compound	ΔG°_{ads}/kJ mol^{-1}
THF	-14.2	CP	-17.1
TFFA	-16.7	FFA	-16.3
THFA	-18.8	FA	-18.8

From the Table it can be seen that there is a surprisingly small difference in ΔG°_{ads} between saturated and unsaturated molecules (say, between THFA and FA), a fact that in this case probably indicates the lack of involvement of the double bonds in the adsorption process.

Gaunitz and Lorenz[56] studied the adsorption of hydroquinone, phenol, biphenyl, and thiophen on mercury from their solutions in water, methanol, and DMF. The purpose of this work was to extend the ideas of partial charge transfer of adsorbed ions to a group of compounds that is known to interact strongly with the metal. For the adsorption of hydroquinone, the partial charge-transfer coefficient (λ) was 0.25 and was independent of the potential. Phenol was found to change its orientation in the adsorbed state from vertical to flat at anodic potentials. In the cathodic desorption region,

---
[53] V. Sh. Tsveniashvili, S. I. Zhdanov, and Z. V. Todres, *Z. analyt. Chem.*, 1967, **224**, 389.
[54] R. G. Barradas and J. M. Sedlak, *Electrochim. Acta*, 1971, **16**, 2091.
[55] D. J. Schiffrin, unpublished results.
[56] U. Gaunitz and W. Lorenz, *Coll. Czech. Chem. Comm.*, 1971, **36**, 796.

evidence was presented for a significant degree of association in the adsorbed state. The capacitance of mercury in contact with methanolic solutions of phenol did not show marked desorption peaks, and these were absent in DMF solutions; also, the degree of adsorption decreased in the series $H_2O > MeOH > DMF$, and it was suggested that this could be due to the stronger π-interactions of phenol with the solvents in the order shown. A similar effect was noted for biphenyl solutions. The adsorption of thiophen from methanolic solutions presents some peculiar features: in solutions more dilute than 2×10^{-2}M, the adsorption at anodic potentials resembles that of an anion. The value of λ under these conditions was estimated at 0.4, which indicates a strong interaction between the sulphur atom and the metal surface. At higher concentrations, greater than 0.1M, thiophen behaves like phenol or biphenyl, and this was attributed to a change to a flat orientation of the adsorbed molecules on the surface, leading to a weakening of the S–Hg interactions.

4 Adsorption Kinetics

Whitehouse[57] has proposed that the series equivalent circuit for representing the interfacial impedance leads to a simplification in the analysis of impedance data in adsorption kinetics studies. For the series combination it was found that

$$1/\omega C_s = [(\omega \tau_D/2)^{\frac{1}{2}} + 1]/\omega(C_0 - C_\infty) \qquad (10)$$

and

$$R_s = [(\omega \tau_D/2)^{\frac{1}{2}} + \omega \tau_H]/\omega(C_0 - C_\infty) \qquad (11)$$

where C_s, R_s are the series capacitance and resistance of the interfacial impedance, ω is the angular frequency, τ_D and τ_H are the relaxation times characteristic of the diffusion and adsorption steps, and C_0 and C_∞ are the zero- and infinite-frequency capacitances. The physical meaning of the impedance components is now similar to that considered in electrode kinetics impedance analysis. Equations (10) and (11) were tested for n-amyl alcohol and sulphide-ion adsorption on mercury.

The use of matrix formalisms has been suggested by Cachet, Cachet, and Lestrade[58] to handle electrode impedance data.

5 The Interface between Platinum (and Platinum-group Metals) and Aqueous Solutions

Thermodynamics and Double-layer Capacitance.—The extension of classical interfacial thermodynamics to adsorption studies on solids has been reviewed by Damaskin *et al.*[3] Interfacial tension variations of solids are not easily measured, but, as shown by Frumkin and Petrii,[59] they can be estimated from the integration of the corresponding Lippmann equation. For the case

---

[57] D. R. Whitehouse, *J. Electroanalyt. Chem. Interfacial Electrochem.*, 1971, **32**, 265.
[58] C. Cachet, H. Cachet, and J. C. Lestrade, *J. Electroanalyt. Chem. Interfacial Electrochem.*, 1971, **32**, App. 5—8.
[59] A. N. Frumkin and O. A. Petrii, *Proc. Acad. Sci. (U.S.S.R.)*, 1971, **196**, 194.

Ionic Double Layers and Adsorption

of an interface where a charge-transfer equilibrium prevails [*i.e.* H˙(Pt) ⇌ H_{aq}^+ +$e^-_{(Pt)}$], two Lippmann equations can be defined according to which chemical potential is held constant (see refs. 15 and 16 in ref. 2). In the present case, integrations were performed under conditions of constant μ_{H^+} and constant $\mu_{H^{\cdot}}$, where μ_{H^+} and $\mu_{H^{\cdot}}$ are the chemical potentials of the proton in solution and of an adsorbed hydrogen atom. The systems studied were 0.005M-H_2SO_4+1M-Na_2SO_4; 0.01M-HCl+1M-KCl; 0.01M-HBr+1M-KBr; 0.01M-NaOH+1.5M-Na_2SO_4; 0.01M-KOH+1M-KCl; 0.01M-KOH+1M-KBr; and 0.01M-KOH+1M-KI. The surface free energy curves for different salts coincided at the cathodic end, close to the reversible hydrogen electrode potential. These curves were reminiscent of the electrocapillary curves for mercury, both in their shape and in the influence of the anion, but those obtained at constant μ_{H^+} varied strongly with potential owing to both hydrogen and oxygen adsorption.

Luk'yanycheva, Strochkova, and Knots[60] studied the double-layer capacitance of the platinum–0.5M-H_2SO_4 interface. The equivalent circuit analysed consisted in a series combination of a Warburg impedance, the adsorption capacitance, and the charge-transfer resistance, shunted by the double-layer capacitance. Measurements were made in the frequency range 15—50 kHz. The capacitance was found to have a characteristic maximum at 0.35 V *vs.* N.H.E., the value of which was about 42 µF cm^{-2}. Similar results have been reported by Flinn, Rosen, and Schuldiner.[61] In continuation with previous galvanostatic investigations, these authors used a charge-injection method. A capacitance maximum of ~33 µF cm^{-2} was found at 0.17 V *vs.* N.H.E. for a 1M solution. For the more dilute solutions studied, a capacitance minimum was observed at about 0.23 V, which was attributed to the occurrence at this potential of the p.z.c. It should be noted here that no reference is made in these papers to previous work by Payne,[62] who obtained results in complete disagreement with those reported, regarding the existence of a capacitance maximum.

Hydrogen and Oxygen Adsorption.—Gokshtein[63] has briefly discussed the evidence for the existence of two types of adsorbed hydrogen on platinum, obtained from small-potential induced stress experiments. It was proposed that these two types of H_{ads} correspond to hydrogen atoms included in the lattice and absorbed on the surface. No relaxation for the exchange process could be found at frequencies of up to 200 kHz.

The stoicheiometry of an adsorbed oxygen film is conveniently expressed as a ratio of monolayer hydrogen coverage, which gives a measure of adsorption independent of surface roughness complications. Biegler, Rand, and

---

[60] V. I. Luk'yanycheva, E. M. Strochkova, V. S. Bagotskii, and J. L. Knots, *Soviet Electrochem.*, 1971, **7**, 252.
[61] D. R. Flinn, M. Rosen, and S. Schuldiner, *Coll. Czech. Chem. Comm.*, 1971, **36**, 454.
[62] R. Payne, *J. Electroanalyt. Chem. Interfacial Electrochem.*, 1968, **19**, 1.
[63] A. Ya. Gokhshtein, *Soviet Electrochem.*, 1971, **7**, 574.

Woods[64] pointed out, however, the precautions that have to be taken in the analysis of voltammetric data, in particular with regard to evaluating the amount of H_{ads} from the integration of the current–potential transients. It was shown that for smooth Pt in 1M-H_2SO_4 solutions, at potentials cathodic to 0.10 V vs. R.H.E., there was a mass-transport-controlled hydrogen-evolution reaction, an effect that must be taken into account when integrating sweep data to perform a realistic separation of hydrogen-adsorption and -evolution currents. It was concluded that the oxygen monolayer contains two oxygen atoms *per* surface platinum atom. A similar study was performed by Rand and Wood[65] for rhodium, palladium, and gold electrodes. For Rh and Pd, a clearly discernable step was found which was assumed to correspond to a ratio of one atom of oxygen *per* surface atom.

Kokoulina, Krasovitskaya, and Krishtalik[66] studied competition effects in the adsorption of oxygen and chlorine on platinum using cyclic voltammetry. The nature of the Pt—O bond was influenced by the potential at which the electrode was oxidized. The reduction of the surface compound occurred at more cathodic potentials when the film was formed at more anodic potentials. No evidence was found for the formation of a chlorine-containing surface film *prior* to chlorine evolution, a result of some importance from an electrosynthetic point of view. Oxygen adsorption increased almost linearly with potential and was little affected by Cl^- ion adsorption.

Similar results regarding the dependence of the strength of the Pd—O bond on the oxidation potential were found for palladium in contact with 0.5M-H_2SO_4 by Gromyko.[67] Also, full coverage corresponded with the formation of two monolayers, probably associated with the formation of the surface compound PdO_2 (see refs. 64 and 65). A Temkin isotherm was found to reproduce the hydrogen-adsorption results. The chemisorption of oxygen on platinum was also studied by Tarasevich and Bogdanovskaya[68] at different values of the pH, using a voltammetric technique. Similar conclusions regarding the strengthening of the Pt—O bond with anodic potential and time as above mentioned were reached. The ageing effect of these surface films of less than a monolayer thickness is an interesting feature of these systems. These effects are to be expected in thick films, where some degree of recrystallization can occur, but in submonolayer films they should be related to slow chemical effects of the adsorbed atoms with the substrate. It is interesting to note that even at high temperatures, in bisulphate melts, these ageing effects have been observed.[69] The possible mechanisms for film formation are extensively discussed in this paper.

[64] T. Biegler, D. A. J. Rand, and R. Woods, *J. Electroanalyt. Chem. Interfacial Electrochem.*, 1971, **29**, 269.
[65] D. A. Rand and R. Woods, *J. Electroanalyt. Chem. Interfacial Electrochem.*, 1971, **31**, 29.
[66] D. V. Kokoulina, Yu. I. Krasovitskaya, and L. I. Krishtalik, *Soviet Electrochem*, 1971, **7**, 1105.
[67] V. A. Gromyko, *Soviet Electrochem.*, 1971, **7**, 854.
[68] M. R. Tarasevich and V. A. Bogdanovskaya, *Soviet Electrochem.*, 1971, **7**, 1469.
[69] A. J. Calandra, personal communication.

The kinetics of oxygen adsorption on palladium were investigated by Tarasevich, Vilinskaya, and Burshtein[70] by cyclic voltammetry. The nature and stability of the surface compounds formed depended on the adsorption potential, in agreement with previous work. The oxidation mechanism proposed involved the primary formation of adsorbed OH˙ radicals, and the kinetics of the reduction of the film depended on the scan rate employed.

Ionic Adsorption.—The surface excess of Na^+ on platinum and rhodium electrodes at different potentials was studied by Kazarinov, Petrii, Topolev, and Losev[71] by the radiotracer method. The solutions studied were $0.5 \times 10^{-3} M\text{-}H_2SO_4 + 10^{-3} M\text{-}Na_2SO_4$; $10^{-3} M\text{-}HCl + 3 \times 10^{-3} M\text{-}NaCl$; $10^{-3} M\text{-}HBr + 3 \times 10^{-3} M\text{-}NaBr$; and $0.5 \times 10^{-3} M\text{-}H_2SO_4 + 3 \times 10^{-3} M\text{-}NaI$. For platinum, Γ_{Na^+} at a potential $E_r = 0$ (E_r is the potential vs. a hydrogen electrode in the same solution under study) was found to be $\sim 8 \ \mu C \ cm^{-2}$ for all the solutions. A significant cationic surface excess was found at potentials anodic to that of the p.z.c., as would be expected from the known adsorptive properties of the anions. Surprisingly, however, the order of adsorption was $Na_2SO_4 < NaI < NaCl < NaBr$, when by comparison with the mercury-solution interface, I^- ion should be more strongly adsorbed than Br^-. Although this is almost certainly the case, if a strong degree of charge transfer on adsorption occurs, it is not necessarily reflected in the Γ_{Na^+} values. Whether this is really the case for mercury is still an open question, but there is strong evidence that some degree of charge delocalization does occur when Cl^- and Br^- ions are adsorbed on to Pt (see refs. 118, 122, and 123 in ref. 1). The comparison of $\Gamma_{SO_4^{2-}}$ values obtained with the radiotracer method was in excellent agreement with the values obtained from charging curves and isoelectrical potential shifts. The results on Rh were similar to those on Pt, but the amount of adsorption appeared to be smaller. The degree of adsorption of sulphate ion found by Schuldiner and Rosen[72] was, however, much greater than the results mentioned above. For a platinum-bead electrode, $q_{SO_4^{2-}}$ was found by these authors to be six times greater than previous estimates by the radiotracer method. The difference in concentrations used in this study was twenty times larger than in the previous work, and this might explain the differences noted. It was also mentioned that the degree of adsorption is altered by the crystallographic features of the metal.

The displacement of adsorbed hydrogen atoms by bivalent ions at a platinum electrode was considered by Petrii, Vasina, and Frumkin.[73] It was found that the adsorption of the cations increased in the order of the increasing radius, that is, $Mg^{2+} < Ca^{2+} < Sr^{2+} < Ba^{2+}$, a result that shows an increasing degree of electrostatic interaction between the cation and the metal surface.

[70] M. R. Tarasevich, V. S. Vilinskaya, and R. Kh. Burshtein, *Soviet Electrochem.*, 1971, **7**, 1152.
[71] V. E. Kazarinov, O. A. Petrii, V. V. Topolev, and A. V. Losev, *Soviet Electrochem.*, 1971, **7**, 1321.
[72] S. Schuldiner and M. Rosen, *J. Electrochem. Soc.*, 1971, **118**, 1138.
[73] O. A. Petrii, S. Ya. Vasina, and A. N. Frumkin, *Soviet Electrochem.*, 1971, **7**, 402.

The effect of Ca^{2+} was found to be equivalent to that of Cs^+. Similar results were obtained by Burshtein, Vilinkaya, and Tarasevich[74] for palladium in alkaline solutions. The true surface area was found to depend on the number of adsorption cycles performed, although a limiting value was reached (see below). The order of adsorption encountered was Na^+, $K^+ < Cs^+ < Ca^{2+} < Ba^{2+}$, and the surface excess decreased with increasing anodic polarization.

The radiotracer method was applied by Balashova[75] to the study of Cl^- and Br^- ion adsorption on palladium. Strong evidence was found for the specific adsorption of the anions, and for Br^- solutions, a limiting coverage of ~ 0.15 of a monolayer was found. A Temkin isotherm described correctly the adsorption behaviour of this ion. As expected, I^- was more strongly adsorbed, and at sufficiently anodic potentials, more than one complete monolayer was formed, suggesting the formation of stable surface compounds. From exchange experiments (where a large amount of inactive salt is added to the solution), it was concluded that the exchange rate close to the p.z.c. was similar for both ions, but when the exchange was performed at fairly anodic potentials (0.6 V vs. N.H.E.), the amount of I^- ion desorbed was very small. This elegant experiment supports the idea that the interaction of I^- with the palladium surface is stronger than that shown by Br^-. Results at variance with those described above were obtained by Vilinskaya and Tarasevich,[76] using both the decrease in oxygen adsorption and potential-jump methods to measure the amount of adsorption. Cl^- and Br^- ions were studied; Br^- adsorption reached a limiting coverage of 0.55 of a monolayer at anodic potentials, three times higher than the radiotracer result. A reason for this discrepancy, besides differences in surface treatment, might be the assumption that the adsorption of one anion results in the displacement of only one adsorbed oxygen atom. The difference appears to be more significant for powdered electrodes than for smooth ones. The kinetics of the adsorption process were further discussed and the Temkin isotherm parameters found were 86, 20, and 14 for HSO_4^-, Cl^-, and Br^- respectively, reflecting the order of ionic character of the adsorbed entity.

The study of the stoicheiometry of displacement of adsorbed hydrogen by I^- ions from the surface of platinum and rhodium electrodes was undertaken by Petrii, Malysheva, and Maksimov,[77] who used charging curves and voltammetric measurements. The amount of adsorbed I^- ion was directly measured by titration after the adsorption equilibrium was reached. The potential in these experiments was controlled both potentiostatically and by the H_2 pressure. An important result in this research is that the amount of H_{ads} displaced by the adsorption of I^- ions is strongly potential-dependent, decreasing from a value of ~ 2.4 for Pt and 2.6 for Rh at $E_r = 0$ to almost

[74] R. Kh. Burshtein, V. S. Vilinskaya, and M. R. Tarasevich, *Soviet Electrochem.*, 1971, **7**, 105.
[75] M. I. Kulezneva and N. A. Balashova, *Soviet Electrochem.*, 1971, **7**, 434.
[76] V. S. Vilinskaya and M. R. Tarasevich, *Soviet Electrochem.*, 1971, **7**, 1465.
[77] O. A. Petrii, Zh. N. Malysheva, and Yu. M. Maksimov, *Soviet Electrochem.*, 1971, **7**, 1010.

zero at $E_r = 0.30$ V. Also, the hydrogen adsorption isotherms reflected this effect. These results throw some doubts on previous estimates of ionic adsorption based only on the decrease of the extent of adsorption.

Wetzel, Müller, and Heckner[78] have measured the adsorption of iodate ion on platinum by a radiotracer method. The amount adsorbed was proportional to $C_{IO_3^-}^{\frac{1}{2}}$, a result which was interpreted assuming dissociative chemisorption of the anion. The claim in this work that, for a 10^{-7}M solution, adsorption equilibrium is reached in two minutes seems to be at variance with a great many adsorption studies of this kind.

The main features of the adsorption of Zn^{2+} on platinum have been extensively studied by Petrii, Malysheva, Kazarinov, and Andreev,[79] using a radiotracer method. ^{65}Zn and ^{35}S were employed in this study. A logarithmic dependence of coverage on time was observed and the kinetics of adsorption could be expressed by

$$v_{ads} = k_{ads} C_{Zn^{2+}} \exp(-K\theta) \qquad (12)$$

where v_{ads} is the adsorption rate, and k_{ads} and K are constants for a given potential. The value of k_{ads} decreased from 5×10^{-7} to 2.7×10^{-5} cm s^{-1} when the potential was altered from 0 to 0.4 (vs. a hydrogen electrode at the same solution concentration) for 10^{-3}M-ZnSO$_4$ + 0.5×10^{-2}M-H$_2$SO$_4$ solutions. This decrease reflected the stronger bonding of Zn^{2+} ion at more cathodic potentials, a result which was verified by exchange experiments with an excess of non-radioactive Zn^{2+} solution: at 0.4 V the exchange reached 85%, compared with a value of 60% at 0 V. Another interesting aspect of the nature of the adsorbed state of zinc on platinum is brought about in this work by the study of temperature effects. It was found that Zn^{2+} is more firmly held to the surface when it is adsorbed at 40 °C than at lower temperatures; exchange occurs under these circumstances only to the extent of 10—15%, but if the adsorption takes place at this temperature and the exchange at lower temperatures, practically no adsorbed charge is lost. This shows conclusively that zinc adsorption on platinum (and very likely other cations) is not an equilibrium process. In the potential region where cation adsorption occurs, SO_4^{2-} ion was also adsorbed to a greater extent than, for instance, for Li$_2$SO$_4$ solutions. It is likely that there is some degree of anion-induced cation adsorption present in this system.

The importance of the state of the surface in this type of work was strongly stressed by the research of Horáni, Solt, and Nagy.[80] The adsorption of Cl$^-$ ions on platinum was studied by the radiotracer technique. It was found that the adsorption curves obtained with freshly prepared electrodes after cathodic regeneration differed significantly from those obtained after several potential excursions across the adsorption region. In the latter case, the potential depen-

[78] R. Wetzel, L. Muller and K. H. Heckner, *Z. phys. Chem. (Leipzig)*, 1971, **246**, 234.
[79] O. A. Petrii, Zh. N. Malysheva, V. E. Kazarinov, and V. N. Andreev, *Soviet Electrochem.*, 1971, **7**, 1630.
[80] G. Horányi, J. Solt, and F. Nagy, *J. Electroanalyt. Chem. Interfacial Electrochem.*, 1971, **31**, 95.

dence of adsorption was similar to that previously observed with several other techniques, but for the first potential excursion two well-defined arrests were noted. This effect was observed in quite a wide concentration range, from 1.3×10^{-5} to 1.3×10^{-3}M solutions. It was suggested that the isotherm describing the whole adsorption curve was a combination of two Langmuir-type isotherms. This work implies that the primary adsorption process, at least in the case of Cl^- ions, is not reverible.

Horányi, Solt, and Vertes[81] studied by the radiotracer method the adsorption of cadmium ion on platinum from H_2SO_4 and HCl solutions. These authors observed a significant increase in the amount of adsorbed HSO_4^- and Cl^- ions at potentials close to zero, where there is a substantial degree of Cd^{2+} ion adsorption. Furthermore, anion adsorption in this potential region was enhanced by an increase in Cd^{2+} ion concentration and it can be concluded that the cation is not in competition for the same surface-active sites as the anions. However, at potentials anodic to 0.5 V, an actual decrease of adsorption of HSO_4^- was noticed when the Cd^{2+} ion concentration was increased. This was attributed to extensive ion-pairing in the solution, thus reducing the activity of the adsorbing anion. Another interesting observation in this work is the absence of any significant effect of cadmium adsorption on the adsorption of hydrogen in the presence of ClO_4^- ions, whereas the charging curves are changed when Cd^{2+}, HSO_4^-, and SO_4^{2-} ions are all present simultaneously. It must be concluded from this work that the anions play some part in the mechanism of catalytic poisoning.

In contrast with its behaviour on platinum, the adsorption of Zn^{2+} on rhodium was found by Petrii, Mahysheva, and Kazarinov[82] to be a reversible process. The radiotracer method was employed and time-independent values of Zn^{2+} adsorption were reached in a short time, and a negligible temperature effect on adsorption was found. The range of adsorption was narrower than that noted for Pt, a fact which was considered to be caused by the more ionic character of the bond in the former case. From charging-curve experiments, it was concluded that adsorbed zinc does not alter the total amount of H_{ads} at $E_r = 0$, and at potentials anodic to this there was even some evidence for an increase in the strength of the Pt—H bond. A complete rationalization of these effects is not available at present.

The increase in the amount of H_{ads} more firmly held to the metal caused by Zn^{2+}_{ads} was also shown by Petrii, Malysheva, and Kazarinov[83] to occur for a platinum electrode. The irreversible character of Zn^{2+} adsorption was also apparent[79] in the large initial hysteresis effects noted on fresh electrodes. Reproducible results were obtained after three potential excursions, a situation resembling the observations by Horányi et al.[80] for chloride adsorption on this metal. The increase in the quantity of H_{ads} with high bond energy was

[81] G. Horányi, J. Solt, and G. Vertes, *J. Electroanalyt. Chem. Interfacial Electrochem.*, 1971, **32**, 271.
[82] O. A. Petrii, Zh. N. Malysheva, and V. E. Kazarinov, *Soviet Electrochem.*, 1971, **7**, 1778.
[83] O. A. Petrii, Zh. N. Malysheva, and V. E. Kazarinov, *Soviet Electrochem.*, 1971, **7**, 1524.

evident from the increase in the second anodic peak in the voltammetric reduction of the hydrogen monolayer.

Although the adsorption of Zn^{2+} does not decrease to any great extent the H_{ads} coverage at $E_r = 0$, this quantity was shown by Cadle and Bruckenstein to be drastically reduced by the adsorption of Cu^{2+}, Ag^+, Pb^{2+}, and $AuCl_4^-$.[84] A linear decrease of H_{ads} coverage with increasing coverage by Cu^{2+} was observed. It was concluded that for each adsorbed copper ion one adsorbed hydrogen atom was displaced. In the case of Ag^+, in contrast with the Cu^{2+} results, there was not a 1:1 relationship between surface coverage by the metal and decrease in hydrogen adsorption, but both ions appeared to inhibit preferentially the weakly bonded hydrogen adsorption sites at intermediate coverages. Adsorbed gold and lead acted as H_{ads} inhibitors, but both forms of H_{ads} were displaced. The difference in behaviour between zinc and the other metals is perhaps somewhat surprising and requires further ellucidation.

The adsorption of Cd^{2+} ion from 1M-KOH solutions was studied by Gilman[85] using a voltammetric technique. The amount of charge adsorbed was calculated by comparison of the anodic sweep of the voltagram in the presence and absence of added Cd^{2+} ions. Significant adsorption was noted, reaching values of ~ 120 μC cm^{-2} at potentials close to the hydrogen reversible potential. Cd^{2+} ion was also found to produce a decrease in hydrogen adsorption, attributable to the blocking of surface sites.

Organic Adsorption.—The study of the adsorption of organic compounds on platinum is complicated by its catalytic activity. Horányi and Nagy[86] set out to establish to what extent it is possible to talk about a true adsorption equilibrium in the case of aromatic compounds, and whether some previously reported potential dependence of adsorption of organic compounds[1] could not be simply related to hydrogenation and oxidation reactions. The adsorption of benzoic, benzenesulphonic, and phenylacetic acids was studied by the radiotracer method. In these three cases, the degree of coverage was independent of the potential in the interval 0.2—0.6 V. The apparent degree of coverage decreased at potentials approaching hydrogen evolution, but it was shown that this effect is simply due to the onset of the hydrogenation reaction. Exchange experiments showed that the adsorbed molecules are very firmly held to the electrode, and very little exchange could thus be detected. This is in sharp contrast with the behaviour of Cl$^-$ and acetic acid. The stability of the bond in the adsorbed state between the aromatic molecules and platinum makes it likely that no adsorption equilibrium exists, and that the adsorption process involves the formation of a stable surface compound. Similar conclusions were obtained by Horányi[87] in relation to the adsorption of *p*-nitrophenol on platinum.

[84] S. H. Cadle and S. Bruckenstein, *Analyt. Chem.*, 1971, **43**, 1858.
[85] S. Gilman, *J. Electrochem. Soc.*, 1971, **118**, 1953.
[86] G. Horányi and F. Nagy, *J. Electroanalyt. Chem. Interfacial Electrochem.*, 1971, **32**, 275.
[87] G. Horányi, *J. Electroanalyt. Chem. Interfacial Electrohem.*, 1971, **31**, App. 1—2.

Macdougall, Conway, and Kozlowska[88] have shown that the strong bonding of some organic compounds with platinum results in the displacement of adsorbed hydrogen. The decrease in H_{ads} coverage on adsorption of thiourea, acetonitrile, and benzonitrile was evaluated both from the integration of the current–potential voltammetric curves and by the measurement of the total anodic charge in the current transient caused by adsorption. The three substances studied induced electrochemical desorption of adsorbed hydrogen, but the analysis was complicated by the possibility of partial hydrogenation of the chemisorbed species. Thiourea was the strongest inhibitor and it did not appear to be electroactive in this potential region.

6 Experimental

An interesting application of modern data acquisition techniques to electrocapillary measurements has been extensively described by Lawrence and Mohilner.[89] The maximum bubble-pressure method was fully automated by coupling an on-line mini-computer to the experimental set up. In this method, the pressure applied to the mercury meniscus in the capillary was controlled by means of solenoid valves. When the maximum pressure was attained, and hence a mercury drop began to form, an electronic birth-detector system[7] operated the digital acquisition of the reading of an electronic pressure transducer. In order to allow for non linearities in the analogue voltage generated by the pressure sensor, its output was calibrated and analytically stored in the computer in the form of a polynomial. Three different rates of pressure increase were used. First, a rough run through the electrocapillary curve was made, which served as reference for the fine measurements. For the latter, the pressure was increased in stages, using the slowest flow rate for the last pressure increase near the maximum bubble pressure. Good accuracy is claimed for this system, and the speeding up and automation of the measurements will undoubtedly allow for a substantial increase in the accuracy of the surface-excess measurements. Full details of the interface hardware and the necessary software are given in this paper.

The chronocoulometric technique for studying adsorption and electrode kinetics has been improved[17] by recording the integrated current transients directly on a time basis proportional to (time)$^{\frac{1}{2}}$. This time basis was generated with the appropriate operational amplifier modules, and allows for a direct calculation of the quantity of adsorbed ion.

Devanathan and Tilak have described[90] a simple transformer bridge for electrode impedance measurements, in which the usual resistive ratio arm (for the Wien configuration), is replaced by two matched 50 mH coils. The off-balance signal of the bridge is taken from the secondary of the ratio coils. No

---

[88] B. Macdougall, B. E. Conway, and H. A. Kozlowska, *J. Electroanalyt. Chem. Interfacial Electrochem.*, 1971, **32**, App. 15—20.
[89] J. Lawrence and D. M. Mohilner, *J. Electrochem. Soc.*, 1971, **118**, 1596.
[90] M. A. V. Devanathan and B. V. K. S. A. Tilak, *Electrochim. Acta*, 1971, **16**, 2121.

Ionic Double Layers and Adsorption

results on the frequency dispersion of this arrangement are given, but the capacitance results agree with Grahame's data.

Jakuszewski, Kozowski, Pszasnyski, and Romanovski[91] have revived an old idea by Frumkin regarding the feasibility of p.z.c. potential determinations by measuring the transient electrical potential appearing on immersion of a metal in a solution. By opposing a suitable potential, the null-charge point was determined as the potential at which no signal could be detected on an oscilloscope when the metal came into contact with a solution. Reasonably good agreement with literature values was found in the case of mercury. Antropov, Gerasimenko, and Gerasimenko[92] preferred to vibrate their electrodes. They used a frequency of vibration of 70 Hz and obtained good agreement with literature values. This method appeared to be capable of working properly in very dilute solutions. Another variation on the same theme is to rush mercury down a capillary partially blocked by a platinum corkscrew,[14] a method that appeared to give p.z.c. potentials independent of the flow rate. It would seem advisable for a critical, comparative study of the various methods proposed so far to be made.

Horányi, Solt, and Nagy[93] have described in some detail the experimental arrangement and background corrections for the radiotracer technique applied to adsorption studies on platinum. The Pt electrode was prepared by electrodeposition on to a thin P.V.C. sheet previously covered with a layer of gold. The latter was applied by vacuum plating and this electrode formed the base of the cell which was mounted directly on the radiation detector.

Morcos[94] has discussed the relationship between liquid meniscus height at a partially immersed electrode and potential, on the basis of the theory of wetting and electrocapillarity.

A graphical method for integrating non-equilibrium capacitance curves has been proposed by Damaskin.[95] This method is applicable to substances undergoing desorption at high anodic and cathodic potentials.

[91] B. Jakuszewski, Z. Kozowoski, S. Partyka, M. Pszasnyski, and S. Romanovski, *Soviet Electrochem.*, 1971, **7**, 772.
[92] L. I. Antropov, M. A. Gerasimenko, and Yu. S. Gerasimenko, *Soviet Electrochem.*, 1971, **7**, 1474.
[93] G. Horányi, J. Solt, and F. Nagy, *J. Electroanalyt. Chem. Interfacial Electrochem.*, 1971, **31**, 87.
[94] I. Morcos, *Coll. Czech. Chem. Comm.*, 1971, **36**, 689.
[95] B. B. Damaskin and S. L. Dyatkina, *Soviet Electrochem.*, 1971, **7**, 244.

5
Organic Electrochemistry - Synthetic Aspects

BY P. M. ROBERTSON

1 Introduction

The success of organic electro-synthesis will ultimately be determined by its degree of involvement in the chemical industry. Although organic electro-synthesis has been studied for over one hundred and thirty years, it is disappointing that it is still regarded as a novelty by industry. It is difficult to see exactly why organic electro-synthesis has failed to make more of an impact. Perhaps the reason is that the technology needed to make the transition from bench-top chemistry to commercial operations has just not developed. It is therefore regrettable that very few institutes have had experience in the operation of electrochemical pilot-plants. However, the interest now developing in electrochemical engineering must be welcomed. A conference was devoted to this subject recently[1] and much interesting work has resulted from the Newcastle and Southampton schools exploring new cell configurations. However, much research in organic electro-synthesis is done without an awareness of the economic requirements of the industry. This does little to improve an image of a subject which to many is an academic curiosity.

Bewick and Pletcher, introducing the previous Report,[2] emphasized the need for organic electrochemists to be equally familiar with the two disciplines of organic chemistry and electrochemistry. In addition, it is probable that more motivated research would result if a knowledge of electrochemical engineering was also included.

There have been many interesting conferences[3-5] and publications during the period of this Report. Unfortunately, the proceedings of the conference most dedicated to organic electrochemistry[3] are not collated in one volume,

[1] Electrochemical Engineering Symposium, Newcastle upon Tyne, England, March, 1971.
[2] A. Bewick and D. Pletcher, in 'Electrochemistry', ed. G. J. Hills (Specialist Periodical Reports), The Chemical Society, London, 1972, vol. 2, p. 1.
[3] Euchem Conference on Organic Electrochemistry, Ronneby, Sweden, June, 1971.
[4] Gesellschaft Deutscher Chemiker, Elektrochemisches Symposium, Jülich, B.R.D., October, 1971.
[5] Dechema-Arbeitsausschusses Elektrochemische Prozesse, Frankfurt, B.R.D., February, 1971.

though most of the work has been published by the individual authors. Eberson and Schäfer have written a useful volume in the series 'Topics in Current Chemistry',[6] and the first of a new series 'Progress in Electrochemistry of Organic Compounds' made its debut.[7] Edited by Frumkin and Ershler and translated into English, the review articles contained in the latter should draw attention to some of the less readily accessible Russian work. Reviews on industrial aspects of electrochemistry[8] and organic electrosynthesis[9-11] have also been published. It should be mentioned that there also exists an 'Analytical Chemistry Review' which overlaps the contents of this chapter to some extent.[12] Of general interest is a comparison of tetraalkylammonium salts as supporting electrolytes for organic electrochemical reactions.[13]

2 Reductions

Compounds containing the Carbonyl Group.—As in previous years, the electrochemistry of carbonyl-containing compounds has been a popular field of study. Much of the work has involved carbonyls containing an activating or a second functional group. In general these compounds have more reaction paths open to them and, in the case of unsaturated ketones, can undergo intramolecular cyclization. Interest in the stereochemical aspects of the reduction of ketones has continued.

Grimshaw, Grimshaw, and Rea[14] investigated the reduction of 2-acetylnaphthalenes and compared the stereoselectivity of the electrochemical and chemical routes. A high degree of stereoselectivity in favour of the ($\pm$)-pinacol was observed in alkaline solutions. This was attributed to the involvement of a dimerization step between one molecule of the radical anion (2) and one molecule of the conjugate acid (3). A significant degree of hydrogen-

$$C_{10}H_7Ac \rightarrow C_{10}H_7-\overset{O}{\underset{\cdot}{C}}-Me \underset{}{\overset{H^+}{\rightleftharpoons}} C_{10}H_7-\overset{OH}{\underset{\cdot}{C}}-Me$$

(1)   (2)   (3)

bonding between the two species just prior to dimerization was assumed, so that the stereochemical preference in this latter step was controlled by the need to minimize non-bonding interactions, while at the same time preserving

---

[6] L. Eberson and H. Schäfer, *Topics Current Chem.*, 1971, **21**, 1.
[7] A. N. Frumkin and A. B. Ershler, 'Progress in the Electrochemistry of Organic Compounds', Plenum Press, London, 1971.
[8] A. Kuhn, 'Industrial Electrochemical Processes', Elsevier, Amsterdam, 1971.
[9] G. Faita, *Chimica e Industria*, 1971, **53**, 472.
[10] J. Chang, R. F. Large, and G. Popp, 'Techniques of Chemistry, Volume 1 Part IIB: Electrochemical Methods', Wiley-Interscience, New York, 1971.
[11] S. Wawzonek, *Synthesis*, 1971, **1**, 285.
[12] D. J. Pietrzyk, *Analyt. Chem.*, 1971, **44**, 457R.
[13] H. O. House, E. Feng, and N. P. Peet, *J. Org. Chem.*, 1971, **36**, 2371.
[14] J. Grimshaw, J. T. Grimshaw, and E. J. F. Rea, *J. Chem. Soc. (C)*, 1971, 683.

the hydrogen bond between two oxygen atoms. The influence of pH on the stereoselectivity was attributed to its effect on the proportion of (2) and (3) present in the solution. Thus in acidic solution insignificant amounts of the radical anion are present, so that dimerization probably occurs by a combination of two radicals (3). If the hydrogen bond between two radicals is much weaker than that between (2) and (3), then the need to minimize non-bonding interactions during dimerization will lead to an increased proportion of *meso*-pinacol through dimerization of non-hydrogen-bonded radicals.

Horner and Degner have investigated the conditions leading to the preferential formation of the alcohol during the cathodic reduction of alkyl aryl ketones.[15,16] They found that when the support electrolyte was tetramethylammonium chloride, increases in the bulkiness of the alkyl group of the ketone reduced the ratio of pinacol to alcohol. The same was true when the support electrolyte was one of the ephedrines (5) or (6). However, in the

$$\begin{array}{cc} \text{H H} & \text{H H} \\ | \; | & | \; | \\ \text{Ph—C—C—Me} & \text{Ph—C—C—Me} \\ | \; | \quad \text{Cl}^- & | \; | \quad \text{Cl}^- \\ \text{HO NH}_2\text{Me} & \text{HO NMe}_3 \\ + & + \\ (1R,2S)\text{-}(4) & (1R,2S)\text{-}(5) \end{array}$$

$$\begin{array}{c} \text{H} \\ | \\ \text{PhCH}_2\text{—C—Me} \\ | \quad \text{Cl}^- \\ \text{NMe}_3 \\ + \\ (S)\text{-}(6) \end{array}$$

presence of (4) there was no change in the ratio of pinacol to alcohol as the alkyl group increased in size. The support electrolyte was also found to have a marked effect on the ratio of *meso*- to *dl*-pinacols produced, and a change in the absolute configuration of the supporting salt also induced a change in the absolute configuration of the alcohol. It is apparent that the configuration of the adsorbed support electrolyte has a great influence on the preferred orientation of the adsorbed depolarizer.

The effect of adjacent groups on the reduction of the carbonyl function was further exemplified in work by Daver and Guillanton[17] and Fleury and Tohier.[18] For example, the β-ketonitriles ArCOCH$_2$CN (Ar = Ph, p-BrC$_6$H$_4$, or furan) were found to reduce to the corresponding pinacols and alcohols,[17] whereas reduction of β-mercaptopyruvic acid[18] at pH < 7.0 gave only the alcohol β-mercaptolactic acid. At higher pH the mercapto-group was eliminated.

[15] L. Horner and D. Degner, *Tetrahedron Letters*, 1971, 1241.
[16] L. Horner and D. Degner, *Tetrahedron Letters*, 1971, 1245.
[17] A. Daver and G. L. Guillanton, *Bull. Soc. chim. France*, 1971, 312.
[18] M.-B. Fleury and J. Tohier, *Bull. Soc. chim. France*, 1971, 2760.

A series of papers on the electroreduction of various diketones was presented by Kariv, Hermolin, Cohen, and Gileadi.[19—22] They observed that the ease of reduction of ketones was affected by the degree of saturation of the molecule and by its ability to enolize. The order of reducibility was αβ-unsaturated > 2-unsubstituted-1,3-diketone > 2,2-disubstituted-1,3-diketone > saturated ketone. The ketones studied had zero-order kinetics, indicating saturation coverage by electroactive species, and the data were consistent with the rate-determining step being the dimerization of adsorbed species. Only one of the carbonyl groups was found to reduce under the conditions employed. Thus the diketone pinacols were the predominant products. In the case of 2,2,5,5-tetramethylcyclohexane-1,3-dione (7) the keto-alcohol was obtained together with a rearrangement product (Scheme 1).

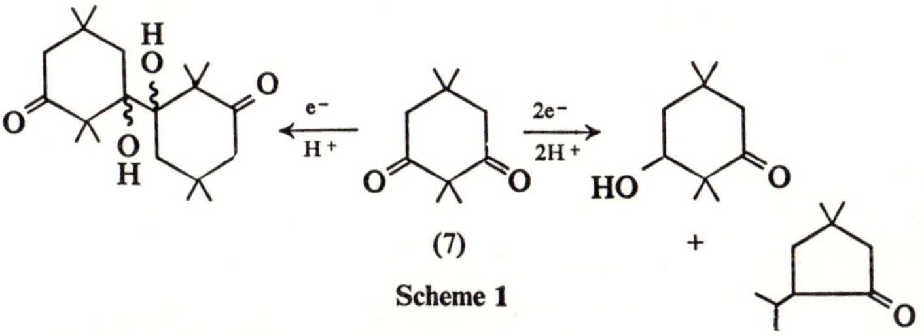

Scheme 1

(7)

Ryvolová-Kejharová and Zuman have observed the formation of the keto-alcohol oxyphthalimidine when phthalimide was reduced at the potential corresponding to a two-electron transfer.[23] The electrochemical reduction of acetylacetone (8) was reported by Thomsen and Lund.[24] At a potential of −1.25 V (S.C.E.) in 0.5M-HCl, four main products were obtained. The proposed reduction path is shown in Scheme 2.

The behaviour of αβ-unsaturated ketones or aldehydes undergoing an electroreduction is somewhat different to that of their saturated counterparts. The conjugated C=C and C=O double bonds lead to an increased stability of the free radical. Dimerization can occur between two β-carbon atoms as well as between two carbonyl carbon atoms, and the cross-coupled product has also been considered. The exact dimerization route is a function of the electron-donating or -withdrawing properties of the substituent groups and especially of steric hindrance effects due to their bulkiness. Pinacol formation was in fact only observed when unsaturated aldehydes or methyl ketones were reduced. Miller, Mandell, and Day[25] obtained 42% of dimeric

---

[19] E. Kariv and E. Gileadi, *Coll. Czech. Chem. Comm.*, 1971, **36**, 476.
[20] E. Kariv, J. Hermolin, and E. Gileadi, *Electrochim. Acta*, 1971, **16**, 1437.
[21] E. Kariv, J. Hermolin, I. Rubinstein, and E. Gileadi, *Tetrahedron*, 1971, **27**, 1303.
[22] E. Kariv, B J. Cohen, and E. Gileadi, *Tetrahedron*, 1971, **27**, 805.
[23] A. Ryvolová-Kejharová and P. Zuman, *Coll. Czech. Chem. Comm.*, 1971, **36**, 1019.
[24] A. D. Thomsen and H. Lund, *Acta Chem. Scand.*, 1971, **25**, 1576.
[25] D. Miller, L. Mandell, and R. A. Day, jun., *J. Org. Chem.*, 1971, **36**, 1683.

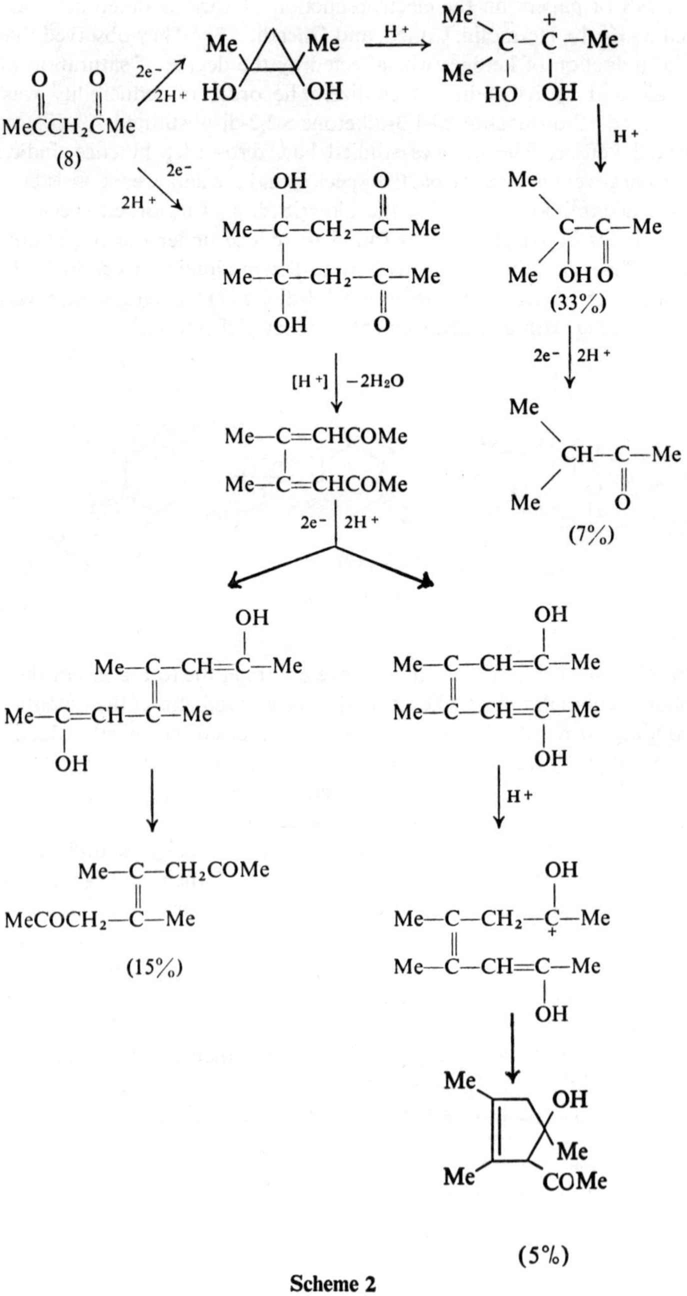

Scheme 2

product when they reduced 3-methylcrotonaldehyde at a mercury electrode [−1.3 V (S.C.E.), pH 5.0]. The dimer composition was 67% of the cis- and trans-isomers of (9) and 24% of the dl- and meso-forms of the diol (10).

$$\begin{array}{cc}
\text{Me} \\
\diagdown \\
C=C \\
\text{Me}\diagup \quad \text{Me} \\
\text{Me} \\
(9)
\end{array} \qquad \begin{array}{c}
\text{Me} \quad \text{OH} \\
\diagdown \quad | \\
C=CH-C-H \\
\diagup \quad | \\
\text{Me} \\
\text{Me} \\
\diagdown \\
C=CH-C-H \\
\diagup \quad | \\
\text{Me} \quad \text{OH} \\
(10)
\end{array}$$

Similarly, Barnes and Zuman[26] have found that crotonaldehyde also gives a pinacol. Reduction at the second wave reduced the aldehyde to crotyl alcohol. The cis- and trans-isomers were obtained in a ratio of 1:5.

The main dimeric species produced by electrolysis at the first polarographic wave of the compounds represented by (11) [except (11f)] were the corresponding diketones (see Table 1). The reductions of (11a) and (11g) at the second

Table 1 *The electroreduction of compound* (11)

$$\begin{array}{c}
O \\
\| \\
R^1 \quad C-R^3 \\
\diagdown \diagup \\
C=C \\
\diagup \diagdown \\
H R^2 \\
(11)
\end{array}$$

	R^1	R^2	R^3	Ref.
(11a)	Ph	H	Me, Ph, or Bu^t	27
(11b)	Ph	H	Me	28
(11c)	Ph	H	Et	28
(11d)	Ph	Me	Me	28
(11e)	Ph	H	Ph	29
(11f)	Ph	H	2-, 3-, or 4-pyridyl	30
(11g)	X	H	Me, Pr^n, Pr^i, or Bu^t	31

polarographic wave were reported to form the corresponding saturated ketones. Albisson and Simonet[31] found that the percentage yield of diketone formed by the electroreduction of the compounds (11g) in the presence of a tetrabutylammonium iodide increased when the alkyl group R^3 increased in size from Me (0%) to Bu^t (95%). However, the presence of lithium ions enhanced the yield of diketone. The same authors,[29] studying the reduction of (11e), also obtained a product formed by cyclization of the diketone (12). The electrochemical reduction selectively formed the trans-isomer, in contrast to the Zn–HOAc reduction which gave the cis.

[26] D. Barnes and P. Zuman, *J. Chem. Soc.* (*B*), 1971, 1118.
[27] J. P. Zimmer, J. A. Richards, J. C. Turner, and D. H. Evans, *Analyt. Chem.*, 1971, **43**, 1000
[28] I. Traore, Y. L. Pascal, and J. Wiemann, *Ann. Chim.* (*France*), 1971, **6**, 27.
[29] J. Simonet and A. Albisson, *Bull. Soc. chim. France*, 1971, 1125.
[30] J.-M. Meunier and E. Laviron, *Bull. Soc. chim. France*, 1971, 3793.
[31] A. Albisson and J. Simonet, *Bull. Soc. chim. France*, 1971, 4213.

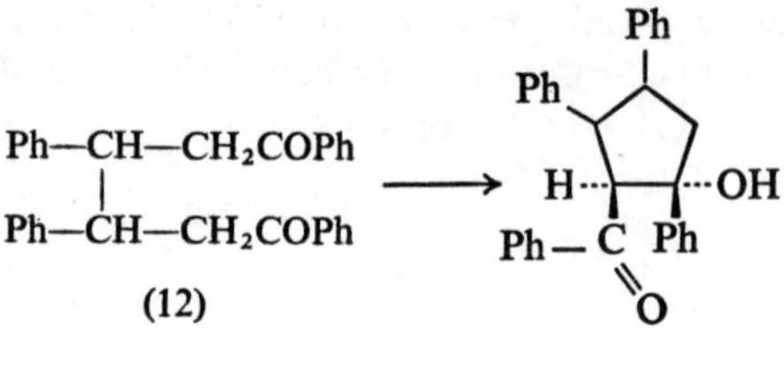

Ph—CH—CH$_2$COPh
|
Ph—CH—CH$_2$COPh

(12)

⟶

(structure 13: cyclopentane with Ph, Ph, H, OH, Ph–C(=O)–Ph substituents)

(13)

Substituted cyclopentanes corresponding to (13) were also obtained by Traore, Pascal, and Wiemann when they electrolytically reduced (11b), (11c), or (11d).[28] An extensive analysis of the stereochemistry of the reduction products was made. The reduction was also found to result in the stereospecific formation of the *meso*-diketones.

Meunier and Laviron reported that compound (11f) gave three polarographic reduction waves.[30] The first two each corresponded to the transfer of one electron, while the third wave increased with pH until at about pH 10 it was a two-electron process. Controlled-potential electrolysis at the potential of the third wave resulted in the formation of the saturated alcohol PhCH$_2$CH$_2$CH(OH)Py, where Py = 2-, 3-, or 4-pyridyl.

Shono and Mitani have developed an interesting electrochemical synthesis of cyclic tertiary alcohols, which have been difficult to prepare by chemical cyclization techniques.[32] A non-conjugated olefinic ketone was reduced at a carbon cathode [−2.7 V (S.C.E.)] using a dioxan (80%)–methanol (20%) solvent system and tetraethylammonium toluene-*p*-sulphonate electrolyte. The products of the reduction of H$_2$C=CH(CH$_2$)$_3$COR were the corresponding *cis*-derivatives of cyclopropane (14). The exclusive formation of the *cis*-

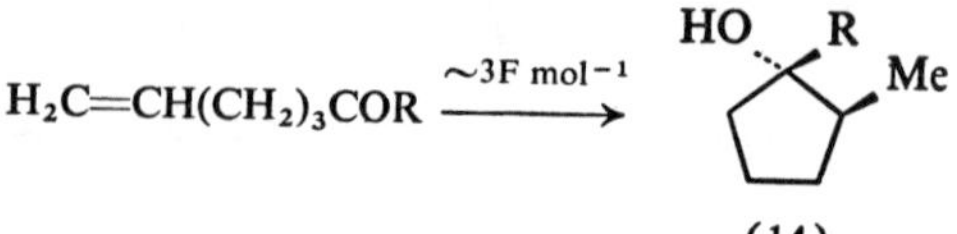

H$_2$C=CH(CH$_2$)$_3$COR $\xrightarrow{\sim 3F\ mol^{-1}}$ (cyclopentane with HO, R, Me substituents)

(14)

R	Yield/%
Me	66
Et	47
Pr^i	45
Bu^n	35
Hex^n	40

Scheme 3

isomer is very interesting in that it contrasts with the chemical route, involving the reaction of a Grignard reagent with the cyclic ketone. This gives mainly the *trans*-isomer. These electrochemically initiated cyclization reactions are analogous to experiments in which one attempts to trap radical intermediates

[32] T. Shono and M. Mitani, *J. Amer. Chem. Soc.*, 1971, **93**, 5284.

Organic Electrochemistry - Synthetic Aspects

with an olefin. The authors found that the carbonyl group always attacked the inner carbon atom of the terminal double bond (Scheme 4). When the substrate was substituted on that carbon atom, cyclization was prevented (Scheme 5). On the other hand, cyclization did occur when a substituent group was located at any other position (Scheme 6). It was not possible by

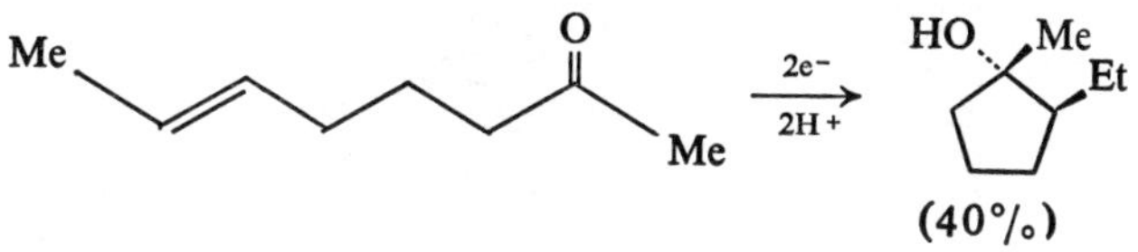

Scheme 4

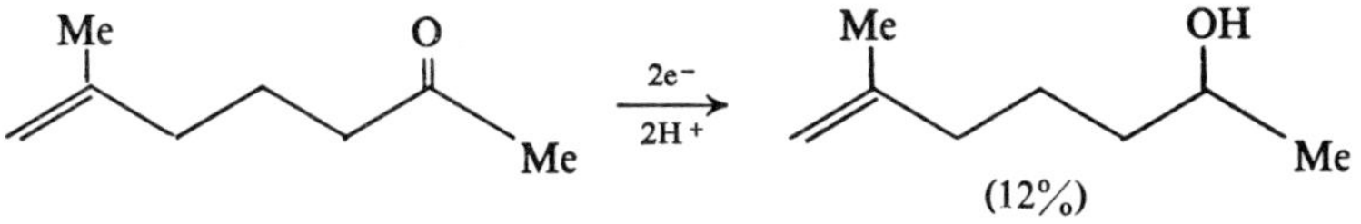

Scheme 5

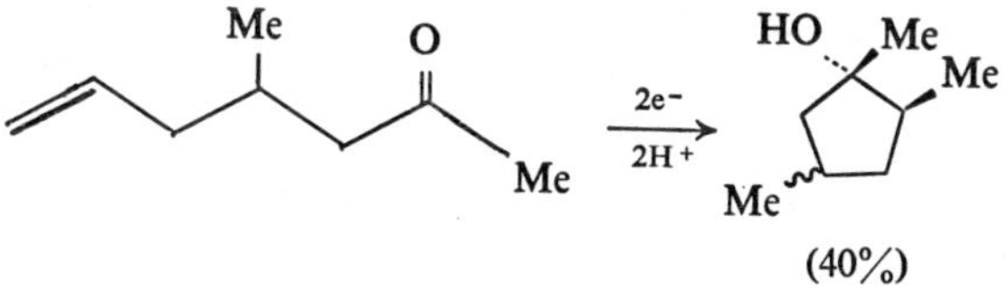

Scheme 6

this technique to prepare strained-ring compounds; only five- or six-membered rings have been successfully prepared (see Table 2).

Barnes, Uden, and Zuman found that electroreduction of glycolaldehyde at the first polarographic wave gives a 90% yield of acetaldehyde.[33] The second polarographic wave was attributed to the reduction of the product. Its limiting current, which was kinetically controlled, was thought to be due to dehydration of the electro-inactive hydrated aldehyde.

A basic consideration in designing an electro-organic synthesis is whether any substituent α to the carbonyl will be eliminated during the course of the reduction. Nadjo and Saveant have extended existing data on this aspect with a study of ring-substituted aryl ketones.[34] Of the compounds investigated (p-phenyl-, p- and m-dimethylamino-, -methoxy-, -methyl-, -trimethylammonium, -fluoro-, -chloro-, and -bromo-benzophenones, and the 2-substituted dimethylamino-, methoxy-, fluoro-, chloro-, and bromo-fluorenones) only the

---

[33] D. Barnes, P. C. Uden, and P. Zuman, *J. Chem. Soc. (B)*, 1971, 1114.
[34] L. Nadjo and J. M. Savéant, *Electroanalyt. Chem.*, 1971, **30**, 41.

Table 2 *Reduction of* $H_2C{=}CH(CH_2)_n COR$

R	n	Product	Yield/%
Me	2	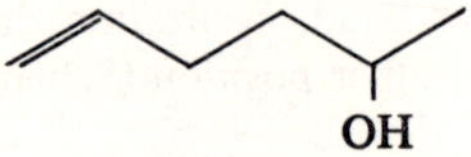 (hex-5-en-2-ol)	33
Me	3	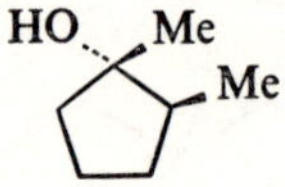 1,2-dimethylcyclopentan-1-ol	66
Me	4	1,2-dimethylcyclohexan-1-ol + oct-7-en-2-ol	50 + 10
Me	5	 non-8-ene-2,3-diol	25
		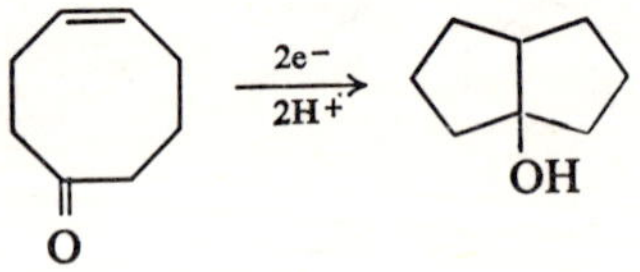 cyclooct-2-enone →(2e⁻, 2H⁺) bicyclic alcohol	64

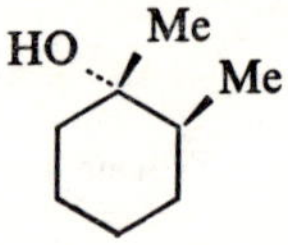

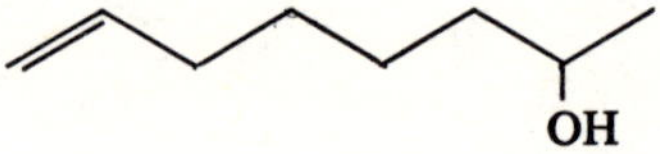

p-chloro- and p- and m-bromo-benzophenones eliminated the substituent (halide). Benzophenone was formed quantitatively (Scheme 7).

Rudd and Conway[35] have investigated the mechanistic aspects of pinacol

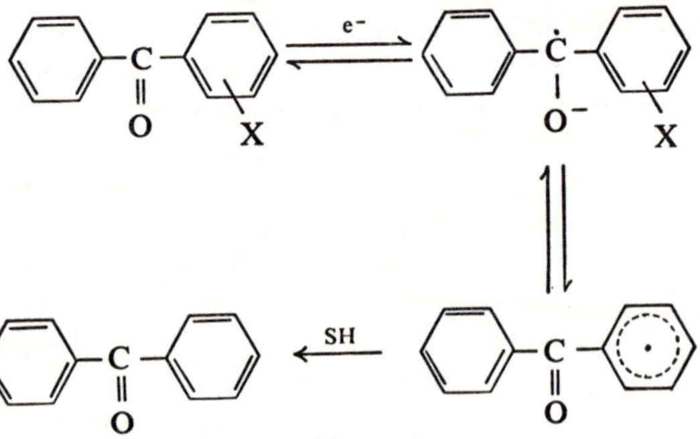

Scheme 7

[35] E. J. Rudd and B. E. Conway, *Trans. Faraday Soc.*, 1971, **67**, 440.

formation. They conclude that the radical anion must be protonated immediately after its formation and then undergo dimerization at the electrode surface. A homogeneous dimerization step was ruled out. Kugatova-Shemyakina, Nikolaeva, Mairanovskii, and Surikova[36] studied the effect of the halfwave potential of interaction between a non-conjugated double bond and a

(15) R = H or Me   (16) R = H or Me

polar substituent. The n.m.r. spectrum of (16) revealed that the conformation of the semicarbazone was axial. Interaction with the double bond in (16) was therefore minimal and, indeed, no difference in half-wave potentials of (15) and (16) was detected. However, the corresponding aldehydes, from which the semicarbazones were derived, had different conformations and the reduction potentials were altered by introducing the double bond.

The polarography and preparative electroreduction of some indanediones was discussed by Zacharová-Kalavská, Perjéssy, and Zelenský.[37–39] At pH values between 8 and 12 these compounds were reduced in a four-electron wave, with saturation of the C=C bond and conversion of one carbonyl group into an alcohol. The reduction of D-erythrose, D-threose, and L-glycerotetrulose was investigated by Fedoroňko, Fülleová, and Linek.[40] They found that the reductions of the aldotetroses were controlled by the kinetics of the dehydration of a hydrated molecule or rupture of the cyclic form. Controlled-potential electrolysis gave the corresponding tetrols. Reduction of L-glycerotetrulose gave successive reductive elimination of first the primary and then the secondary hydroxy-groups. A side reaction involved the reduction of the keto-function to a secondary alcohol. Guillanton[41] reported the reduction of dehydracetic acid (17). In neutral or alkaline solutions the electrolysis gave the diol (18), characteristic of other ketone reductions. In acidic solutions the diol partially decomposed (Scheme 8).

The electroreduction of 3-flavonolosides in DMF[42] and very acidic solu-

---

[36] G. P. Kugatova-Shemyakina, G. M. Nikolaeva, S. G. Mairanovskii, and N. V. Surikova, *Tetrahedron Letters*, 1971, 1725.
[37] D. Zacharová-Kalavská, I. Zelenský, and A. Perjéssy, *Coll. Czech. Chem. Comm.*, 1971, **36**, 2716.
[38] D. Zacharová-Kalavská, A. Perjéssy, and I. Zelenský, *Coll. Czech. Chem. Comm.*, 1971, **36**, 2712.
[39] D. Zacharová-Kalavská and A. Perjéssy, *Coll. Czech. Chem. Comm.*, 1971, **36**, 1406.
[40] M. Fedoroňko, E. Fülleová, and K. Linek, *Coll. Czech. Chem. Comm.*, 1971, **36**, 114.
[41] G. L. Guillanton, *Compt. rend.*, 1971, **272**, C, 11.
[42] J. P. Haluk and M. Metche, *Chim. analyt.*, 1971, **53**, 149.

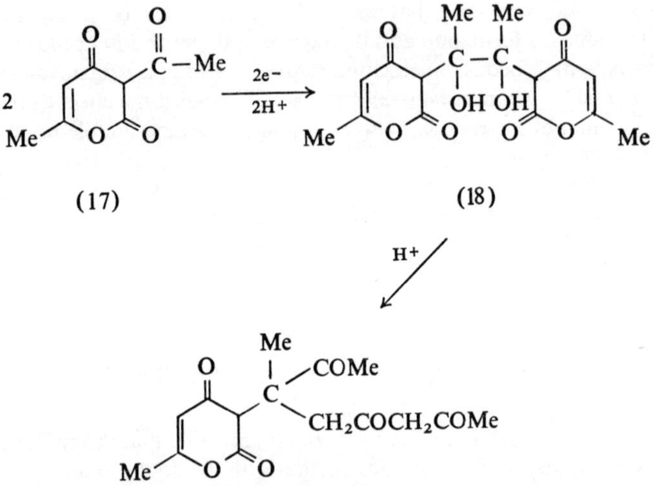

Scheme 8

tions[43] has been studied by Haluk and Metche. The electrochemical reduction was found to be more selective than catalytic hydrogenation since mainly the heterosidic groups linked at non-aromatic structures were attacked. The same authors also studied flavonols, flavanonols, and flavanones.[44] The electrochemical reduction was found to give anthocyanidin pigments with higher purity than those obtained by currently known methods.

Reductive Coupling Reactions.—This section is devoted to hydrodimerizations akin to the Baizer adiponitrile process. Cross-coupling reactions are also considered here. However, attention is drawn to the fact that many of the reductions reviewed in the section on unsaturated ketones could also be included.

Baizer's interest in reductive coupling reactions has continued and together with Chruma he has published on the subject of mixed reductive couplings with acrylonitrile.[45] These reductions can in principle give three products:

$$A + B \xrightarrow[H_2O]{2e^-} HAAH + HABH + HBBH$$

where A and B are:

A (more easily reduced partner)   B

Acrylonitrile (AN) .... { Acetone / Styrene / 1,1-Diphenylethylene

PhCOPh
Diethyl malonate } .... Acrylonitrile
PhCHO

[43] J. P. Haluk and M. Metche, *Chim. analyt.*, 1971, **53**, 375.
[44] J. P. Haluk and M. Metche, *Chim. analyt.*, 1970, **52**, 1245.
[45] M. M. Baizer and J. L. Chruma, *J. Electrochem. Soc.*, 1971, **118**, 450.

The object of the work was to find the conditions for optimum yields of the mixed product HABH. In the case of attempting to couple two compounds whose reduction potentials are close, the normal behaviour of two activated olefins is not to couple. Thus AN does not cross-couple with styrene or 1,1-diphenylethylene at the reduction potential of AN, but it does form the cross-coupled product, 5-hydroxy-5-methylvaleronitrile, with acetone. However, electrolysis at the reduction potential of the more difficultly reduced compound does lead to cross-coupled products in all cases, with improved yields. For example, benzophenone was previously reported not to couple with AN at -1.73 V. The main product was shown to be polyacrylonitrile. However, at potentials more cathodic than -2.00 V the cross-coupled product was formed. At these negative potentials polymer formation was also suppressed.

Difficulty in distinguishing between ECE and EC mechanisms was exemplified by the opposing conclusions of Petrovich and Baizer[46] and Childs, Maloy, Keszthelyi, and Bard.[47] The latter authors studied the mechanism of the electrohydrodimerization of diethyl fumarate (DEF) in DMF by cyclic voltammetry, double-potential step chronoamperometry, and controlled-potential coulometry. The results indicated that the first step of the reduction was a one-electron transfer, giving the anion radical of DEF (in agreement with Baizer). In contrast to earlier views, however, they found that dimerization of these anion radicals was the major path to the hydrodimer:

$$R + e^- \rightarrow R^{\bar{\cdot}} \qquad 2 R^{\bar{\cdot}} \rightarrow R_2^{2-}$$

Petrovich and Baizer[46] studied the effect of counterions on the electrolytic reductive coupling of DEF and PhCH=CRX (where R=H for X=CN or CONMe$_2$ and R=Ph for X=CN). The results were interpreted on the basis of the counterion having a 'directive influence' on the reaction pathway. Those considered were:

(i) ECE mechanism.

$$XCH{=}CHY \xrightarrow{e^-} (XCHCHY)^{\bar{\cdot}} \xrightarrow{XCHCHY} (XCHCHY)_2^{\bar{\cdot}}$$
$$\downarrow e^-$$
$$XCH_2CHYCHXCH_2Y$$
$$+ YCH_2CHXCHXCH_2Y$$

(ii) EC mechanism.

$$2(XCH{=}CHY) \xrightarrow{2e^-} 2(XCHCHY)^{\bar{\cdot}} \longrightarrow (XCHCHY)_2^{2-}$$
$$\downarrow H_2O$$
$$XCH_2CHYCHYCH_2Y$$

They found that the counterion had quite a marked effect on the mechanism, and concluded that it was caused by the effects of ion pairing with the inter-

---

[46] J. P. Petrovich and M. M. Baizer, *J. Electrochem. Soc.* 1971, **118**, 447.
[47] W. V. Childs, J. T. Maloy, C. P. Keszthelyi, and A. J. Bard, *J. Electrochem. Soc.*, 1971, **118**, 874.

mediate anion radical. However, while the experimental results of Petrovich and Baizer and Childs et al. were in agreement, Petrovich and Baizer based their interpretation on rather suspect models and ended up with different mechanistic conclusions. The consensus of opinion would seem to favour the interpretation of Childs et al.

Nonaka and Sekine used a.c. techniques to study the crossed hydro-coupling of acetone with activated olefins.[48] The reaction was deduced to take place *via* the nucleophilic addition of a hydroacetone anion to the olefin (Scheme 9).

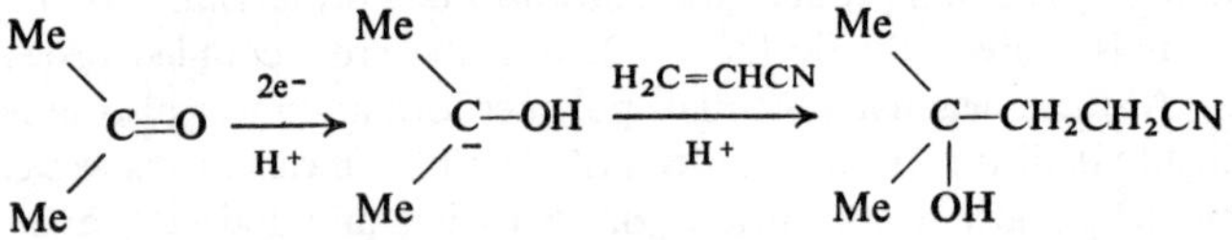

$$\begin{array}{c}Me\\ \diagdown\\ C=O\\ \diagup\\ Me\end{array} \xrightarrow[H^+]{2e^-} \begin{array}{c}Me\\ \diagdown\\ C-OH\\ \diagup\\ Me\end{array}^{-} \xrightarrow[H^+]{H_2C=CHCN} \begin{array}{c}Me\\ \diagdown\\ C-CH_2CH_2CN\\ \diagup |\\ Me\ \ OH\end{array}$$

Scheme 9

Differential capacity measurements on the acrylonitrile–adiponitrile system were used by Tomilov, Fedorova, Klimov, and Tedoradze to investigate of the role of adsorption in electrolytic hydrodimerization.[49] The main effects of tetraethylammonium cations were twofold, firstly to form an anhydrous layer next to the electrode, and secondly to raise the surface concentration of acrylonitrile in the more negative potential region where the electroreduction of acrylonitrile occurs.

Reduction of Halides.—The apparently straightforward electrochemical reduction eliminating a halide atom from a molecule is in fact a reaction of considerable complexity. Even though there has been much activity in this field it is still difficult to make generalizations about the mechanism. Aryl halides seem to give the most complex behaviour: for example, there are reports of the elimination of halide at a polarized mercury cathode prior to any electron transfer.[50] However, the reaction of the radical, formed after elimination of the halide, seems to be normal: hydrogen abstraction, dimerization, formation of an organometallic compound, or an increase in the unsaturation of the substrate, as in the case of dihalogeno-compounds.

The stereochemical aspects of the cathodic reduction of halogenated organic derivatives have recently been reviewed by Butin.[51] A comparison was made between the behaviour of these compounds on mercury and on platinum. Specific interaction between the halogen and the mercury electrode was shown to occur during reduction, *i.e.* direct contact between the halogen derivative and mercury is a condition for the production of the transition state.

[48] T. Nonaka and T. Sekine, *Denki Kagaku*, 1971, **39**, 29.
[49] A. P. Tomilov, L. A. Fedorova, V. A. Klimov, and G. A. Tedoradze, *Soviet Electrochem.*, 1971, **7**, 936.
[50] Y. Matsui, T. Soga, and Y. Date, *Bull. Chem. Soc. Japan*, 1971, **44**, 513.
[51] K. P. Butin, *Russ. Chem. Rev.*, 1971, **40**, 525.

The reduction of 2-halogeno-2-nitropropane has been studied by Bartak and Hawley[52] and also by Armand, Pinson, and Simonet.[53] The latter authors found that in aprotic solvents the dimer 2,3-dimethyl-2,3-dinitrobutane was formed in almost quantitative yield. They assumed that the reaction proceeded by the addition of two electrons, elimination of the halide, and dimerization between the so-formed anion and the original substrate. This mechanism was thought to be unlikely by Bartak and Hawley, who investigated the reaction path in some detail using cyclic voltammetry and the effect of proton and hydrogen-atom donors on the product yields. They concluded from their studies that the most likely path involved the combination of a 2-nitropropyl anion with a 2-nitropropyl radical (Scheme 10).

$$RX + e^- \rightleftharpoons RX^{\overline{\cdot}}$$
$$RX^{\overline{\cdot}} \rightarrow R\cdot + X^-$$
$$R\cdot + e^- \rightleftharpoons R^-$$
$$R\cdot + RX^{\overline{\cdot}} \rightarrow R^- + RX$$
$$R\cdot + R^- \rightarrow RR^{\overline{\cdot}}$$
$$RR\cdot + RX \rightarrow RR + RX^{\overline{\cdot}}$$

Scheme 10

The polarographic reduction of *cis*- and *trans*-3-chlorophenylpropenenitriles was studied by Guillanton and Daver.[54] Two 2-electron waves were obtained in each case. The second wave occurred at the same potential for both isomers and was attributed to the reduction of the phenylpropenenitrile. However, the ease of reduction at the first wave was dependent on the conformation of the molecule. The *cis*-isomer was reduced more easily than was the *trans*-isomer. Controlled-potential electrolysis gave 2-phenylpropanenitrile in excellent yields.

Matsui, Soga, and Date[50] found rather complex behaviour for the electroreduction of benzhydryl bromide in DMF. They observed that the substrate dissociated prior to electron transfer:

$$3Ph_2CHBr + Hg \rightarrow 2Ph_2CH\cdot + Ph_2CH^+ + HgBr$$

The polarogram then showed three waves, which the authors attributed to the processes:

$$2 HgBr_3^- + Hg \rightleftharpoons 3 HgBr_2 + 2e^-$$
$$2 HgBr_3^- + 4e^- \rightleftharpoons 2 Hg + 6 Br^-$$
$$Ph_2CH\cdot + e^- \rightleftharpoons Ph_2CH^-$$

Thus controlled-potential electrolysis at the plateau of the second wave

---

[52] D. E. Bartak and M. D. Hawley, *J. Electroanalyt. Chem. Interfacial Electrochem.*, 1971, **33**, 13.
[53] J. Armand, J. Pinson, and J. Simonet, *Analyt. Letters*, 1971, **4**, 219.
[54] G. L. Guillanton and A. Daver, *Compt. rend.*, 1971, **272**, C. 421.

[−0.60 V (S.C.E.)] resulted in the formation of *sym*-tetraphenylethane in almost quantitative yield (89%) as a result of the reactions:

$$2\ PhCH\cdot \rightarrow Ph_2CHCHPh_2$$
$$Ph_2CH^+Br^- \rightarrow Ph_2CHBr$$

However, when the c.p.e. was performed at the third wave [−1.40 V (S.C.E.)] the products were 38% diphenylmethane and 51% *sym*-tetraphenylethane:

$$2\ Ph_2CH\cdot + 2\ e^- \rightleftharpoons 2\ Ph_2CH^-$$
$$Ph_2CH^- + solvent \rightarrow Ph_2CH_2$$
$$Ph_2CH^- + Ph_2CH^+ \rightarrow Ph_2CHCHPh_2$$

Brown, Thirsk, and Thornton[55] studied the reduction of benzyl bromide and chloride at a mercury cathode. Controlled-potential electrolysis at −1.35 V (*versus* Ag|AgClO$_4$ 0.01 mol l^{-1}), which corresponds to the foot of the polarization curve, gave dibenzylmercury with 60% current efficiency, whereas c.p.e. at −2.1 V gave toluene in almost quantitative yield. Bibenzyl was formed only in very small amounts. An unequivocal explanation of all the results was not attempted, but in several respects there were similarities with the benzhydryl bromide system,[50] especially the formation of monomeric hydrocarbon at more cathodic potentials.

The synthesis of lead tetra-alkyls by the reduction of alkyl halides at a lead cathode was discussed by Fleischmann, Pletcher, and Vance.[56] The use of DMF as solvent was shown to give improved conditions for the synthesis. In fact the effect of solvent on this process was quite striking. In DMF alkali-metal halides could be used as support electrolytes and yields of lead alkyls were as high as 90%. This is in contrast to previous studies which showed that alkali-metal halides were not suitable support electrolytes in propylene carbonate. The use of DMF is noteworthy in that cheap alkali-metal halides may be used which give higher solution conductivities.

Electroreduction of molecules with halide substituted at two or more carbon atoms usually results in an increase in unsaturation of the molecule[57,58] or in cyclization.[59−61] The reductive elimination of halide from perhalogenated molecules was described by Seiber[57] for the preparation of acetylenic compounds. Good yields were obtained when the reductions were performed at lead, rather than mercury, cathodes. For example, octachlorostyrene was reduced to pentachloroethynylbenzene in 77% yield, and the reductions of the

---

[55] O. R. Brown, H. R. Thirsk, and B. Thornton, *Electrochim. Acta*, 1971, **16**, 495.
[56] M. Fleischmann, D. Pletcher, and C. J. Vance, *J. Electroanalyt. Chem. Interfacial Electrochem.*, 1971, **29**, 325.
[57] J. N. Seiber, *J. Org. Chem.*, 1971, **36**, 2000.
[58] H. R. Koch and M. G. McKeon, *J. Electroanalyt. Chem. Interfacial Electrochem.*, 1971, **30**, 331.
[59] M. Rifi, *Coll. Czech. Chem. Comm.*, 1971, **36**, 932.
[60] M. Rifi, *J. Org. Chem.*, 1971, **36**, 2016.
[61] A. J. Fry and W. E. Britton, *Tetrahedron Letters*, 1971, 4363.

heptachloro-2- and -3-vinylpyridines gave the corresponding acetylenes in 63% and 67% yields, respectively. By-products were formed by further reductive elimination or by protonation of the intermediate carbanions (Scheme 11).

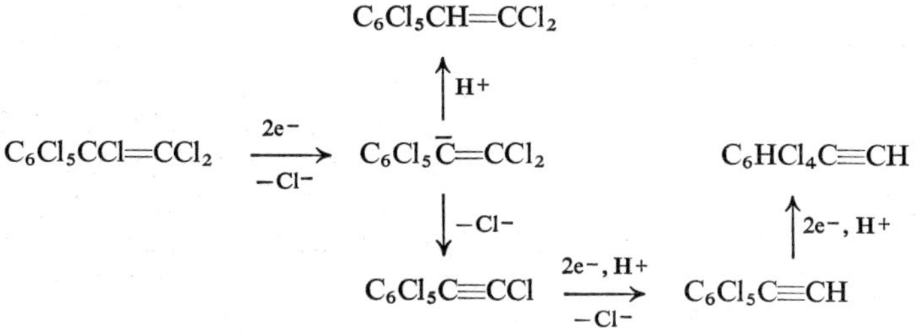

$$C_6Cl_5CCl=CCl_2 \xrightarrow[-Cl^-]{2e^-} C_6Cl_5\bar{C}=CCl_2 \xrightarrow{H^+} C_6Cl_5CH=CCl_2$$

$$C_6Cl_5\bar{C}=CCl_2 \xrightarrow{-Cl^-} C_6Cl_5C\equiv CCl \xrightarrow[-Cl^-]{2e^-, H^+} C_6Cl_5C\equiv CH$$

$$C_6HCl_4C\equiv CH \xleftarrow{2e^-, H^+} C_6Cl_5C\equiv CH$$

Scheme 11

The reduction of *trans*-3-*cis*-4-dibromo-t-butylcyclohexane and $2\beta,3\alpha$-dibromo-9βH,10αH-decalin at mercury cathodes has been re-examined by Koch and McKeon.[58] In the light of their new evidence the reduction is characteristic of other vicinal dibromides, in that it involves a 2-electron step to give an alkene, and does not proceed by a 4-electron step as was previously reported by Zavada.

Simonov, Cemenov, and Manina found that the dimethyl ester of *cis,cis*-tetrachloromuconic acid (19) gave two polarographic waves.[59] In a neutral medium the electrode reaction consumed ten electrons, six protons, and eliminated the four substituent chlorine atoms. Intermediates detected were α,α'-dichloromuconic acid and the dimethyl ester of muconic acid (Scheme 12).

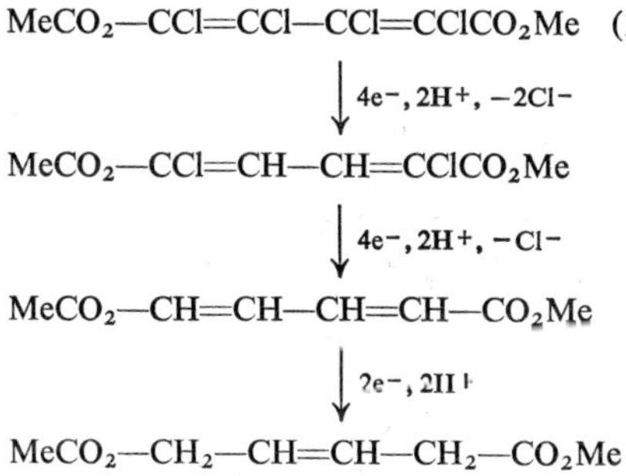

$$MeCO_2-CCl=CCl-CCl=CClCO_2Me \quad (19)$$
$$\downarrow 4e^-, 2H^+, -2Cl^-$$
$$MeCO_2-CCl=CH-CH=CClCO_2Me$$
$$\downarrow 4e^-, 2H^+, -Cl^-$$
$$MeCO_2-CH=CH-CH=CH-CO_2Me$$
$$\downarrow 2e^-, 2H^+$$
$$MeCO_2-CH_2-CH=CH-CH_2-CO_2Me$$

Scheme 12

The half-wave potentials and the greater adsorbability of the *cis,cis* than the *trans,trans* molecule led the authors to believe that elimination of the chlorine was easier from the *trans*-position.

Three papers have dealt with the formation of cyclic compounds.[60-62]

---

[62] V. D. Simonov, V. A. Cemenov, and F. A. Mamina, *Zhur. obshchei Khim.*, 1971, **41**, 980.

Rifi argued that the reduction of vicinal and 1,3-dihalides proceeds by a concerted mechanism.[60, 61] His conclusions were based on half-wave potentials and product analysis. The ease of reduction of *cis*-dihalides, relative to the *trans*-isomers, was thought to be due to the easier formation of the transition state for the concerted mechanism. The absence of propane and chloropropane in the reduction products of 1-bromo-3-chloropropane was thought to indicate that the intermediates were not anionic. However, Fry and Britton think otherwise.[62] They proposed that the formation of cyclopropanes during the electroreduction of 2,4-dibromopentane is in fact a step-wise process which proceeds through a carbanion intermediate. The concerted mechanism was rejected since the reduction of diastereomeric 2,4-dibromopentanes was not stereospecific. In addition, pentane was detected in reduction products, which probably comes from an anionic intermediate. The special affinity of the halide group for mercury electrodes was not mentioned in these works. It

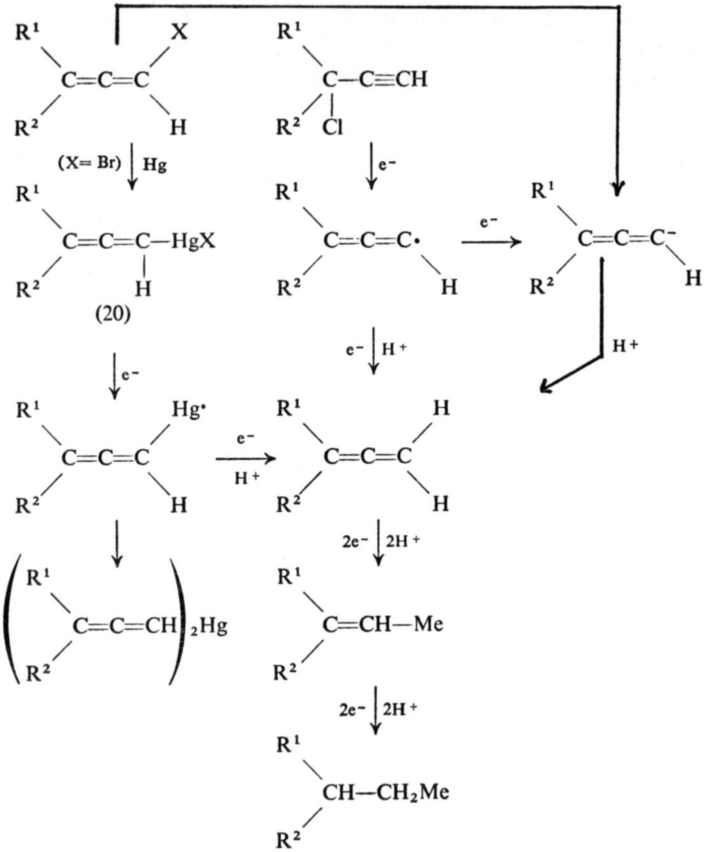

Scheme 13

is possible that the ease of reduction of the *cis*-dihalo geno-compounds is a result of stronger interaction with the electrode than is possible with the *trans*-compound.

Doupeux, Martinet, and Simonet studied the reduction of several unsaturated halogen compounds of the general formula $R^1R^2C=C=CHX$ and $R^1R^2XC-C\equiv CH$.[63] Most of the compounds were reduced in four steps. The first two waves given by 3-bromo-1,1-diphenylpropa-1,2-diene, 4-bromo-2-phenylbuta-2,3-diene, and 3-bromo-1-phenylpropa-1,2-diene were attributed to the reduction of an organometallic compound (20) formed prior to electron transfer. The overall reaction path proposed is shown in Scheme 13.

Aliphatic and Heterocyclic Azomethines.—Pinson and Armand have made a study of both aromatic and aliphatic compounds containing the $-\overset{|}{N}=\overset{|}{C}-\overset{|}{C}=N-$ group.[64, 65] Quinoxalines were found to be reduced at pH 7 in a 2-electron wave having a pH-dependent half-wave potential ($\Delta E/\text{pH} = 60$ mV). The transfer of two electrons and two protons was assigned to this wave, and it was shown by cyclic voltammetry to be pseudo-reversible. The product obtained by controlled-potential electrolysis was a 1,4-dihydroquinoxaline, which could not be further reduced. A re-arrangement of 2- or 3-substituted 1,4-dihydroquinoxalines (21) to either the corresponding 1,2- or 3,4-dihydroquinoxalines (22) was observed. The substituent groups had some directive influence on this rearrangement. Favourable conditions for this step necessitated alkaline conditions. Further reduction was possible only following the rearrangement (Scheme 14).

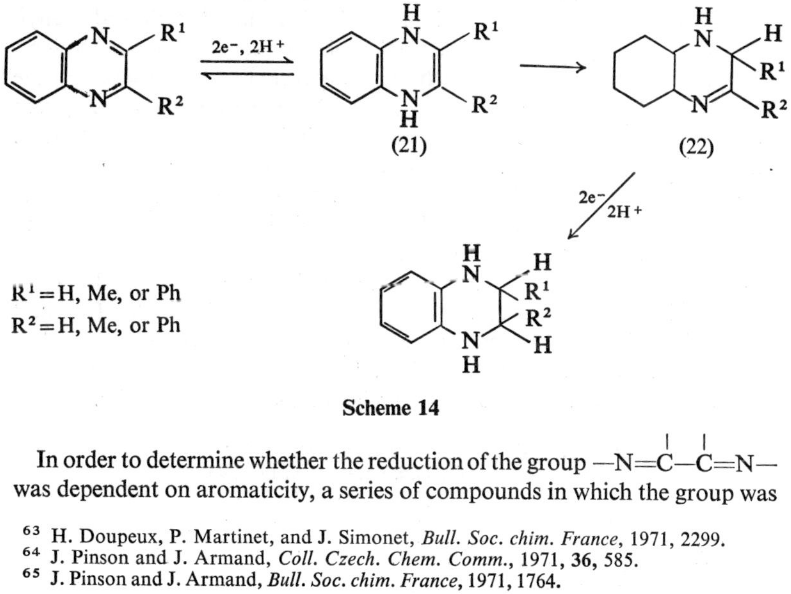

$R^1 = H$, Me, or Ph
$R^2 = H$, Me, or Ph

Scheme 14

In order to determine whether the reduction of the group $-N=\overset{|}{C}-\overset{|}{C}=N-$ was dependent on aromaticity, a series of compounds in which the group was

[63] H. Doupeux, P. Martinet, and J. Simonet, *Bull. Soc. chim. France*, 1971, 2299.
[64] J. Pinson and J. Armand, *Coll. Czech. Chem. Comm.*, 1971, **36**, 585.
[65] J. Pinson and J. Armand, *Bull. Soc. chim. France*, 1971, 1764.

not part of an aromatic ring was studied.[65] The compounds 2,3-dimethyl- and 2,3-diphenyl-5,6-dihydropyrazine were reduced irreversibly to the tetrahydro-derivatives. The reduction was in fact easier than for the pyrazines or quinoxalines. There was no rearrangement of the product, as observed with some quinoxalines. Some non-cyclic di-imines behaved similarly. In this case, however, non-symmetrical products were also formed (Scheme 15). Presumably this product is prefered owing to the increased conjugation of the azomethine group with a phenyl ring.

The reduction of phenazine was reported by Nakamura and Yoshida.[66] As with quinazolines, a 2-electron wave was obtained at high pH. The influence of amino-, carboxy-, hydroxy-, methoxy-, or chloro-groups substituted at positions 1 or 2 was relatively weak.

The kinetics of the reduction of pyrimidine in aqueous and non-aqueous solvents have been discussed by O'Reilly and Elving.[67]

Reductions of various substituted quinazolines (23) by Kwee and Lund resulted in hydrogenation of one C=N bond and elimination of the substituent group.[68] Quinazoline (24) was detected polarographically as an intermediate in the reduction. Cyclic voltammetry showed its presence only after the first sweep, indicating the slow elimination step following the transfer of two electrons. The rate of elimination of R was dependent on its effectiveness as a leaving group and on the pH: higher pH gave a higher rate. At more negative potentials and in alkaline conditions, the dihydroquinazoline (25) was reduced further to tetrahydroquinazoline (26). In addition, a dimerization was observed in the reduction of 4-methoxy- and 4-amino-quinazolines in alkaline solutions, whereby 4,4'-biquinazoline (27) was formed (Scheme 16).

Lund and Jensen found that 4-dimethylamino- and 4-methoxy-1-phenylphthalazines (28) were reduced in a single 2-electron step in alkaline solutions, whereas in acidic solutions a 4-electron step occurred.[69] The products of the 2-electron reduction were the corresponding 1,2-dihydro-derivatives. Fission of the N—N bond occurred during electrolysis in hydrochloric acid solutions. The mechanism was likened to that of aromatic phenylhydrazone reduction, in which cleavage of the N—N bond precedes saturation of the C=N double bonds (Scheme 17). In neutral and alkaline solutions further reductions to 3-imino-1-phenylisoindoline occurred.

The effect of proton donors on the electroreduction of 2-phenyl-3-phenyliminoindolenine has been discussed by Andruzzi, Cardinali, and Trazza.[70] Bogatskii et al. have reported on the polarographic reduction of some 1,4-benzodiazepines and their derivatives.[71] Some correlations were found between their ease of reduction and psychopharmacological activity.

[66] S. Nakamura and T. Yoshida, *Denki Kagaku*, 1971, **39**, 502.
[67] J. E. O'Reilly and P. J. Elving, *J. Amer. Chem. Soc.*, 1971, **93**, 1871.
[68] S. Kwee and H. Lund, *Acta Chem. Scand.*, 1971, **25**, 1813.
[69] H. Lund and E. T. Jensen, *Acta Chem. Scand.*, 1971, **25**, 2727.
[70] R. Andruzzi, M. E. Cardinali, and A. Trazza, *Ann. Chim. (Italy)*, 1971, **61**, 66.
[71] A. V. Bogatskii, S. A. Andronati, V. P. Gultyai, I. Vikhliaev, A. F. Galatin, Z. I. Zhilina, and T. A. Klygul, *Zhur. obshchei Khim.*, 1971, **41**, 1358.

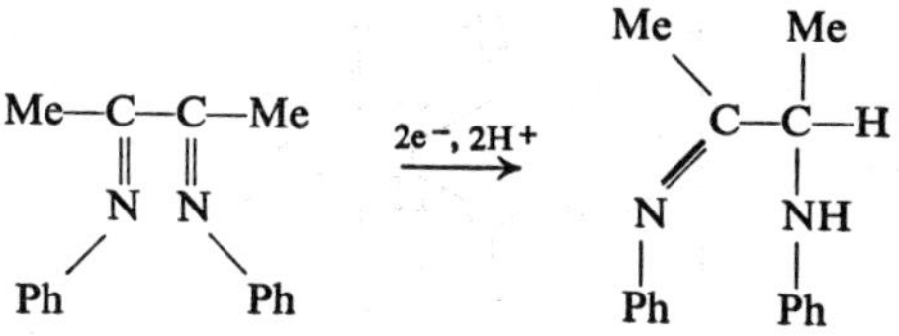

Scheme 15

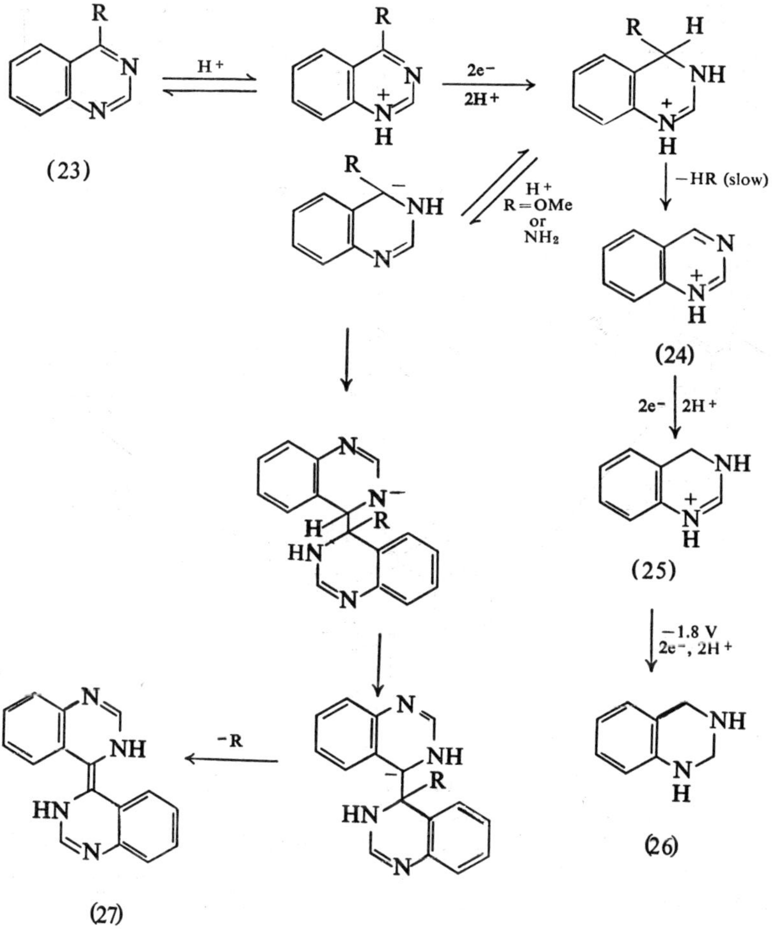

Scheme 16

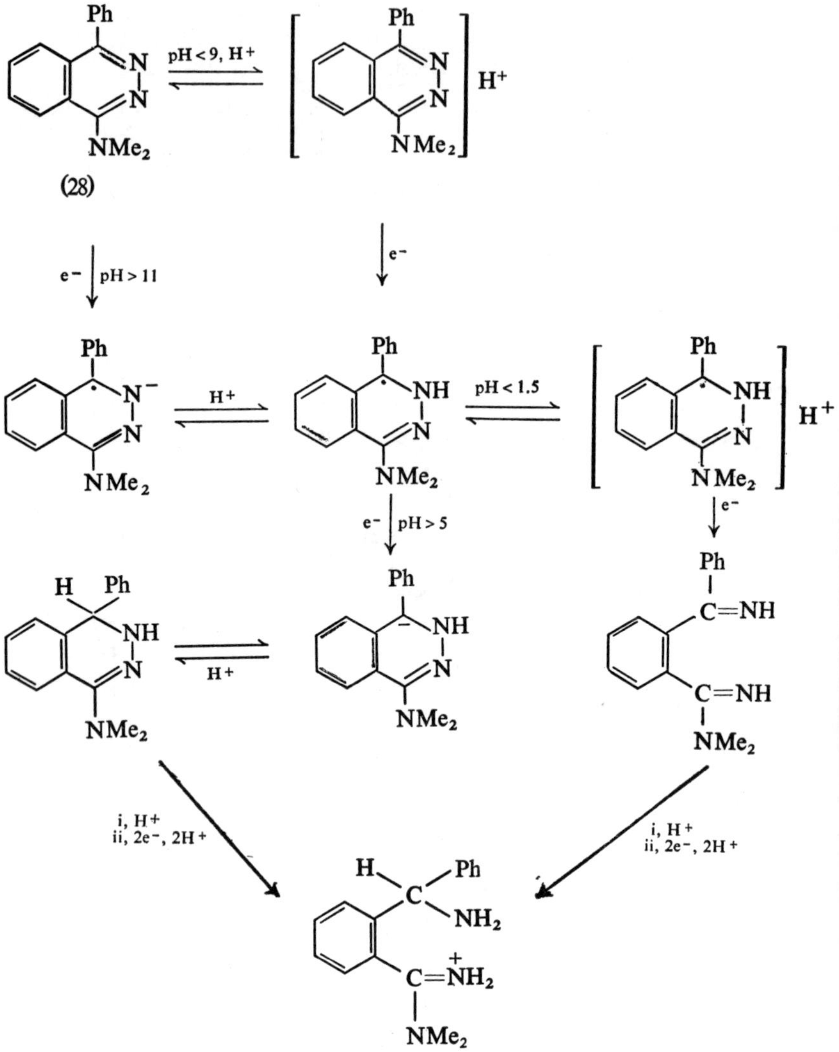

Scheme 17

The polarographic reduction of purine-2,6-disulphonic (29) was shown by McAllister and Dryhurst[72] to be quite involved. No fewer than eight polarographic waves were obtained. At a low pH (0—3.5) the purine was reduced polarographically in three steps, corresponding to those shown in Scheme 18. The nature of the polarographic waves was discussed in some detail.

[72] D. L. McAllister and G. Dryhurst, *J. Electroanalyt. Chem. Interfacial Electrochem.*, 1971, **32**, 387.

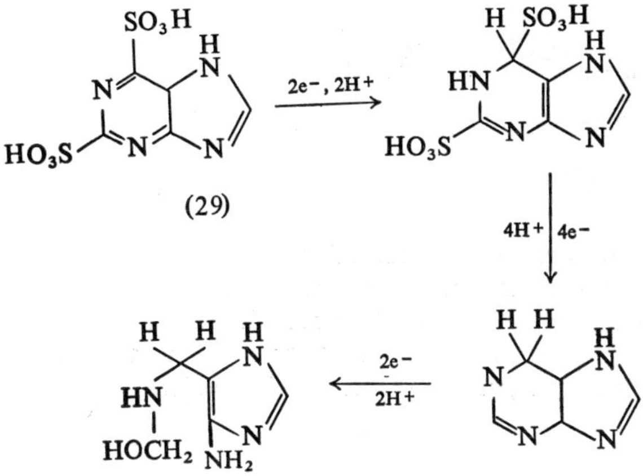

Scheme 18

A short communication by Weinberg, Hoffmann, and Reddy[73] was novel in two aspects, firstly in their use of molten ammonium toluene-p-sulphonate as the solvent support system and secondly in the reaction they studied. Ammonium toluene-p-sulphonate at 140 °C was found to be a good support system for cathodic reactions (the potential range at mercury electrodes was 0.0 V to −2.5 V versus Ag|Ag$^+$). A novel synthesis of amino-acids was then demonstrated when benzalaniline was reduced at −2.0 V (versus Ag|Ag$^+$) in the presence of CO_2 to give dl-α-phenylphenylglycine. The current efficiency was 60%. Reductive carboxylations, which have received little attention, would therefore seem to have useful synthetic applications.

The effect of intermolecular hydrogen-bonding on the reduction behaviour of azomethines was investigated by Mairanovskii, D'yachenko, and Gol'dfarb.[74] It has previously been shown that the proton-donating ability of phenolic groups through an intermolecular hydrogen-bridge bond is too weak to influence the reduction of azomethine groups. However, Mairanovskii et al. have now observed an effect with sulphur homologues of the phenolic compounds, as a result of the stronger electron-donating properties of the thiol group. Compound (30) was reduced in DMF at a potential 200 mV less cathodic than compound (31). Weak proton donors had little effect on the reduction behaviour of (30) but introduced a second wave for (31). Only when a strong acid (HCl) was present did the hydrogen-bridge bond in (30) break and become protonated, resulting in a second wave. The new wave was kinetically controlled by the bridge-breaking step. Evidently the hydrogen-bridge bond gives a very stable structure.

Hansen and Dryhurst re-examined the electroreduction of alloxan and its

---

[73] N. L. Weinberg, A. K. Hoffmann, and T. B. Reddy, *Tetrahedron Letters*, 1971, 2271.
[74] S. G. Mairanovskii, I. A. D'yachenko, and Ya. L. Gol'dfarb, *Soviet Electrochem.*, 1971, **7**, 29.

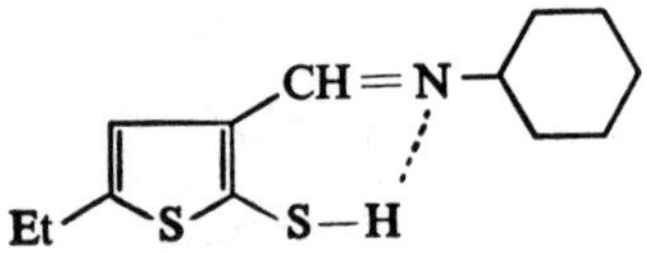

N-methyl derivatives.[75] They found that three distinct routes to the dialuric acid anion existed. They were the reduction of the protonated, unprotonated, and hydrated alloxan. Smith and Rogers studied the c.p.e. of aryl-substituted oxadiazoles.[76] They believed that complete saturation of the heteroaromatic ring together with ring fission occurred (Scheme 19).

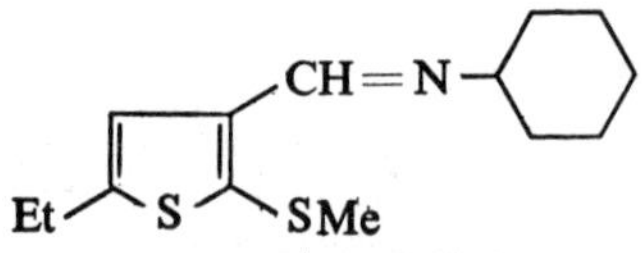

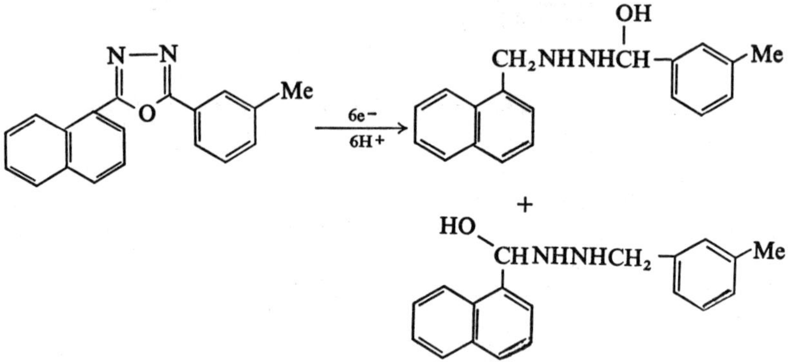

Scheme 19

The reduction of some benzylidine-4-imino-1,2,4(4*H*)-triazoles in DMF was discussed by Kitaev, Skrebkova, Zverev, and Maslova.[77] The half-wave potentials were found to correlate linearly with the Hammett σ-constant.

Armand, Boulares, and Pinson found that monosemicarbazones and anils

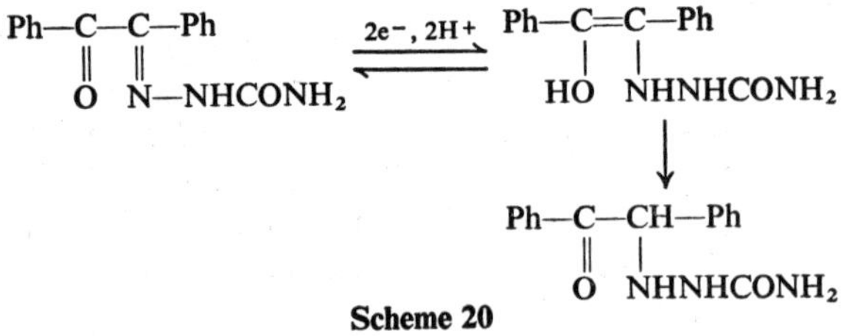

Scheme 20

[75] B. H. Hansen and G. Dryhurst, *J. Electrochem. Soc.*, 1971, **118**, 1747.
[76] G. L. Smith and J. W. Rogers, *J. Electrochem. Soc.*, 1971, **118**, 1089.
[77] Yu. P. Kitaev, I. M. Skrebkova, V. V. Zverev, and L. I. Maslova, *Bull. Acad. Sci., U.S.S.R.*, 1971, **20**, 23.

of diketones ($R^1COC=NR^2R^3$, where $R^2 = Ph$ or C_6H_{11}[78] or $R^2 = NHCONH_2$[79]) were reduced selectively at the azomethine group (Scheme 20). This was in contrast to conclusions by Fleet and Jee, who assumed that the product obtained when the diketone ($R^1 = Ph$, $R^2 = NHCONH_2$) was reduced was the imine PhCOC(=NH)Ph. The hydrazones and phenylhydroxylamines behaved differently even though the electrolysis conditions were similar. The main products resulted from fission of the N—N bond (Scheme 21). When the phenyl groups were replaced by 2-pyridyl groups the product of

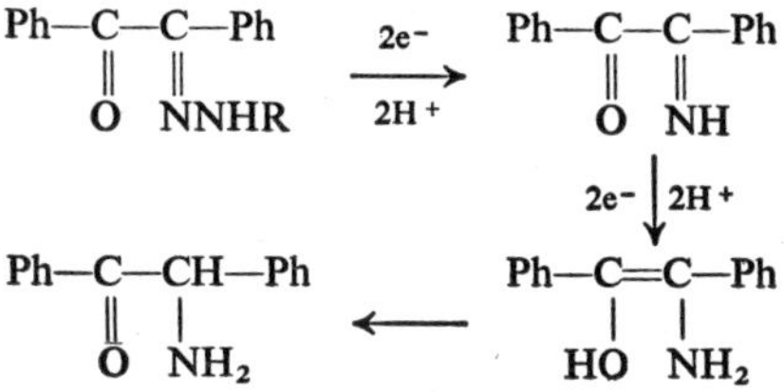

$$\begin{array}{c}
\text{Ph–C–C–Ph} \\
\parallel \ \ \parallel \\
\text{O NNHR}
\end{array} \xrightarrow[2H^+]{2e^-} \begin{array}{c}
\text{Ph–C–C–Ph} \\
\parallel \ \ \parallel \\
\text{O NH}
\end{array}$$

$$\downarrow 2e^-, 2H^+$$

$$\begin{array}{c}
\text{Ph–C–CH–Ph} \\
\parallel \ \ | \\
\text{O NH}_2
\end{array} \longleftarrow \begin{array}{c}
\text{Ph–C=C–Ph} \\
| \ \ \ \ | \\
\text{HO NH}_2
\end{array}$$

Scheme 21

the reduction was the enol. According to the authors, this was the first open-chain enaminol prepared.

Manoušek[80] has shown that the electroreduction of the monophenylhydroxylamine of ascorbic acid (32) also follows the path represented by Scheme 21.

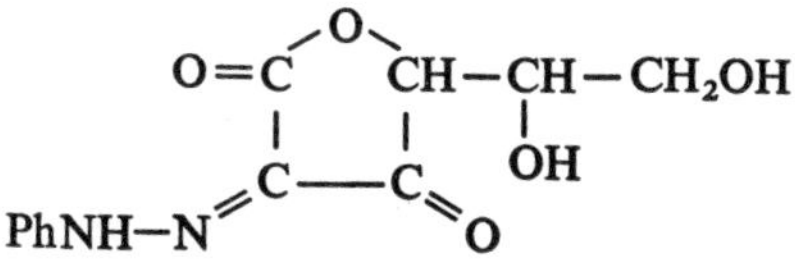

$$\text{PhNH–N} \begin{array}{c} \diagup \text{O} \diagdown \\ \diagup \ \ \ \ \ \ \ \diagdown \\ =\text{C}----\text{C}= \end{array} \begin{array}{c} \text{O=C} \ \ \ \text{CH–CH–CH}_2\text{OH} \\ | \ \ \ \ \ | \\ \ \ \ \ \ \ \ \ \ \ \ \ \ \text{OH} \\ \text{O} \end{array}$$

(32)

Compounds containing a Quaternary Ammonium Centre.—Reports on the reduction of compounds containing a quaternary ammonium group were all in agreement (with one exception[81]) that the first step was the transfer of one electron, with no participation by a proton. Tyssee and Baizer[82] have studied the reduction of NN-diethylaziridinium salts (33) which have reduction potentials accessible enough to make for easier mechanistic investigations. The addition of the first electron was accompanied by ring-opening, producing a triethylamine radical. The reduction potential of this species was less cathodic

---

[78] J. Armand, L. Boulares, J. Pinson, and P. Souchay, *Bull. Soc. chim. France*, 1971, 1918.
[79] J. Armand, L. Boulares, and J. Pinson, *Compt. rend.*, 1971, **273**, C, 120.
[80] O. Manoušek, *Z. Chem.*, 1971, **11**, 18.
[81] H. P. Cleghorn, J. E. Gaskin, and D. Lloyd, *J. Chem. Soc. (B)*, 1971, 1615.
[82] D. A. Tyssee and M. M. Baizer, *J. Electrochem. Soc.*, 1971, **118**, 1420.

Table 3 Reduction of R¹⟨N⁺(Me)(R²)⟩

R¹	R²	Products (yield %)
PhCH₂—Me—	Me	PhMe (90), bibenzyl (trace)
PhCH₂—Me—	Ph	PhMe (90), bibenzyl (3)
o-C₆H₄(CH₂—)(CH₂CH₂—)	Me	o-MeC₆H₄(CH₂CH₂NMe₂) (80), o-CH₂=CH-C₆H₄-CH₂NMe₂ (6)
o-C₆H₄(CH₂—)(CH₂CH₂CH₂—)	Me	PhCH₂CH₂CH₂NMe₂ (98)
PhCH₂CH₂—	Me	PhCH₂Me (40), PhCH=CH₂ (20)

than the aziridinium salt. The nature of the subsequent protonation or hydrogen-abstraction reaction was determined by electrolysing the aziridinium salt in a D_2O–H_2O solvent. Triethylamine was obtained with an isotopic enrichment factor of 1. This was consistent with an anion protonation reaction following the addition of the second electron. The mechanism proposed is shown in Scheme 22. The side reaction giving diethylamine and ethylene was characteristic of there being an anhydrous layer at the electrode surface.

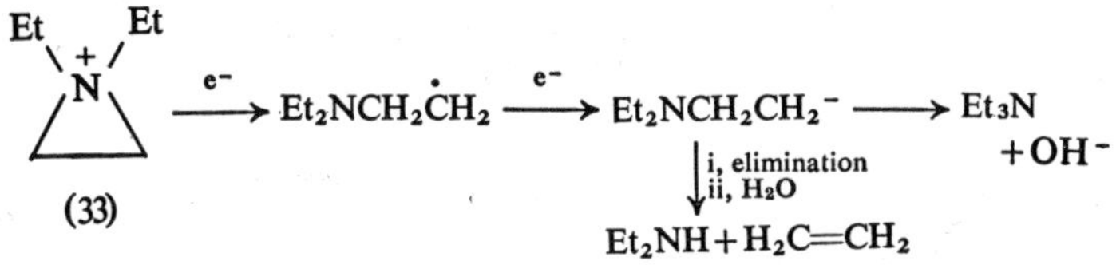

$$\underset{(33)}{\overset{Et}{\underset{}{\overset{+}{N}}}\triangle} \xrightarrow{e^-} Et_2NCH_2\dot{C}H_2 \xrightarrow{e^-} Et_2NCH_2CH_2^- \longrightarrow Et_3N + OH^-$$

$$\downarrow \text{i, elimination} \atop \text{ii, H}_2\text{O}$$

$$Et_2NH + H_2C{=}CH_2$$

Scheme 22

Wróbel and Krawczyk used liquid ammonia as the solvent in their investigations on the electrolysis of quaternary ammonium salts.[83] The optimum conditions were found to be: 20 mmol l^{-1} quaternary ammonium nitrate in NH_3, 20 mmol l^{-1} substrate, and a current density of 4 mA cm^{-2}. All electrolyses were carried out until 2 F mol^{-1} electricity had been consumed. The ease of reduction of the C—N bond was dependent upon the nature of the substituent groups on the quaternary nitrogen. The order found was:

$$E_{N-C_{aryl}} < E_{N-C_{benzyl}} < E_{N-C_{alkyl}}$$

Fission of a C—N bond followed by protonation was the main reaction pathway and gave a product in very high yields with good current efficiencies (see Table 3).

Kato, Nakaya, and Imoto studied the reduction of quaternary ammonium salts of aza-heteroaromatic compounds.[84—86] They found that the stability of the radical formed after the addition of the first electron was enhanced when substituent groups which caused either greater delocalization of the unpaired electron or gave cationic charge were introduced into the molecule. The electrolytic reduction of some 3-hydroxypyridine quaternary cations has been studied by Ferles, Hamid, and Hrubá.[87] Complete saturation of the pyridine ring took place under the conditions used (Pb cathode, 20% H_2SO_4 solvent, no potential control, 7 F mol^{-1}) (Scheme 23). By investigating the electroreduction products of compounds suspected to be intermediates, a mechanism of the overall reaction was suggested (Scheme 24). However, the presence of some products was difficult to explain.

[83] J. T. Wróbel and A. R. Krawczyk, *Roczniki Chem.*, 1971, **45**, 1465.
[84] S. Kato, J. Nakaya, and E. Imoto, *Rev. Polarog. (Japan)*, 1971, **17**, 56.
[85] S. Kato, J. Nakaya, and E. Imoto, *Rev. Polarog. (Japan)*, 1971, **17**, 46.
[86] S. Kato, J. Nakaya, and E. Imoto, *Bull. Chem. Soc. Japan*, 1971, **44**, 1928.
[87] M. Ferles, A. H. Attia, and H. Hrubá, *Coll. Czech. Chem. Comm.*, 1971, **36**, 2057.

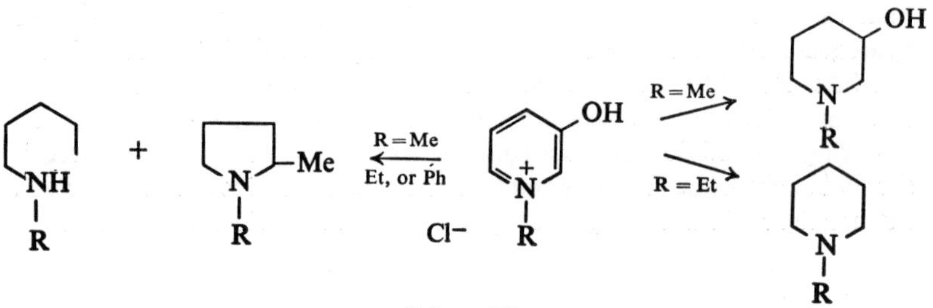

Scheme 23

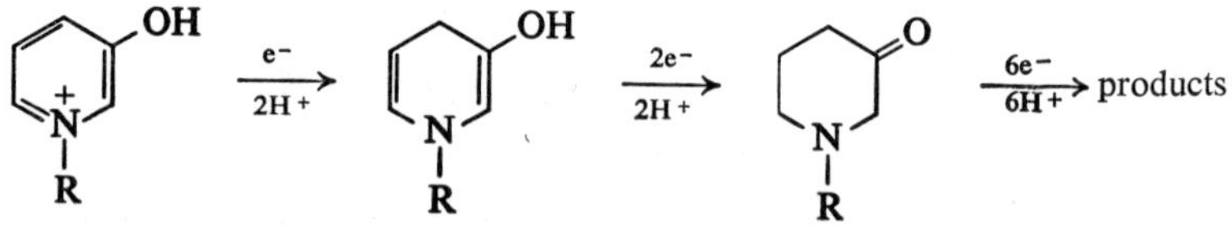

Scheme 24

The exception to the 'normal' reduction path of quaternary nitrogen-containing compounds was observed by Cleghorn, Gaskin, and Lloyd[81] in their study of the diazepinium perchlorates (34). They noticed a pH dependence of half-wave potential and concluded that a proton was involved before or during the transfer of the first electron. The protonated molecule containing two quaternary nitrogen atoms and a conjugated double-bond system was reduced quite easily (Scheme 25).

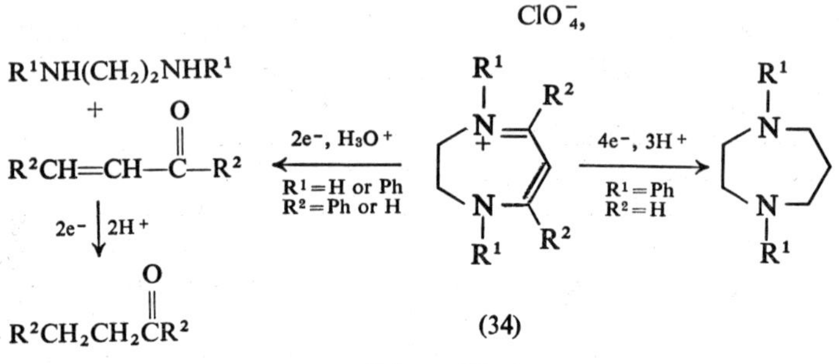

Scheme 25

Timofeeva, Lizogub, Muravich-Alexander, and Elczov found that the quaternary ammonium compounds of the type (35) were reduced at a dropping mercury electrode in a single two-electron step to the iminazoline derivative (Scheme 26).[88] The sulphur homologue or the methylthio-derivative (X = —SMe) were found to eliminate the sulphur function quite readily.

[88] E. N. Timofeeva, A. V. Lizogub, H. L. Muravich-Alexander, and E. V. Elczov, *Zhur. obshchei Khim.*, 1971, **41**, 2539.

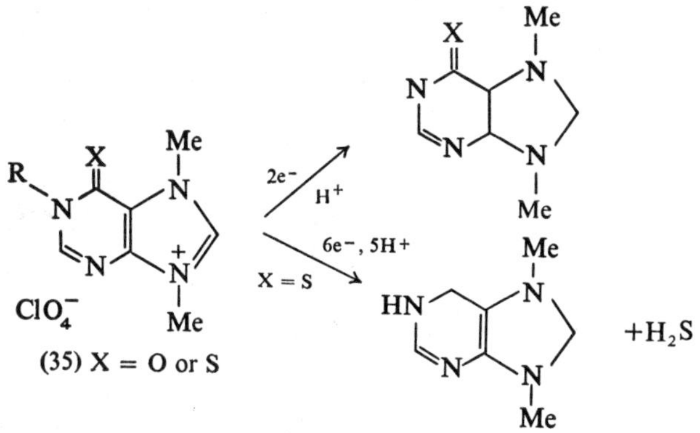

Scheme 26

Nitro-compounds.—Hazard and Tallec[89] have investigated the electroreduction of 4-nitro- and 2-nitro-4'-X-azobenzenes (X = Cl, OH, or OMe). These molecules gave two polarographic waves corresponding to the uptake of 2 and between 4 and 8 electrons, respectively. Controlled-potential electrolysis of the 4-nitro-4'-X-azobenzenes at the first wave [−500 mV (S.C.E.)] in alkaline solution gave a quantitative yield of the 4-hydroxylamine-4'-X-azobenzenes in an electrode reaction consuming 4 electrons mol^{-1}. The discrepancy between the polarographic and preparative run 'n' values was accounted for by assuming a slow disproportionation step in the reaction path (Scheme 27).

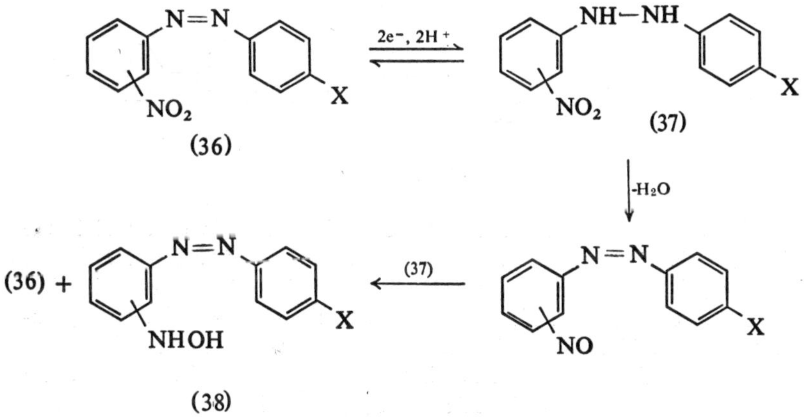

Scheme 27

On decreasing the pH the formation of (38) went less readily. In acetate buffer the chloro-derivative reduced only to (37) and in chloroacetate buffer only

[89] R. Hazard and A. Tallec, *Compt. rend.*, 1971, **273**, C, 1114.

the methoxy-derivative was reduced further. The results with the 2-nitro-4'-X-azobenzenes were similar, except that (38) was not obtained owing to a very rapid intramolecular cyclization: (Scheme 27a).

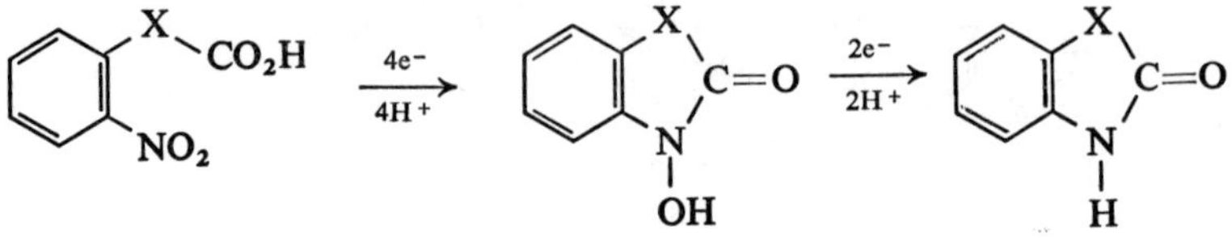

$$\underset{NO_2}{\underset{|}{C_6H_4}}\text{-X-}CO_2H \xrightarrow[4H^+]{4e^-} \text{[benzisoxazolinone-OH]} \xrightarrow[2H^+]{2e^-} \text{[benzisoxazolinone-H]}$$

X = CH_2, CHOH, or CO

Scheme 27a

The dehydration of both the *o*- and *p*-nitrohydrazobenzenes was found to be favoured by basic conditions and the presence of an electron-donor substituent at the 4'-position. Electrolysis at more negative potentials [−1.0 V (S.C.E.)] resulted in complete reduction of the nitrogen functions, giving *o*- and *p*-phenylenediamines and the 4-X-aniline. Tallec and his collaborators, Mannereau and Bobic, also reported the electrosynthesis of heterocyclic compounds by reduction of *o*-nitroaromatic carboxylic acids.[90] The chemical reduction is known to give lactams in good yields but the isolation of hydroxamic acids is more difficult. In contrast, the electrochemical route preferentially gives the hydroxamic acids. Polarography of the acids in $1N\text{-}H_2SO_4$–EtOH (50/50) showed two waves of 4 and 2 electrons, respectively. C.p.e. at the first wave [−400 mV (S.C.E.)] gave the hydroxamic acids in 90% yield (Scheme 28).

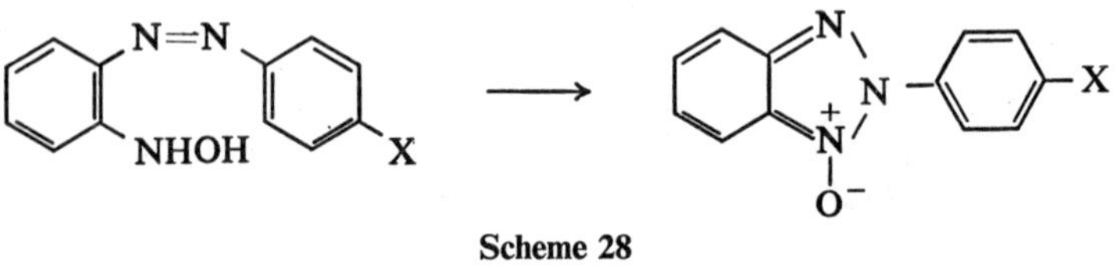

Scheme 28

Electrolysis at the potential of the second wave again gave mainly the hydroxamic acid, together with no more than 20% of lactam. Reikhman, Stradyn, Gavar, and Giller reported the mechanism of the electrochemical reduction of two ethylenic nitro-compounds: α-phenyl- and α-(2-furyl)-β-nitroethylene.[91] The effect of substituent groups on the reduction of this type of ethylenic nitro-compound was investigated by Armand and Convert.[92] Although the compound $PhCH:CRNO_2$ (R=H) is known to reduce to phenylacetaldoxime and β-phenylethylhydroxylamine, it was found that when

[90] A. Tallec, G. Mennereau, and G. Robic, *Compt. rend.*, 1971, **273**, C, 1378.
[91] G. O. Reikhman, Ya P. Stradyn, P. A. Gavar, and S. A. Giller, *Zhur. obshchei Khim.*, 1971, **41**, 906.
[92] J. Armand and O. Convert, *Coll. Czech. Chem. Comm.*, 1971, **36**, 351.

R = Br or Cl the behaviour was quite different. For instance, controlled-potential electrolysis of the bromo-derivative (39) at -0.4 V, pH 1, and in H_2O–dioxan gave a product composed of 80% phenylacetonitrile and 10% benzaldehyde. Chemical reduction with Zn–HOAc is also known to give phenylacetonitrile. The reaction mechanism proposed is shown in Scheme 29.

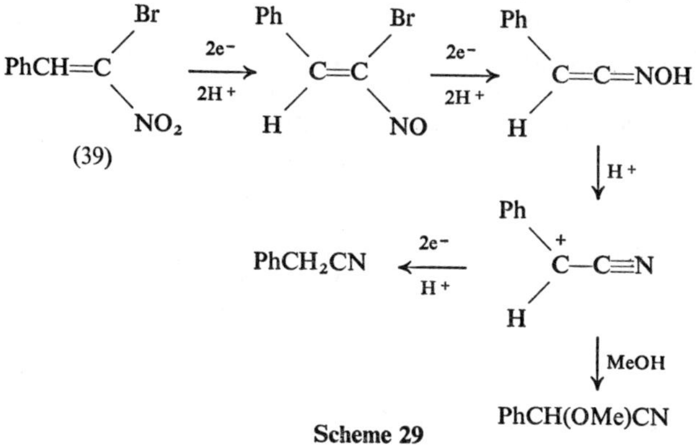

$$PhCH=C\begin{matrix}Br\\ \\NO_2\end{matrix} \xrightarrow[2H^+]{2e^-} \begin{matrix}Ph\\ \\ \\H\end{matrix}C=C\begin{matrix}Br\\ \\NO\end{matrix} \xrightarrow[2H^+]{2e^-} \begin{matrix}Ph\\ \\H\end{matrix}C=C=NOH$$

$$\downarrow H^+$$

$$PhCH_2CN \xleftarrow[H^+]{2e^-} \begin{matrix}Ph\\ \\H\end{matrix}C\!-\!C\!\equiv\!N$$

$$\downarrow MeOH$$

$$PhCH(OMe)CN$$

Scheme 29

The benzaldehyde was thought to be formed by the partial hydrolysis of (39). When the solvent was methanol, methoxylation of the intermediate carbonium ion gave some phenylmethoxyacetonitrile. It is interesting that under such mild conditions a nitro-compound may be reduced to a nitrile. On account of the known decomposition paths of nitronic acids derived from secondary nitro-compounds, the reduction of *cis*-2,3-dinitrobut-2-ene was proposed to go *via* the dinitronic acid. This could then decompose to give the observed products: diacetal (40), diacetal mono-oxime (41), and dimethyl-furoxan (42) (Scheme 30).

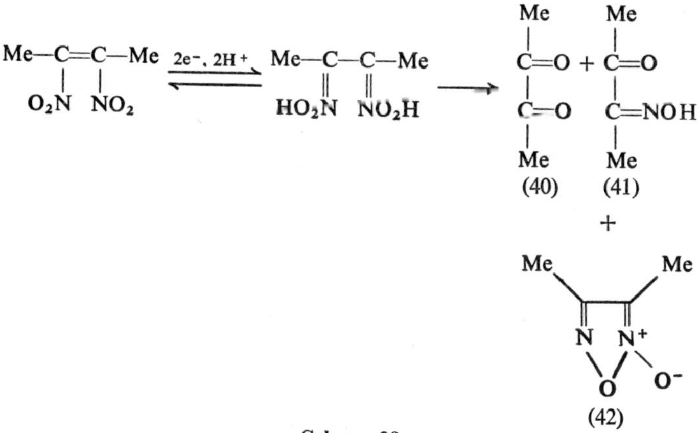

$$\underset{\underset{O_2N\ \ NO_2}{|\ \ \ \ |}}{Me\!-\!C\!=\!C\!-\!Me} \xrightleftharpoons{2e^-,\ 2H^+} \underset{\underset{HO_2N\ \ NO_2H}{\|\ \ \ \ \|}}{Me\!-\!C\!-\!C\!-\!Me} \longrightarrow \begin{matrix}Me\\|\\C=O\\|\\C=O\\|\\Me\\(40)\end{matrix} + \begin{matrix}Me\\|\\C=O\\|\\C=NOH\\|\\Me\\(41)\end{matrix}$$

$$+$$

$$\underset{(42)}{\begin{matrix}Me & & Me\\ \diagdown & & \diagup\\ & C=C &\\ \| & & \|\\ N & & N^+\\ \diagdown & & \diagup\\ & O & O^-\end{matrix}}$$

Scheme 30

Lovreček, Vajtner, and Hranilović[93] re-investigated the electroreduction of 2-nitrobutanol in the hope of finding suitable conditions for, and explaining the mechanism of, the formation of 2-aminobutanol. The reduction followed the normal path of aliphatic nitro-compounds in that the hydroxylamine was the main product. However, when strongly acidic (10% HCl) solutions were electrolysed at elevated temperatures (70 °C), low but significant amounts of 2-aminobutanol were obtained.

The effect of the solvent on the kinetics of the polarographic reduction of simple nitroalkanes was discussed by Gavioli, Grandi, and Andreoli.[94] In aqueous solution the electrochemically active form was the protonated nitroalkane, while at higher pH the reduction of the neutral species was prevalent.

Intramolecular coupling of two hydroxylamine groups was thought by Rogers and Watson to occur during the electrolysis of 2,2′-dinitrobiphenyl.[95] The resulting compound was 9,10-diazaphenanthrene. However, its identification was based on only u.v. and half-wave potential data. A detailed study by Heyrovský, Vavřička, and Holleck[96] of the mechanism of reduction of *p*-dinitrobenzene to *p*-phenylenediamine showed the presence of *pp*′-dinitroazoxybenzene as an intermediate at all pH values. Darchen[97] has also studied this process and has proposed reaction intermediates that were not considered by Heyrovský *et al.* The presence of *p*-nitrosoaniline was observed in all cases. A disproportionation was considered to be the reaction path leading to

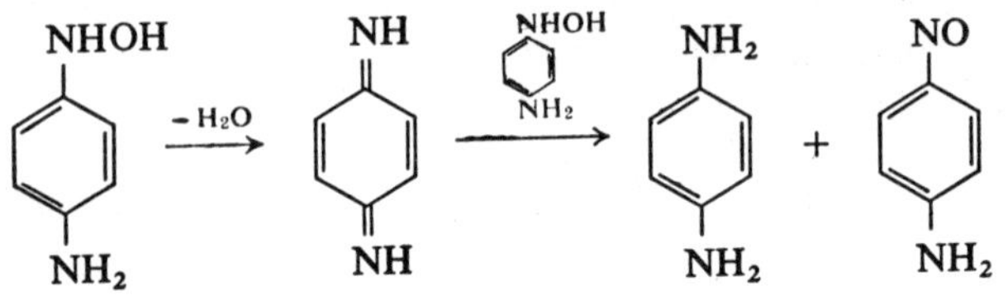

Scheme 31

its formation (Scheme 31). The *p*-nitrosoaniline was more easily reduced than the *p*-dinitrobenzene.

The reduction of *p*-nitroaniline at a ring-disc electrode has been studied by Podlibner and Nekrasov.[98] In very alkaline solutions (5.7 M-KOH) it was possible to stop the reduction at the *p*-hydroxylamineaniline stage. The

[93] B. Lovreček, Z. Vajtner, and J. Hranilović, *Tetrahedron Letters*, 1971, 3319.
[94] G. B. Gavioli, G. Grandi, and R. Andreoli, *Coll. Czech. Chem. Comm.*, 1971, **36**, 730.
[95] J. W. Rogers and W. M. H. Watson, *Analyt. Chim. Acta*, 1971, **54**, 41.
[96] M. Heyrovský, S. Vavřička, and L. Holleck, *Coll. Czech. Chem. Comm.*, 1971, **36**, 971.
[97] A. Darchen, *Compt. rend.*, 1971, **272**, C, 2193.
[98] B. G. Podlibner and L. N. Nekrasov, *Soviet Electrochem.*, 1971, **7**, 1143.

mechanism of the electroreduction of some aminoazobenzenes was discussed by Ladányi, Vajda, and Vámos.[99] The compounds 4-aminobenzene, 2,4-diaminobenzene, and 2,4-diamino-4′-ethoxy-azobenzene were reduced in an overall 4-electron step to the corresponding amines. It was thought that a disproportionation reaction was involved. Cyclic voltammetry showed that after the first sweep, a new wave was formed owing to the redox behaviour of the product.

The optimization of electrolysis conditions for the synthesis of metanilic acid[100] and p-aminobenzoic acid[101] from m-nitrobenzenesulphonic acid and p-nitrobenzoic acid, respectively, was discussed by Udupa et al. The metanilic acid synthesis worked with highest current efficiencies at copper or tin cathodes, low current densities, and a solution temperature of 40 °C. A suitable electrolyte was 15% sulphuric acid. At lower concentrations the current efficiency was reduced slightly, whereas at higher values appreciable amounts of phenol were produced. The concentration of depolarizer had little effect on the current efficiency, but it was found to be advantageous to work in strong solutions (150 g l^{-1}) since the metanilic acid precipitated out. It is interesting that whereas the reduction p-nitrobenzoic acid at tin electrodes gave the highest current efficiency (97%), in contrast to the previous discussion, copper electrodes were bad for this process. High operating temperatures (65—70 °C) were necessary to obtain the excellent efficiencies quoted.

The polarography of 5-nitro-orotic acid was reported by Gupta, Kishore, and Raghavan.[102] A mechanism was proposed but no product analysis was attempted.

Other Nitrogen Compounds.—A simple and efficient laboratory electrosynthesis of 1,1-dialkylhydrazines from the corresponding N-nitrosamines was developed by Iversen.[103] The electrolysis was performed at a potential less negative than -0.9 V (vs. Ag/Ag$^+$ in 4N-HCl) in a solvent–support electrolyte system of 4N-HCl in EtOH. In general, very high yields of hydrazines were obtained from the 20 symmetric, unsymmetric, and cyclic N-nitrosamines investigated. A feature of the reaction system was the harmless anodic reaction selected, in which the elementary chlorine produced was scavenged by the ethanol to give mainly chloroacetaldehyde diethylacetal. Ginsburg et al. found that the ease of reduction of some fluoronitrosoalkanes went in the order CFCl$_2$NO < CF$_2$ClNO < CF$_3$NO < aromatic nitroso-compounds.[104] The electrolysis of α-amino-ketones in acidic medium has previously been

---

[99] L. Ladányi, M. Vajda, and Gy. Vámos, Acta Chim. Acad. Sci. Hung., 1971, **68**, 47.
[100] S. Thangavelu, G. S. Subramanian, and H. V. K. Udupa, Denki Kagaku, 1971, **39**, 5.
[101] R. Kanakam, A. P. Shakunthala, S. Chidambaram, M. S. V. Pathy, and H. V. K. Udupa, Electrochim. Acta, 1971, **16**, 423.
[102] S. L. Gupta, N. Kishore, and P. S. Raghavan, Electrochim Acta, 1971, **16**, 2135.
[103] P. E. Iversen, Acta Chem. Scand., 1971, **25**, 2337.
[104] V. A. Ginsburg, V. V. Smolianizkaia, A. N. Medvedev, V. Z. Phaermark, and A. P. Tomilov, Zhur. obshchei Khim., 1971, **41**, 2284.

reported to result in the reductive cleavage of the C—N bond. An exception to this rule has been found by Kariv, Hermolin, and Rubinstein[105] for the reduction of quinone and dihydrocinchoninone. Cleavage of the C—N bond was indeed observed, but in very acidic solutions (35% H_2SO_4) protonation of the ring-nitrogen atom and the resulting mesomeric structures had the effect of steering the reaction along a path to the alcohol (Scheme 32).

Andruzzi, Cardinali, and Carelli[106] reported that isatin 3-oxime was re-

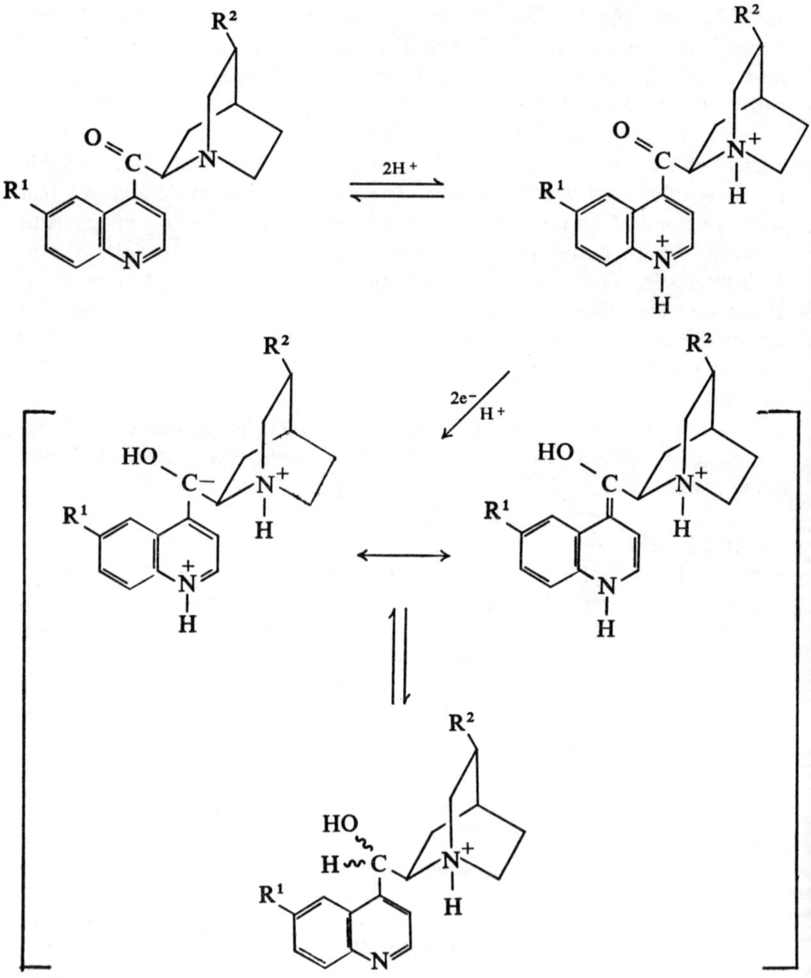

Scheme 32

[105] E. Kariv, J. Hermolin, and I. Rubinstein, *Tetrahedron*, 1971, **27**, 3707.
[106] R. Andruzzi, M. E. Cardinali, I. Carelli, and A. Trazza, *Ann. Chim.* (*Italy*), 1971, **61**, 415.

duced in a single pH-dependent four-electron wave at the hydroxyazomethine group to the corresponding keto-amine. The first step was thought to be rupture of the N—O bond. The polarographic and e.s.r. properties of some aromatic N-oxides have been studied by Miyazaki, Nishikida, and Kubota.[107, 108] Leibson, Mairanovski, and Kounik[109] suggested that the shifting of the half-wave potential of 3-methoxycarbonylisoxazoline N-oxide and the anomalous change of slope of the reduction wave of 3-phenylisoxazoline with increasing pH was a result of the closeness of the reduction potentials of the $\overset{+}{N}-\overset{-}{O}$ group and the isoxazoline ring.

Carboxylic Acids and Derivatives.— The use of highly reducing conditions to effect the conversion of an aliphatic amide (acetamide) into an amine has been investigated by Bewick and Avaca.[110] Hexamethylphosphoramide was found to be a good solvent for the process. It facilitated the electrogeneration of solvated electrons in the presence of the highly acidic conditions prevailing when acetic acid was added as a proton donor. High acidities and low temperatures and current densities favoured the formation of the amine (Scheme 33, path B).

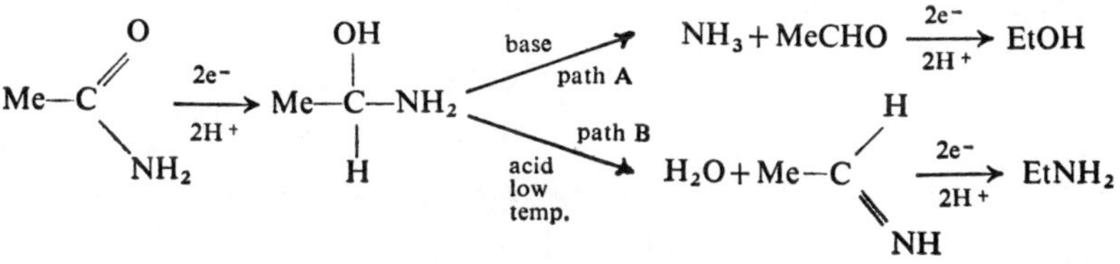

$$\text{Me}-\overset{O}{\underset{NH_2}{C}} \xrightarrow[2H^+]{2e^-} \text{Me}-\overset{OH}{\underset{H}{\underset{|}{C}}}-NH_2 \begin{matrix} \xrightarrow{\text{base, path A}} & NH_3 + MeCHO & \xrightarrow[2H^+]{2e^-} & EtOH \\ \xrightarrow[\text{low temp.}]{\text{acid, path B}} & H_2O + Me-\overset{H}{\underset{NH}{C}} & \xrightarrow[2H^+]{2e^-} & EtNH_2 \end{matrix}$$

Scheme 33

The polarographic reduction of m-substituted benzamides m-X·C_6H_4-$CONH_2$ (X=H, Me, CO_2H, $CONH_2$, or CN) has been reported by Vesheva, Ovchinnikova, and Reishakhrit.[111] The half-wave potentials of the amides were found to correlate well with the Hammett coefficients. At pH < 3 the protonated amide was the electroactive species. Electrolysis in ethanol produced the corresponding alcohols. Some naphthoic acid esters have been studied polarographically by Przhiyalgovskaya, Kheifets, Dmitrievskaya, and Bezuglyi,[112] who observed an overall two-electron process. In aprotic media

---

[107] H. Miyazaki, K. Nishikida, and T. Kubota, *Bull. Chem. Soc. Japan*, 1971, **44**, 277.
[108] H. Miyazaki and T. Kubota, *Bull. Chem. Soc. Japan*, 1971, **44**, 279.
[109] V. N. Leibson, S. G. Mairanovskii, and E. I. Konnik, *Izvest. Akad. Nauk S.S.S.R., Ser. khim.*, 1971, 1429.
[110] L. A. Avaca and A. Bewick, presented at the Electrochemical Engineering Symposium, Newcastle upon Tyne, March, 1971.
[111] L. V. Vesheva, R. A. Ovchinnikova, and L. S. Reishakhrit, *Zhur. obshchei Khim.*, 1971, **41**, 975.
[112] N. M. Przhiyalgovskaya, L. Ya. Kheifets, L. I. Dmitrievskaya, and V. D. Bezuglyi, *Zhur. obshchei Khim.*, 1971, **41**, 1110.

(HMPA) two one-electron steps occurred, whereas in 70% methanol the two waves merged into a single two-electron step. The ease of reduction of the anion radical was discussed in terms of resonance structures and the delocalization of charge in the dianion.

Harrison and Shoesmith have studied the electroreduction of some aromatic carboxylic acids.[113] No other products besides the aldehyde or alcohol were obtained. The rate of reduction was found to be controlled by the electrode reaction. The observed order of rates (acetylsalicylic, m-chlorobenzoic > o- and p-chlorobenzoic > salicylate–borate, benzoic > p-methylbenzoic > o-methyl-benzoic > p-methoxybenzoic > p-hydroxybenzoic) was as expected from inductive and resonance effects.

Mikhailov, Mairanovskii, and Gul'tyai thought that the main product of the electroreduction of esters of α-thiophencarboxylic acid on a dropping mercury electrode was a ring-hydrogenated compound.[114] The reduction of the ester group to the corresponding aldehyde was a side-reaction that occurred only to a negligible extent.

The electroreduction of some unsaturated esters was reported by Klemm, Olson, and White.[115] Compound (43) gave two polarographic waves at -1.94 V and -2.26 V (S.C.E.). Reduction at the plateau of the first wave gave a dimeric product (44). On the other hand, electrolysis at the second

$$\text{MeO-}\underset{\text{MeO}}{\text{C}_6\text{H}_3}\text{-CH=CHCO}_2\text{Et} \xrightarrow[\text{H}^+]{2e^-} \text{MeO-}\underset{\text{MeO}}{\text{C}_6\text{H}_3}\text{-}\overset{\overset{\text{O}}{\|}}{\text{C}}\text{-CH(CO}_2\text{Et)-}\underset{\text{OMe}}{\text{C}_6\text{H}_3}\text{-OMe} + \text{EtO}^-$$

(43) → (44)

wave resulted in hydrogenation of the double bond of (43a). These results

$$\left(\text{MeO-C}_6\text{H}_4\right)_2\text{C=CHCO}_2\text{Et} \xrightarrow[2\text{H}^+]{2e^-} \left(\text{MeO-C}_6\text{H}_4\right)_2\text{CHCH}_2\text{CO}_2\text{Et}$$

(43a)

were the basis for formulating the conditions for hydrogenation of the αβ carbon–carbon double bond in compound (45) to give racemic deoxypicropodophyllin (46).

---

[113] J. A. Harrison and D. W. Shoesmith, *J. Electroanalyt. Chem. Interfacial Electrochem.*, 1971, **32**, 125.
[114] V. S. Mikhailov, S. G. Mairanovskii, and V. P. Gul'tyai, *Soviet Electrochem.*, 1971, **7**, 1159.
[115] L. H. Klemm, D. R. Olson, and D. V. White, *J. Org. Chem.*, 1971, **36**, 3740.

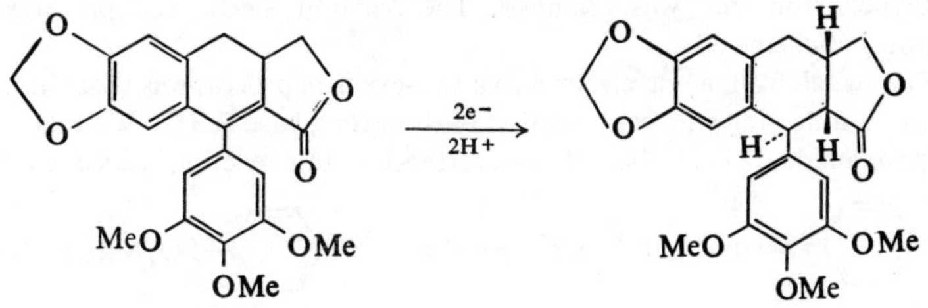

(45) → (46)

Organosulphur Compounds.—The electrochemical cleavage of the arlysulphonamide group is a potentially important method for removing a protective arylsulphonyl group from a peptide or amino-acid. Okumura, Iwasaki, Matsuoka, and Matsumoto[116] have tested the method on fourteen N-tosylamino-acids and peptides. Using a lead cathode and methanol solvent, the electrolysis was carried out galvanostatically at a current density of 2—5 A dm^{-2}. The current efficiency was low, but the product resulting from the S—N bond cleavage was obtained in high yield. The rate of reaction was strongly dependent on the solvent. Dioxan, DMF, EtOH, and H$_2$O gave lower reaction rates than methanol.

A more detailed study of arylsulphonamides was made by Cottrell and Mann.[117] The cleavage of the S—N bond was found to take place in a single irreversible reduction step. Primary and secondary amino-compounds in an aprotic solvent (MeCN) gave a one-electron step, whereas for tertiary amino-compounds or primary and secondary ones in the presence of a proton donor,

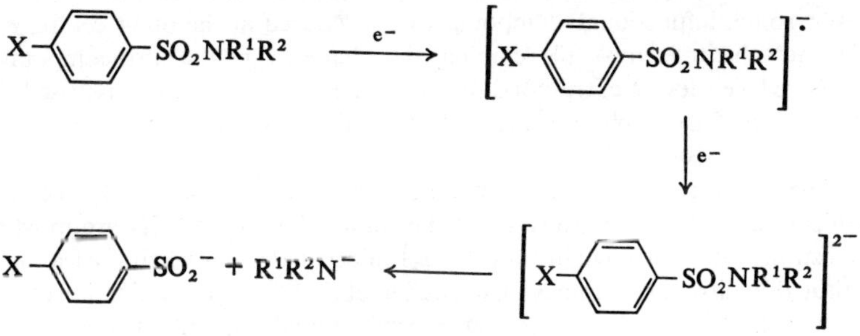

$$X\text{–}C_6H_4\text{–}SO_2NR^1R^2 \xrightarrow{e^-} [X\text{–}C_6H_4\text{–}SO_2NR^1R^2]^{\cdot -}$$

$$\downarrow e^-$$

$$X\text{–}C_6H_4\text{–}SO_2^- + R^1R^2N^- \longleftarrow [X\text{–}C_6H_4\text{–}SO_2NR^1R^2]^{2-}$$

X = Me, CN, or H
R^1 = H, Me, Et, or Pr
R^2 = H, Me, or Et

Scheme 34

[116] K. Okumura, T. Iwasaki, M. Matsuoka, and K. Matsumoto, *Chem. and Ind.*, 1971, 929.
[117] P. T. Cottrell and C. K. Mann, *J. Amer. Chem. Soc.*, 1971, **93**, 3579.

a two-electron step was obtained. The reaction mechanism proposed is shown in Scheme 34. The switch from a one-electron to a two-electron process was thought to be due to a following chemical reaction of the strong base RHN^- with the arylsulphonamide (Scheme 34a). The occurrence of this reaction in aprotic media

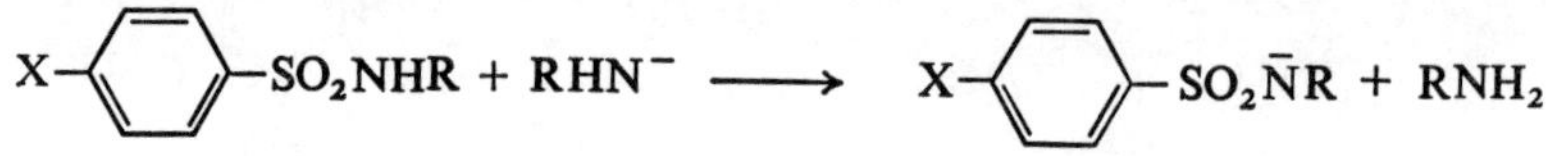

$$X-\langle\bigcirc\rangle-SO_2NHR + RHN^- \longrightarrow X-\langle\bigcirc\rangle-SO_2\bar{N}R + RNH_2$$

Scheme 34a

makes an overall one-electron process. Controlled-potential electrolysis gave the amine and sulphinate in almost quantitative yields. The authors pointed out an inherent advantage in the electrochemical elimination: the ability by selecting a suitably substituted tosyl protective group (*e.g.* with electron-withdrawing substituents such as CN) to perform the reaction at lower potentials. This would be advantageous in the case of a peptide containing another electrochemically reducible centre. The reported formation of products resulting from an S—C bond fission was substantiated only by the results obtained with *NN*-diethyl-*p*-cyanobenzenesulphonamide. Even in this case only a low yield of the nitrile was obtained. The other cyanobenzenesulphonamides studied gave no detectable S—C fission products.

Simonet and Jeminet have also investigated the reduction of substituted aromatic sulphones, with the interest being in the factors influencing the position of the bond fission.[118] Their study of phenyl methyl sulphones substituted by NH_2, OMe, Me, F, or H groups in the phenyl ring showed only reductive cleavage to the corresponding phenylsulphinic compounds. However, some 'bifunctional' compounds were reduced at the other centre, *e.g.* bromo-substituted molecules eliminated the halogen and the nitro-compounds were reduced to the corresponding anilines. The same authors discussed the formation of arylsulphinic acids by the reduction of some other sulphones at a mercury cathode.[119]

The electroreduction of aromatic thiocyanates was shown by Bartak, Shields, and Hawley[120] to result in the elimination of the SCN group when substituted on a side chain (nitrobenzyl thiocyanate) but when nuclear substituted (*p*-nitrophenyl thiocyanate), fission of the SCN with elimination of CN occurred. Using cyclic voltammetry, product analysis, and pH-dependence results, the elimination of SCN from nitrobenzyl thiocanate was shown not to involve a carbanion intermediate. The products 4,4-dinitrobibenzyl (*ca.* 90% yield) and *p*-nitrotoluene (*ca.* 10% yield) were formed exclusively by hydrogen-atom abstraction and radical dimerization of the *p*-nitrobenzyl radical.

[118] J. Simonet and G. Jeminet, *Bull. Soc. chim. France*, 1971, 2754.
[119] G. Jeminet and J. Simonet, *Compt. rend.*, 1971, **272**, *C*, 661.
[120] D. E. Bartak, T. M. Shields, and M. D. Hawley, *J. Electroanalyt. Chem. Interfacial Electrochem.*, 1971, **30**, 289.

Hojman, Stefanović, Stanković, and Zuman[121] report that, contrary to the present literature, the reduction of rubeanic acid at pH 3.6—5.7 takes place in two steps. Firstly, a four-electron transfer at -0.6 V and then a two-electron transfer at -1.0 V. The only product identified was hydrogen sulphide. Little evidence was therefore presented to substantiate the mechanism proposed.

Gladkova, Mairanovskii, and Stoyanovich[122] studied the polarographic reduction of $\alpha\alpha$-, $\alpha\beta$-, and $\beta\beta$-dithienyl sulphides. A two-electron reduction step was observed which, despite the extremely negative potentials involved, was appreciably influenced by adsorption behaviour of the reactant and products. The reduction resulted in the cleavage of the C—S bond to give products that derived from the less basic intermediate carbanion (Scheme 35).

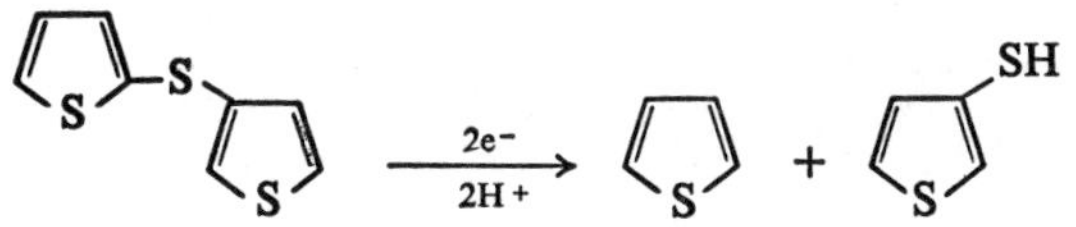

Scheme 35

Miscellaneous Reductions.—Bewick and Avaca have used electrochemically generated solvated electrons to reduce anthracene.[110] With acetic acid as a proton donor in hexamethylphosphoramide the reduction products were di-, tetra-, hexa-, and octa-hydroanthracenes. By suitable choice of reaction conditions (anthracene concentration and current density) it was possible to obtain any of these products selectively, in high yields. The electroreduction of the trityl cation was reported by Plesch, Stasko, and Robson to form the trimer 4-(diphenylmethyl)phenyltriphenylmethane.[123] Bitropyl was formed by electrolysis of the tropylium cation.[124]

Colchester and Entwisle[125] have described the conditions for the synthesis of NN-disubstituted bipyridylium salts by electrolysis of an N-substituted pyridinium salt. High yields were obtained when the pyridinium salt was substituted in the 4-position by a good leaving group that was also capable of forming a stable anion (*e.g.* CN or Cl). It was claimed that electrolysis of 4-cyano-1-methyl-pyridinium methyl sulphate at -0.80 V (S.C.E.) gave a 70% yield of 1,1-dimethyl-4,4′-bipyridylium methyl sulphate.

Weinberg has discussed the uses of mercury olefin complexes for the introduction of functional groups into olefins.[126] Both reductive and oxidative electrosynthesis procedures were considered (Scheme 36).

The reduction of olefin-mercury complexes was also studied by Esikova

---

[121] J. Hojman, A. Stefanović, B. Stanković, and P. Zuman, *J. Electroanalyt. Chem. Interfacial Electrochem.*, 1971, **30**, 469.
[122] L. K. Gladkova, S. G. Mairanovskii, and F. M. Stoyanovich, *Soviet Electrochem.*, 1971, **7**, 306.
[123] P. H. Plesch, A. Stasko, and D. Robson, *J. Chem. Soc.* (*B*), 1971, 1634.
[124] P. H. Plesch and A. Stasko, *J. Chem. Soc.* (*B*), 1971, 2052.
[125] J. E. Colchester and J. H. Entwisle, U.S.P. 3 627 651.
[126] N. L. Weinberg, U.S.P. 3 629 080.

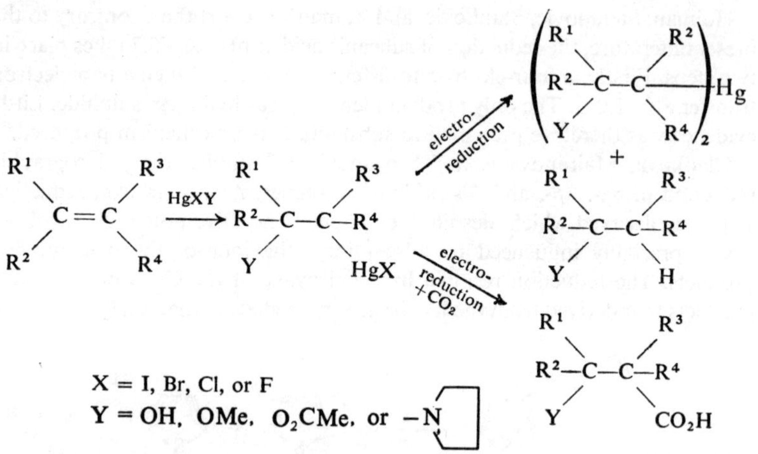

Scheme 36

X = I, Br, Cl, or F
Y = OH, OMe, O₂CMe, or —N⟩

et al.[127] The predominant reaction below pH 14 was the reductive substitution of both R and Hg by hydrogen in compounds of the type RC_2H_4Hg (R = OH, OEt, or $MeCO_2$). Only when the pH was greater than 14 or the solvent aprotic was the substituted alkane RC_2H_5 obtained. The electrochemical reduction of the organomercury compounds bis-(2-acetyl-5-thienyl)mercury and 2-acetyl-5-bromomercurithiophen has also been studied by Kashin, Levinson, Butin, and Ershler.[128] The reduction of a functional group within a benzyl-substituted phosphonium salt without benzyl-group cleavage has apparently not been reported. Stocker et al.[129] now report that the electro-

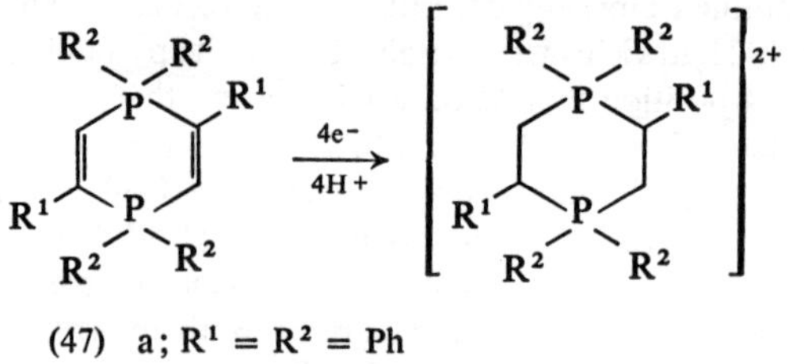

(47) a; $R^1 = R^2 = Ph$
b; $R^1 = H, R^2 = Ph$
c; $R^1 = Me, R^2 = PhCH_2$

Scheme 36a

[127] I. A. Esikova, O. N. Temkin, A. P. Tomilov, R. M. Flid, and G. P. Pavlikova, *Soviet Electrochem.*, 1971, **7**, 133.
[128] A. N. Kashin, I. M. Levinson, K. P. Butin, and A. B. Ershler, *Soviet Electrochem.*, 1971, **7**, 940.
[129] J. H. Stocker, R. M. Jenevein, A. Aguiar, G. W. Prejean, and N. A. Portnoy, *Chem. Comm.*, 1971, 1478.

chemical reduction proceeds without cleavage. Several 1,4-diphosphoniacyclohexa-2,5-dienes (47) were investigated. In general, good yields of the saturated products were obtained (Scheme 36a).

The formation of ion-pair complexes between support electrolyte and radical anions or dianions can exert considerable effects on the electroreduction behaviour. Indeed, their effect may be used to lower the potential necessary for an electrolysis. Fuginago, Izutsu, and Nomura have systematically investigated the effect of alkali metals on the reduction potentials of 1,2- and 1,4-naphthoquinones.[130] The positive shift in the reduction potential increased in the order $K^+ < Na^+ < Li^+ < Mg^{2+} < Zn^{2+}$ and was greater in poor cation-solvating solvents. Similar trends were observed by Kalinowski[131] and by Kryszczynska and Kalinowski.[132] Thus, whereas the cations Et_4N^+ and K^+ had no effect on the half-wave potential of azobenzene, 5-phenylazopyridine,[132] or several aromatic ketones,[131] the ions Ba^{2+}, Sr^{2+}, and Li^+ shifted the waves to more anodic potentials.

3 Oxidations

Aromatic Hydrocarbons.—Nyberg has studied the anodic coupling of some aromatic hydrocarbons.[133-135] Bimesityl was shown to be formed in good yields (49%) when mesitylene was oxidized at a platinum anode in acetonitrile.[133] Optimum conditions involved the use of high substrate concentrations (2 mol l^{-1}) and tetrabutylammonium tetrafluoroborate as support electrolyte. This process compared very favourably with a chemical route using ferric chloride as oxidant, especially with respect to efficiency and product purity. The reaction worked less well on carbon anodes, and the presence of water also had a detrimental effect. Some methyl-substituted benzenes were also investigated.[134] It was found that nitromethane was a better solvent for the preparation of biphenyls; other main products were diphenylmethanes. The product distribution was found to correlate with the unpaired electron charge density on the cation radical. A large unpaired electron charge density located on a free ring position corresponded to a high percentage of biphenyl in the product. Their formation was suggested to occur by electrophilic attack of cation radicals on the starting compound:

$$ArCH_3 \xrightarrow{-e^-} ArCH_3^{+\cdot} \xrightarrow{ArCH_3} \text{biphenyls}$$

while for the formation of diphenylmethanes the following reaction path was proposed:

---

[130] T. Fujinaga, K. Izutsu, and T. Nomura, *J. Electroanalyt. Chem. Interfacial Electrochem.*, 1971, **29**, 203.
[131] M. K. Kalinowski, *Roczniki Chem.*, 1971, **45**, 469.
[132] H. Kryszczyńska and M. K. Kalinowski, *Roczniki Chem.*, 1971, **45**, 1747.
[133] K. Nyberg, *Acta Chem. Scand.*, 1971, **25**, 534.
[134] K. Nyberg, *Acta Chem. Scand.*, 1971, **25**, 2499.
[135] K. Nyberg, *Acta Chem. Scand.*, 1971, **25**, 2983.

$$ArCH_3^{+\cdot} \xrightarrow{-H^+} ArCH\cdot \xrightarrow{-e^-} ArCH_2^+$$

$$ArCH_2^+ + ArCH_3 \xrightarrow{-H^+} \text{diphenylmethanes}$$

The anodic coupling of some 5-alkyl-substituted m-xylenes[135] was also found to give biphenyls as major product, but with very low current efficiencies.

Intermolecular coupling reactions were the subject of two papers by Stuart and Ohnesorge.[136,137] Tetra-anisylethylene was oxidized in a 2-electron process to give initially a dication species.[137] This was found to undergo a slow (*ca.* 2 days) intramolecular cyclization reaction to form 3,6-dimethoxy-9,10-bis-(p-anisyl)phenanthrene. A more detailed study of the kinetics of a similar reaction, *viz.* the oxidation of tetraphenylethylene (TPE) to 9,10-diphenylphenanthrene (DPP), was also attempted.[136] The reaction was found to be rather complex and involved inhibition at more anodic potentials, probably as a result of oxidation of the product. The overall reaction to the phenanthroline involved the loss of two electrons, and cyclic voltammetry indicated that the first transfer was more reversible than the second. It was not possible to distinguish between EEC or ECE mechanisms (Scheme 37).

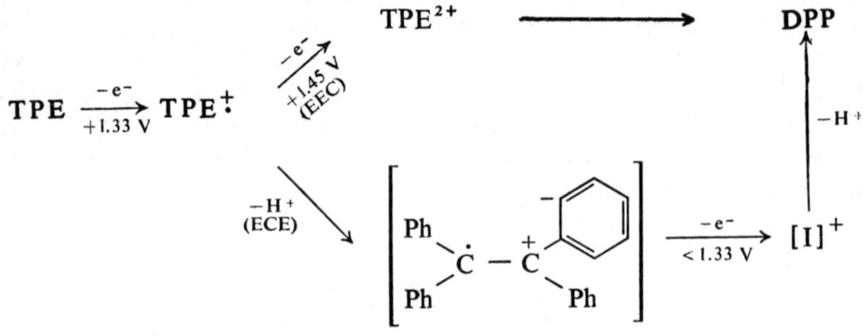

Scheme 37

When hydrocarbons are oxidized in a medium containing nucleophilic species the tendency to couple is decreased and substitution reactions are favoured. For example, Parker, Dirlam, and Eberson[138] reported the anodic methoxylation of 9,10-disubstituted anthracenes (Scheme 37a). The electrolysis was performed in methanol containing sodium methoxide and the *cis*-isomer was the major product. Substitution on the side chain rather than on the aromatic nucleus usually occurs to a greater extent when species of lower nucleophilicity are present, *e.g.* acetic acid instead of HOAc–AcO$^-$. Magnusson, Olofsson, and Nyberg[139] have investigated the parameters affecting the side-chain acetoxylation of methylbenzenes (toluene, p-xylene, mesitylene,

---

[136] J. D. Stuart and W. E. Ohnesorge, *J. Amer. Chem. Soc.*, 1971, **93**, 4531.
[137] J. D. Stuart and W. E. Ohnesorge, *J. Electroanalyt. Chem. Interfacial Electrochem.*, 1971, **30**, App. 11.
[138] V. D. Parker, J. P. Dirlam, and L. Eberson, *Acta Chem. Scand.*, 1971, **25**, 341.
[139] C. Magnusson, B. Olofsson, and K. Nyberg, *Chemica Scripta*, 1971, **1**, 57.

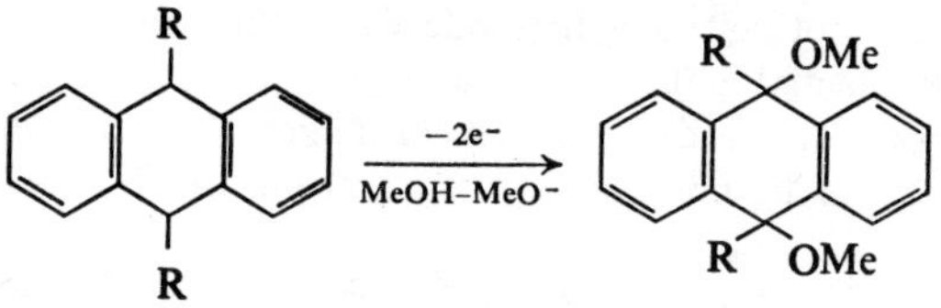

Scheme 37a

R = Me or Et

durene, pentamethylbenzene, and hexamethylbenzene). The highest yields of side-chain-substituted products were obtained at carbon anodes and with tetrabutylammonium perchlorate as support electrolyte. The formation of side products, e.g. bibenzyls and diphenylmethanes, was interpreted, as Nyberg had done previously,[134] in terms of the charge-density distribution of the free electron in the cation radical. Using the optimum conditions found, current yields of the acetates were greater than 60%. It was thought that this method would be advantageous for the synthesis of less accessible side-chain acetates.

Anodic substitution reactions have been postulated to take place *via* either an EC_BE or an EC_NE mechanism:

$$EC_BE: ArCH_3 \xrightarrow{-e^-} ArCH_3^{+\cdot} \xrightarrow{AcO^-} ArCH_2^{\cdot} \xrightarrow{-e^-} ArCH_2^+$$

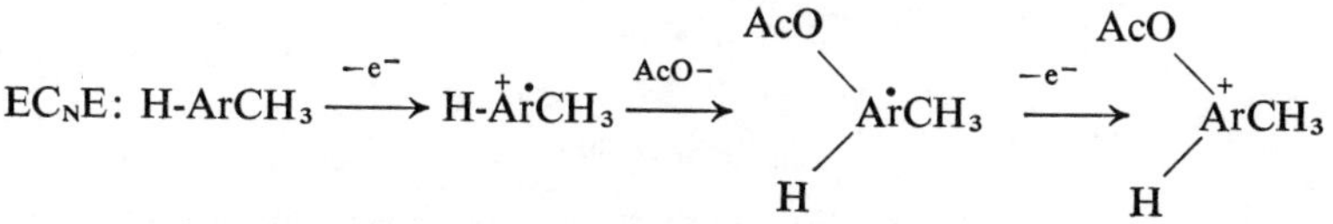

$$EC_NE: H\text{-}ArCH_3 \xrightarrow{-e^-} H\text{-}\overset{+\cdot}{Ar}CH_3 \xrightarrow{AcO^-} \underset{H}{\overset{AcO}{\diagdown}}\!\!\overset{\cdot}{Ar}CH_3 \xrightarrow{-e^-} \underset{H}{\overset{AcO}{\diagdown}}\!\!\overset{+}{Ar}CH_3$$

Bernhardsson, Eberson, Nyberg, and Rietz have considered the possibility of β-hydrogen elimination occurring in the benzylic carbonium ion formed as an intermediate in the EC_BE mechanism.[140] They chose to study the anodic acetoxylation of indane and to search for indene in the reaction products. The normal acetoxylated products were formed with 35% current efficiency

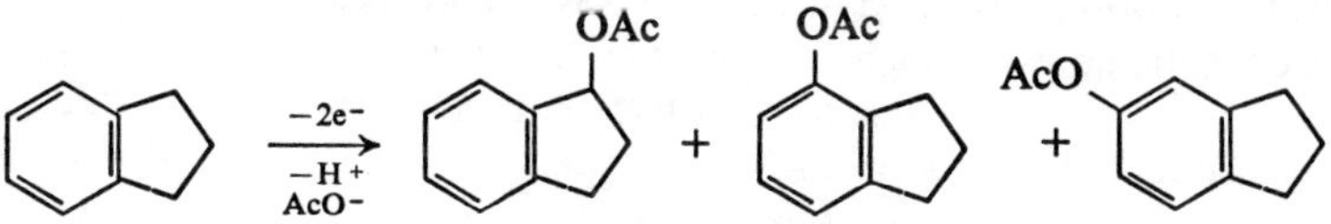

Scheme 38

79 : 9 : 12

(Scheme 38). No indene was detected, but there was evidence of its presence as an intermediate (48), in that disubstituted products were obtained (Scheme

[140] E. Bernhardsson, L. Eberson, K. Nyberg, and B. Rietz, *Acta Chem. Scand.*, 1971, **25**, 1224.

38a). In fact, it is not surprising that indene undergoes further oxidation since its oxidation potential is 0.36 V less anodic than that of indane. However, since no indene at all was detected, it seemed likely that it was produced in an adsorbed state and did not escape from the electrode.

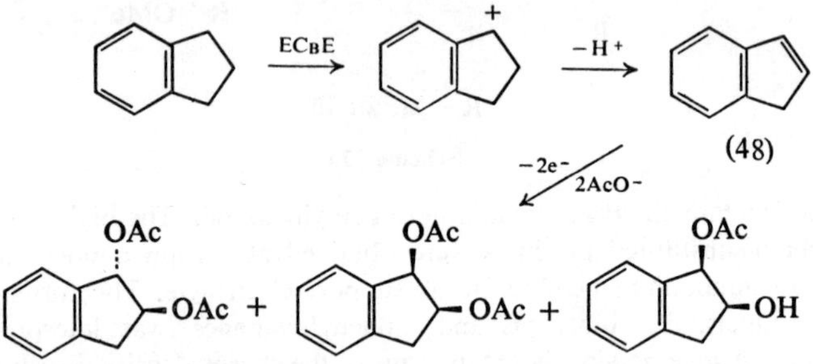

$$\text{indane} \xrightarrow{EC_BE} \text{[cation]} \xrightarrow{-H^+} \text{indene} \quad (48)$$

$$\xrightarrow[2AcO^-]{-2e^-}$$

OAc, OAc (37)	OAc, OAc (35)	OAc, OH (28)
37 :	35 :	28

Scheme 38a

The exact mechanism of anodic substitution of aromatic compounds is in fact not completely clear-cut. A challenger to the ECE mechanism is one involving disproportionation (Scheme 39). At the recent Euchem conference

$$Ar \underset{}{\overset{-e^-}{\rightleftharpoons}} Ar^{+\cdot}$$
$$2Ar^{+\cdot} \rightleftharpoons Ar + Ar^{2+}$$
$$Ar^{2+} + Py \rightarrow ArPy^{2+}$$
$$ArPy^{2+} \rightarrow \text{further reaction}$$

Scheme 39

on electrochemistry[3] the acceptance of one mechanism or the other was contested in discussion sessions by Parker and Shine. Marcoux has now presented a comparison of the ECE and disproportionation mechanisms.[141] Digital simulation provided 'working-curves' of the apparent n value (n_{app}) plotted against log[nucleophile]t for both mechanisms. Experimental values of n_{app} for the substitution of 9,10-diphenylanthracene by pyridine suggested that the disproportionation mechanism was a more likely route.

Completing this section is the report by Usanovich, Solomin, and Kryuchkova[142] on the anodic oxidation of toluene emulsions with aqueous solutions. The oxidation at platinum in 30% sulphuric acid gave benzaldehyde and m-cresol with current efficiencies of 27—30% and 15—20%, respectively. In 15% sodium hydroxide solution the major product was benzaldehyde, but with a decreased current efficiency (10%).

Alkenes and Alkanes.—The oxidation of alkenes in aqueous solutions is

[141] L. Marcoux, *J. Amer. Chem. Soc.*, 1971, **93**, 537.
[142] M. I. Usanovich, A. V. Solomin, and E. I. Kryuchkova, *Soviet Electrochem.*, 1971, **7**, 885.

strongly dependent on the electrode material. Only in the case of activated alkenes are the oxidation potentials on carbon or platinum electrodes below that for oxygen evolution. However, gold anodes behave somewhat differently and partial oxidation of simple alkenes is possible several hundred millivolts below the oxygen evolution potential. Johnson, Lai, and James have examined the role of gold by using anodes of Au–Pt alloys over a range of composition.[143] Cwiklinski and Périchon[144] investigated the oxidation of ethylene on gold anodes in 1N-H_2SO_4 at potentials before the onset of passivation by a gold oxide film. At 0.95 V (N.H.E.) a typical product distribution was: acetaldehyde 76%, acetic acid 0%, glyoxal 5%, glycolaldehyde 1%. At more anodic potentials [1.2 V (N.H.E.)] the yield of acetic acid increased: acetaldehyde 24%, acetic acid 40%, glyoxal 8%, glycolaldehyde 4%, formic acid 3%. Huang and Hsu[145] postulated the participation of a vinyl cation intermediate in the electro-oxidation of cis-1-(4-morpholino)-1,2-diphenylethane. The main products were benzoin and benzil.

Oxidation of alkenes in non-aqueous media leads to other products, in many cases derived from attack of the intermediate carbonium ion on the solvent. Torii et al. have investigated the oxidation of α-methoxy-γγ-dimethylaconic acid (49) in methanol.[146] The molecule was oxidized most easily at the double bond to give methoxylated products. Similarly, Sainsbury found

that oxidation of 1,2-dimethoxy-4-prop-1-enylbenzene in the presence of pyridine took place at the olefinic double bond, with subsequent nucleophilic attack by the pyridine.[147]

The bifunctionalization of olefins, making use of a halogen and nucleophile, has been discussed by Weinberg and Hofmann.[148] Their reaction,

---

[143] J. W. Johnson, S. C. Lai, and W. J. James, *Electrochim. Acta*, 1971, **16**, 1763.
[144] C. Cwiklinski and J. Périchon, *Compt. rend.*, 1971, **272**, C, 1930.
[145] S. J. Huang and E. T. Hsu, *Tetrahedron Letters*, 1971, 1385.
[146] S. Torii, T. Furuta, Y. Miyaoka, H. Sako, H. Tanaka, and K. Uneyama, *Bull. Chem. Soc. Japan*, 1971, **44**, 2258.
[147] M. Sainsbury, *J. Chem. Soc. (C)*, 1971, 2888.
[148] N. L. Weinberg and A. K. Hoffman, *Canad. J. Chem.*, 1971, **49**, 740.

which was analogous to the Prévost reaction, generated the halogen *in situ* by oxidation of the corresponding halide ion. The classical method is shown in Scheme 40 for comparison. The reaction is normally driven to the right by the presence of the silver salt, which forms an insoluble halide. The adaptation of this process to an electrochemical approach eliminates the need for the

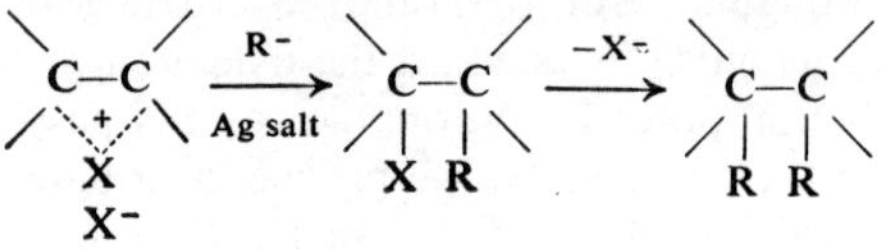

$$\underset{\underset{X^-}{X}}{\overset{}{\diagdown C - C \diagup}} \xrightarrow[\text{Ag salt}]{R^-} \underset{X \quad R}{\diagdown C - C \diagup} \xrightarrow{-X^-} \underset{R \quad R}{\diagdown C - C \diagup}$$

X = I, Br, or Cl
R = OH, OAlk, OAc, NO_2, N_3, NCO, or $\overset{+}{N}\equiv CMe$

Scheme 40

silver salt, since the reaction can now be displaced to the right by the electro-oxidative removal of the halide (Scheme 41). Using bromide or chloride

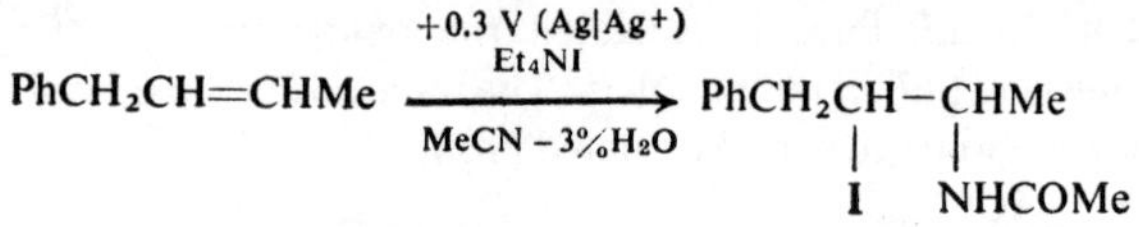

$$PhCH_2CH=CHMe \xrightarrow[\text{MeCN} - 3\% H_2O]{+0.3 \text{ V (Ag|Ag}^+) \atop Et_4NI} PhCH_2\underset{I}{CH}-\underset{NHCOMe}{CHMe}$$

(51% current efficiency)

Scheme 41

electrolytes, vicinal dihalides were formed in appreciable amounts (Scheme 42).

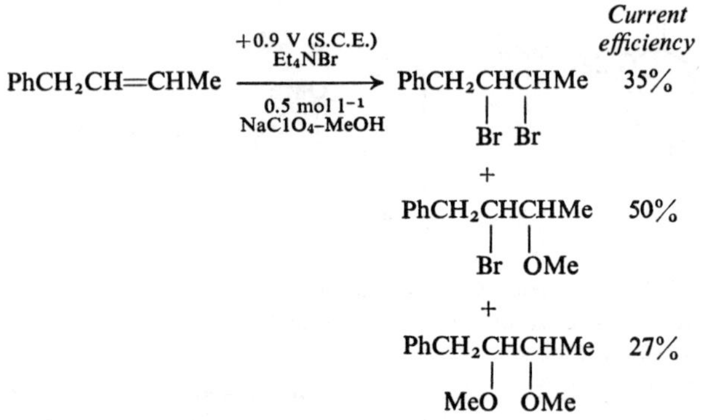

$$PhCH_2CH=CHMe \xrightarrow[\substack{0.5 \text{ mol l}^{-1} \\ NaClO_4-MeOH}]{+0.9 \text{ V (S.C.E.)} \atop Et_4NBr}$$

	Current efficiency
$PhCH_2\underset{Br \;\; Br}{CHCHMe}$	35%
+	
$PhCH_2\underset{Br \;\; OMe}{CHCHMe}$	50%
+	
$PhCH_2\underset{MeO \;\; OMe}{CHCHMe}$	27%

Scheme 42

Intramolecular cyclization was found to occur when certain olefinic alcohols were anodically oxidized. Shono, Ikeda, and Kimura[149] reported that oxidation of citroneroll (50) in acetonitrile at +1.8 V (S.C.E.) gave rose oxide (51)

[149] T. Shono, A. Ikeda, and Y. Kimura, *Tetrahedron Letters*, 1971, 3599.

Organic Electrochemistry – Synthetic Aspects 165

in 26% yield. Similarly, anodic oxidation of *endo*-norbornenemethanol (52) gave a 13.8% yield of the 2-oxatetracyclononane (53).

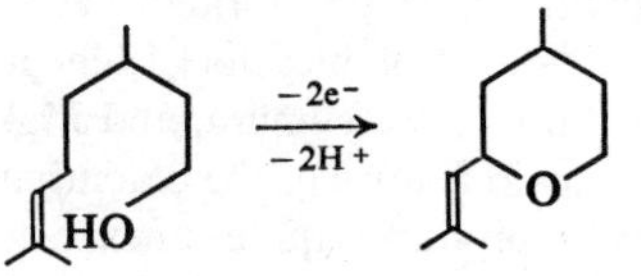

(50)   (51)

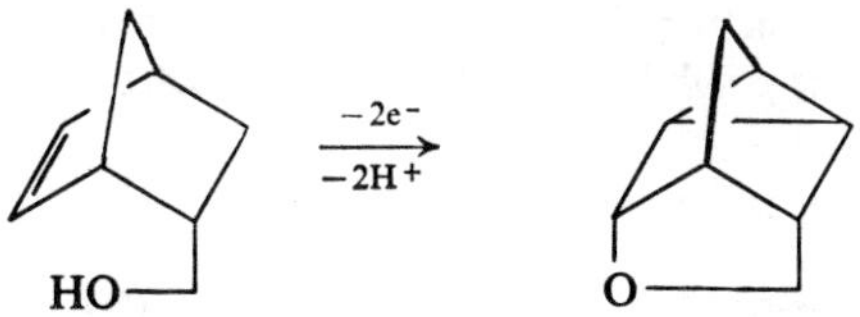

(52)   (53)

The formation of a carbonium ion as an intermediate in the oxidation of alkanes leads one to anticipate products with a rearranged carbon skeleton. This has indeed been reported in the case of cyclic[150] and bicyclic[151] alkanes. The Southampton school have continued their study of extremely oxidizing conditions with an interesting report on the use of fluorosulphonic acid as

$$MeCO_2H + FSO_3H \rightleftharpoons MeCO_2H_2^+ + FSO_3^- \rightleftharpoons MeCO^+ + FSO_3^- + H_2O$$

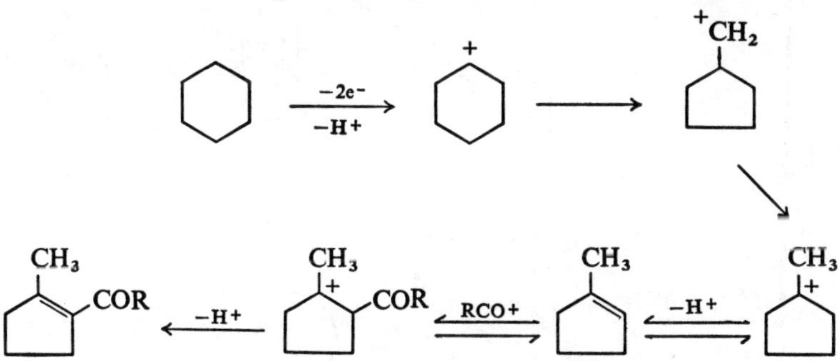

For R = Me, current efficiency = 30%

Scheme 43

solvent. Bertram, Fleischmann and Pletcher[150] showed it was possible to oxidize alkanes containing five or more carbon atoms in this medium. The main product of the reaction was an $\alpha\beta$-unsaturated ketone (Scheme 43). Using propionic instead of acetic acid to provide the support electrolyte, the

[150] J. Bertram, M. Fleischmann, and D. Pletcher, *Tetrahedron Letters*, 1971, 349.
[151] T. Shono, Y. Matsumura, and Y. Nakagawa, *J. Org. Chem.*, 1971, **36**, 1771.

ethyl derivative was obtained (R = Et). A similar reaction was obtained with the alkanes n-hexane, n-heptane, n-pentane, isopentane, and cyclo-octane, showing that the formation of αβ-unsaturated ketones was a quite general reaction. The anodic oxidations of bicyclo[4,1,0]heptane and bicyclo[3,1,0]hexane were studied by Shono, Matsumura, and Makagawa.[151] Opening of the smaller ring was the main feature of the oxidation. Interestingly, electrochemical oxidation gave products quite unexpected for the equivalent chemical reactions. For example, protonation or acidic solvolysis of these compounds proceeds by rupture of an external bond of the cyclopropane ring, whereas the anodic oxidation takes place at the internal bond. The products obtained on electrolysis at carbon anodes in methanol with tetraethylammonium toluene-p-sulphonate as support electrolyte, together with the

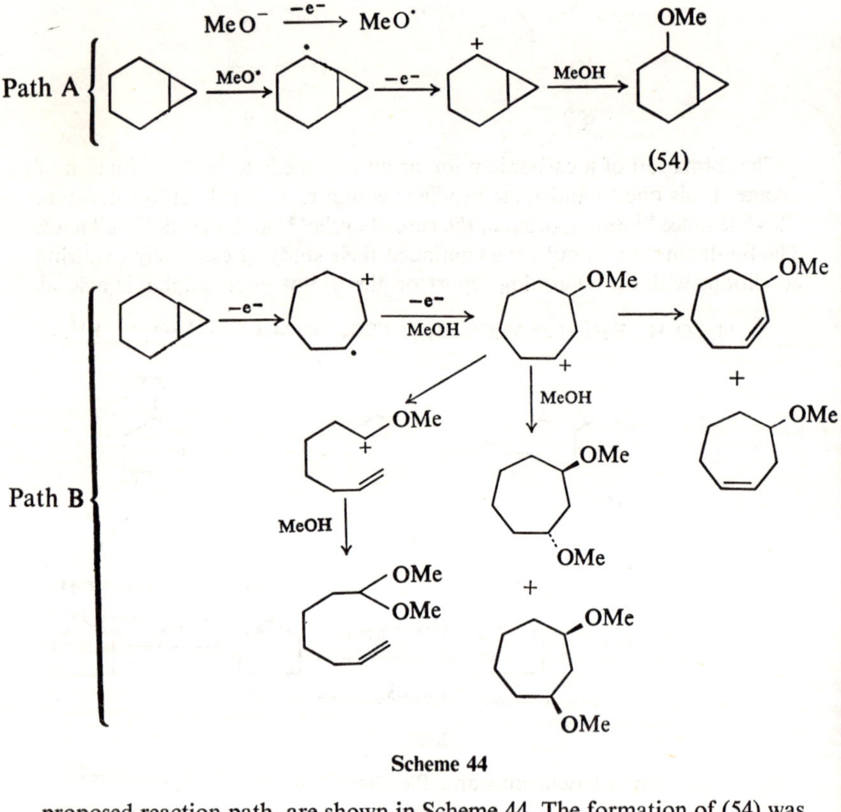

Scheme 44

proposed reaction path, are shown in Scheme 44. The formation of (54) was thought to be due to attack of a methoxyl radical on the substrate, since on electrolysis at 1.3 V in MeOH–NaOMe it was produced exclusively. At this potential the only anodic reaction possible is the oxidation of the methoxide anion. It should be mentioned that rearrangements of cyclic alkenes have also been observed at a polarized fuel-cell electrode.[152,153] In many cases the electrode material may have sufficient catalytic activity to promote this without necessarily forming a carbonium ion.

Carboxylic Acids.—It is now nearly 140 years since the discovery of the Kolbe reaction. The fact that there is still considerable interest in it is continuing proof of its usefulness and versatility in organic electrosynthesis.

Colman, Utley, and Weedon[154] have compared the relative importance of radical and carbonium ion pathways for the oxidation of phenyl acetate ions in methanol. The relative yields of Kolbe dimer and the benzyl methyl ether (carbonium ion product) were dependent on the nature of substituents in the phenyl ring and the presence of other ions. Electron-donating substituents biased the route to the carbonium ion product, and with the p-methoxy-substituted substrate the product distribution was 99% $ArCH_2OMe$ and less than 1% dimer. Highest dimer yields (74%) were obtained with the perfluorinated phenyl compound. The addition of small amounts of sodium perchlorate was found to block the formation of the dimer. Coleman and Eberson[155] have considered another variation of the Kolbe mechanism, to which they have given the name pseudo-Kolbe reaction. By this reaction the oxidation of an aryl acetate ion proceeds by the initial removal of an electron from the aryl group. However, little evidence was found to support the proposal of this added complication in the Kolbe mechanism. Arora and Woolford[156] demonstrated the importance of steric hindrance imparted by the substituent on the oxidation path of α-halogenocarboxylic acids. Electrolysis of chloro-valeric, -caproic and -isobutyric acids in methanol gave none of the Kolbe dimer. Instead, the main product was the chloro-ester. In water, the chloro-acids behaved differently from their bromo-counterparts in that a carbonium ion formed by the Kolbe oxidation of the acid combined with the original carboxylate anion to give a new ester (55) (Scheme 45). Only straight-chain α-fluorocarboxylic acids gave any Kolbe dimer. Thus the branched-chain acid 2-ethyl-2-fluorobutanoic acid gave pentan-3-one as major product

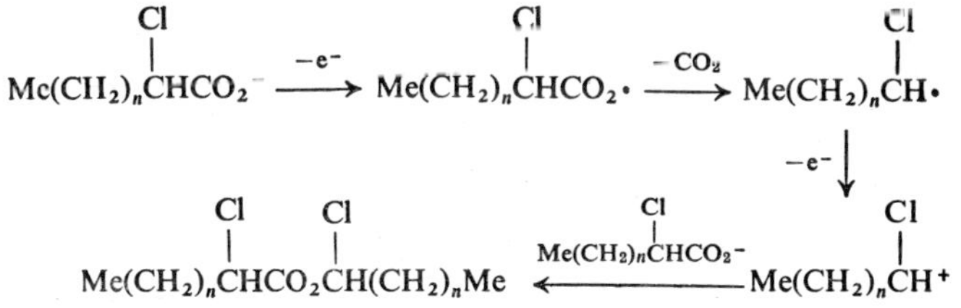

$$Me(CH_2)_n\overset{Cl}{\underset{|}{C}}HCO_2^- \xrightarrow{-e^-} Me(CH_2)_n\overset{Cl}{\underset{|}{C}}HCO_2\cdot \xrightarrow{-CO_2} Me(CH_2)_n\overset{Cl}{\underset{|}{C}}H\cdot$$

$$\Big\downarrow -e^-$$

$$Me(CH_2)_n\overset{Cl}{\underset{|}{C}}HCO_2CH(CH_2)_nMe \xleftarrow{Me(CH_2)_n\overset{Cl}{\underset{|}{C}}HCO_2^-} Me(CH_2)_n\overset{Cl}{\underset{|}{C}}H^+$$
(55)

Scheme 45

[152] H. J. Barger, G. W. Walker, and R. J. York, *J. Amer. Chem. Soc.*, 1971, **93**, 2800.
[153] H. J. Barger, G. W. Walker, and R. J. York, *J. Electrochem. Soc.*, 1971, **118**, 1713.
[154] J. P. Coleman, J. H. P. Utley, and B. C. L. Weedon, *Chem. Comm.*, 1971, 438.
[155] J. P. Coleman and L. Eberson, *Chem. Comm.*, 1971, 1300.
[156] P. C. Arora and R. G. Woolford, *Canad. J. Chem.*, 1971, **49**, 2681.

(24—32%). The important factor in determining whether these acids gave the dimer was thought to be the magnitude of the steric hindrance given by the α-halogeno-group.

The anodic oxidation of the monoesters of dicarboxylic acids to the corresponding diester (Brown–Walker reaction) has also received attention. Tarkhanov et al.[157] studied the dependence of the current yields on the length of the carbon chain of the monoester. As with the Kolbe reaction, the diester yields were virtually independent of the number of CH_2 units in the chain, over the range of two to eight. A patent by Dennison and Holland[158] on the continuous production of diesters has also appeared in the literature. They gave results on the oxidation of methyl hydrogen glutarate. The current efficiencies were very similar to those obtained by Tarkhanov (~60%). A direct extraction using a hydrocarbon solvent was used to make the process continuous. Osman et al.[159] studied the oxidation of monoethyl malonate at a platinum electrode in the presence of naphthalene. Although the major product was diethyl succinate (Brown–Walker reaction) a certain amount of the radical addition product with naphthalene was observed (Scheme 46).

$$EtO_2CCH_2CO_2^- \xrightarrow[-CO_2]{-e^-} EtO_2CCH_2\cdot \xrightarrow[-H^+]{C_{10}H_8} EtO_2CCH_2C_{10}H_7$$

$$\downarrow -e^-$$

$$EtO_2\overset{+}{C}CH_2CO_2\cdot \qquad EtO_2CCH_2CO_2C_{10}H_7$$

Scheme 46

Several other papers were concerned with the behaviour and trapping of radical or carbonium ion intermediates. For example, Nagamori and Sekine[160] reported the oxidation of acetate and the subsequent coupling of the methyl radicals with chloroform. Konstantinov, Chelepin, and Koloskova obtained an interesting ester when they oxidized furan-2-carboxylic acid (and also the sulphur and selenium homologues) in DMF.[161] The preparative runs were performed galvanostatically and the potential was in the range 2.6—2.9 V (S.C.E.). The formation of an ester (56) with the solvent was thought to be due to the simultaneous oxidation of both the carboxylic acid and the DMF with subsequent coupling of the radicals (Scheme 47).

Chkir and Lelandais[162] investigated the coupling of Kolbe oxidation intermediates with the trapping agents CH_2=CR^2Y (Y = CHO, COMe, CO_2Et, or CN). Oxidation of some simple carboxylic acids (R^1CO_2H, R^1 = Me, Et, Pr^n, or Bu^i) in the presence of these compounds gave the dimeric product

---

[157] G. A. Tarkhanov, L. I. Solov'eva, L. P. Kovsman, and G. N. Freidlin, Soviet Electrochem., 1971, 7, 272.
[158] R. W. Dennison and F. S. Holland, U.S.P. 3 582 484.
[159] A. Osman, A. A. Sueliam, F. M. El Sheikh, and A. I. Khudair, Soviet Electrochem., 1971, 7, 575.
[160] H. Nagamori and T. Sekine, Denki Kagaku, 1971, 39, 584.
[161] P. A. Konstantinov, I. V. Chelepin, and N. H. Koloskova, Khim. geterotsikl. Soedinenii, 1971, 915.
[162] M. Chkir and D. Lelandais, Chem. Comm., 1971, 1369.

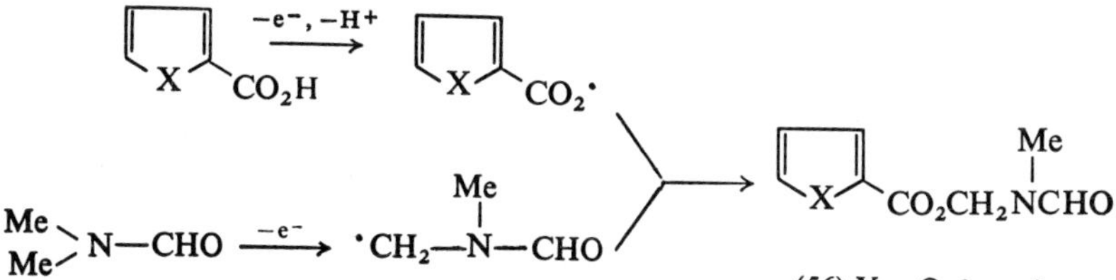

Scheme 47

$(R^1CH_2CHR^2Y)_2$ in good yield. Thomas[163] has studied the coupling of the carbonium ion produced by the anodic oxidation of isobutyric acid with the solvent acetonitrile. The formation of N-acetyl-N-isobutyrylisopropylamine (58) and isobutyric anhydride (59) as well as the expected N-isopropylacetamide (60) led to the postulation of intermediate (57). Non-cyclic compounds of this type have not been synthesized before (Scheme 48).

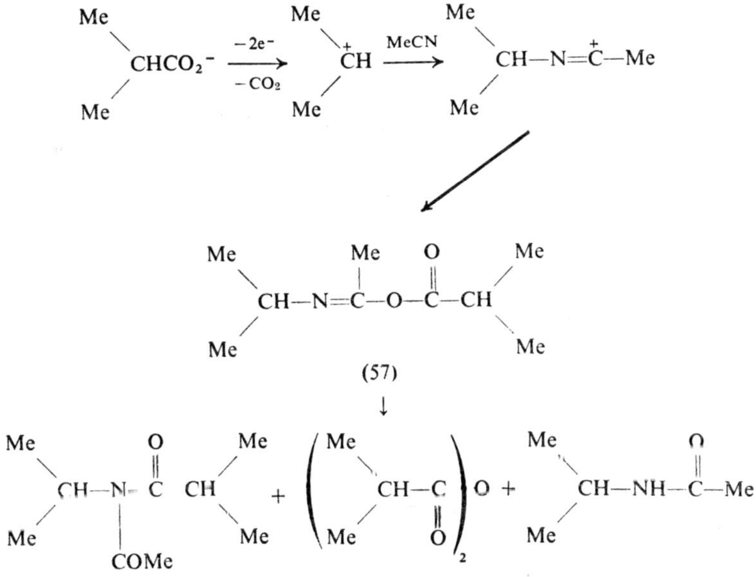

Scheme 48

Uneyama, Torii, and Oae[164] found that the anodic oxidation of sodium phenylthioacetate gave a very low yield of the Kolbe dimer $PhSCH_2CH_2SPh$ (2.3%). This was thought to be due to the low oxidation potential of the

---

[163] H. G. Thomas, *Angew. Chem.*, 1971, **83**, 579.
[164] K. Uneyama, S. Torii, and S. Oae, *Bull. Chem. Soc. Japan*, 1971, **44**, 815.

phenylthiomethyl cation and the formation of a stable thiophenoxyl radical. The main products were derived from these two species. The formation of the thiophenoxyl radical (61) by α-elimination of a methylenecarbene moiety was demonstrated by trapping experiments. The reaction pathways are shown in Scheme 49.

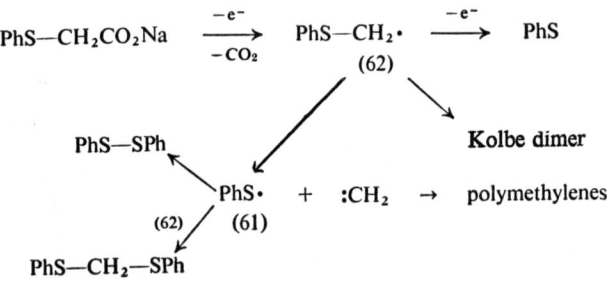

Scheme 49

Bodennee, Laurent, and Laurent[165] studied the oxidation of four isomers of pentanoic acid with a special interest in the behaviour of the radical intermediate. They observed that hydrogen-atom abstraction to give alkanes occurred only in the case of acids yielding primary radicals. Rodewald and Lewis[166] tried to identify the oxidation path of cyclopropanecarboxylic acids. In contrast to previous reports they found that the Kolbe dimer (with no ring-opening) was the major product. Regarding the formation of ring-opened side-products, the reaction paths in Scheme 50 were considered. It was

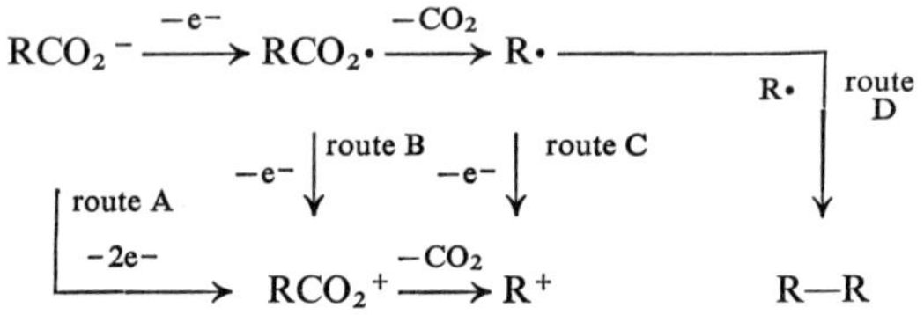

$$RCO_2^- \xrightarrow{-e^-} RCO_2\cdot \xrightarrow{-CO_2} R\cdot$$

route A: $-2e-$
route B: $-e-$
route C: $-e-$
route D: $R\cdot \rightarrow R-R$

$RCO_2^+ \xrightarrow{-CO_2} R^+$

Scheme 50

pointed out that the concerted and stereospecific opening of cyclopropyl cations upon departure of the leaving group was well documented. Thus a loss of CO_2 from the *cis* and *trans* carboxonium ions (63) and (64) would be

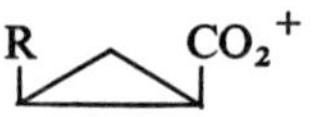

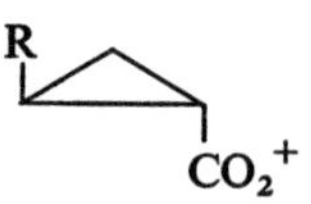

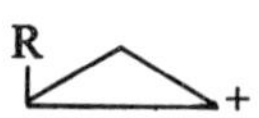

(63)   (64)   (65)

expected to give different isomer distributions. In fact, oxidation of the corre-

[165] G. Bodennec, A. Laurent, and E. Laurent, *Bull. Soc. chim. France*, 1971, 1691.
[166] L. B. Rodewald and M. C. Lewis, *Tetrahedron*, 1971, **27**, 5273.

sponding acids gave identical isomer distributions, indicating a common intermediate (65). The formation of ring-opened products by route C (Scheme 50) therefore seemed more likely. In contrast to this work, Takeda, Wada, and Murakami[167] found that no Kolbe dimer or product retaining the cyclopropyl ring were obtained from the electro-oxidation of 2,2-dichloro-3-phenylcyclopropanecarboxylic acid (66) (Scheme 51).

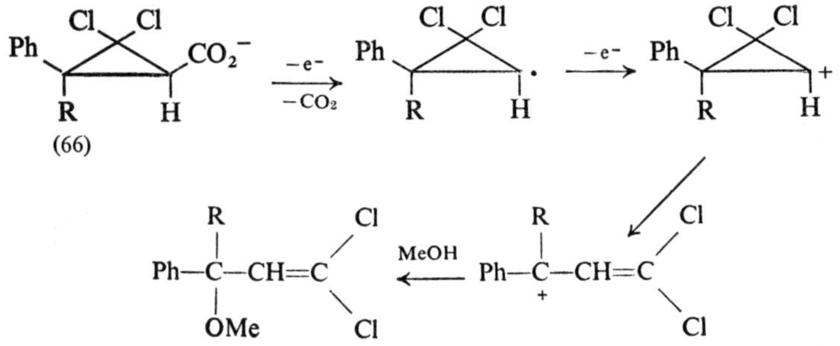

Scheme 51

The anodic decarboxylation of glycidic acids has been observed to give different products from those of the classical methods. Waters and Witkop[168] found that the main products resulting from the electrolysis of some glycidic

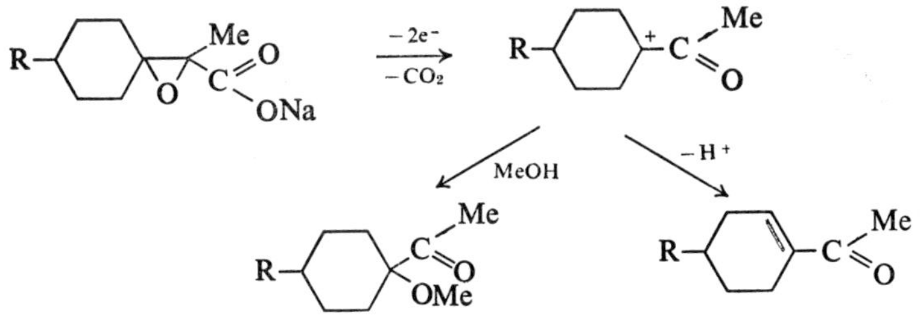

Scheme 52

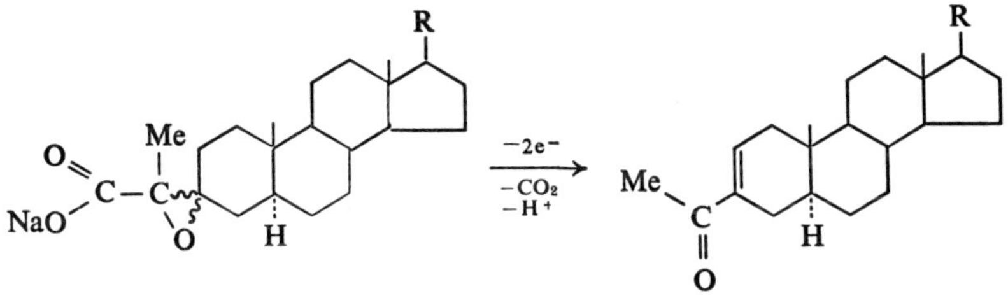

Scheme 52a

[167] A. Takeda, S. Wada, and Y. Murakami, *Bull. Chem. Soc. Japan*, 1971, **44**, 2729.
[168] J. A. Waters and B. Witkop, *J. Org. Chem.*, 1971, **36**, 3232.

acids in methanol at a platinum anode were the α-methoxy and αβ-unsaturated ketones (Scheme 52). The technique was applied to the decarboxylation of some steroids. In this case the β-elimination occured specifically in the Δ²-position (Scheme 52a).

Aromatic Amines.—The anodic oxidation of aniline in alkaline solutions has received relatively little attention in comparison with work in acidic media. The results presented in two new papers on this topic are in reasonable agreement. Giorgio, Lepri, and Heimler[169] found that the oxidation of aniline at platinum and carbon anodes was favoured on increasing the pH. An average of two electrons was consumed in c.p.e. experiments and the rate-determining

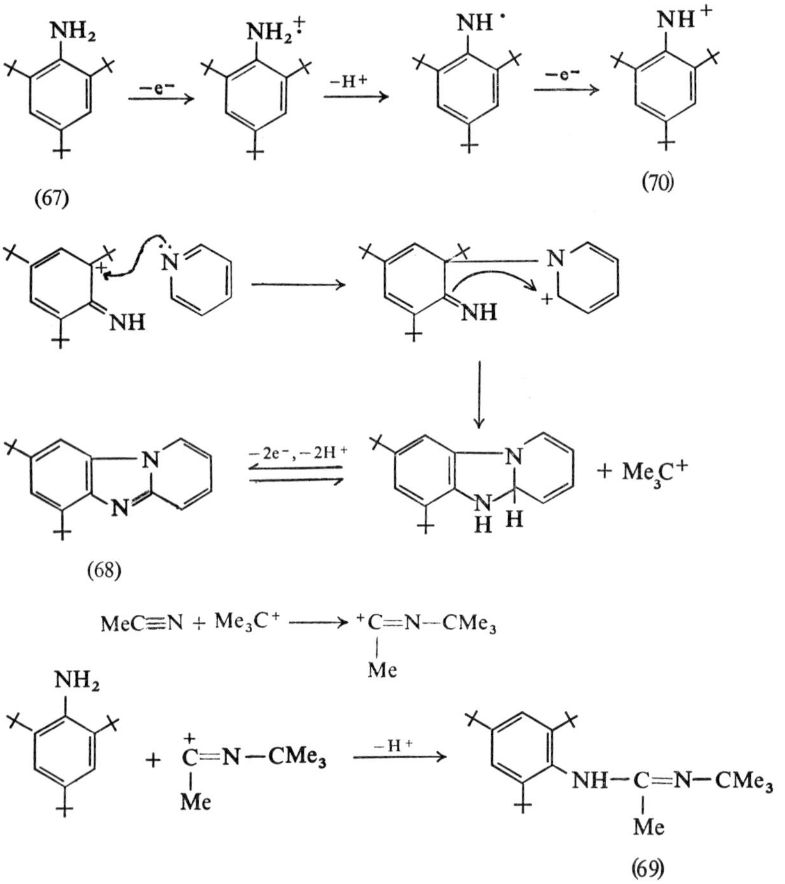

Scheme 53

[169] P. G. Desideri, L. Lepri, and D. Heimler, *J. Electroanalyt. Chem. Interfacial Electrochem.*, 1971, **32**, 225.

step was the transfer of the first electron. Matsuda et al.[170] came to the same conclusions. Both reported the main product to be azobenzene but did not agree on the second major product. Matsuda proposed head-to-tail coupling to give p-aminodiphenylamine, whereas Giorgio et al. proposed tail-to-tail coupling resulting in N-phenylquinone di-imine. Nickel was found to be the most active electrode material for the oxidation.

The oxidation of 2,4,6-tri-t-butylaniline (67) in acetonitrile with pyridine present, as reported by Cauquis, Cross, and Geniès[171,172] was considerably more complex. The main products formed by reaction between the aniline, pyridine, and acetonitrile were 6,8-di-t-butylpyrido[1,2-a]benzimidazole (68) and an amidine (69). The mechanism proposed to explain the formation of these exotic products is shown in Scheme 53. The presence of a methyl group substituted on the 2-position of pyridine was sufficient to prevent the intramolecular cyclization to (68). Thus on adding 2-picoline or 2,4- or 2,6-lutidine, the imidazole (68) was not formed, but hexa-t-butylazobenzene and a quinone imine resulted instead. In this case it is possible that the substituted pyridine acts as a base which deprotonates the cation (70), resulting in dimerization. At the same time, residual water in the solvent gives the quinone monoimine (70a). The absence of (68) when using 2-substituted pyridines

(70a)

supported the view that the formation was not by a radical pathway. This work was also discussed in a short review of some aspects of the electrochemistry of aromatic nitrogen compounds.[173]

Cauquis, Gognard, and Serve have studied the oxidation of diphenylamine (DPA) in acetonitrile.[174] Voltammetry and cyclic voltammetry showed the occurrence of two kinetically controlled one-electron transfers (E_p 0.67 and 0.77 V) corresponding to the oxidation of a deprotonated DPA radical and the free DPA, respectively. On repeated sweeping the build-up of NN'-diphenylbenzidine (DPB) was indicated by two additional peaks coming at potentials less anodic than those given by DPA. In agreement with previous authors the radical cation of DPA could not be detected. However, the presence of the DPB radical cation was indicated by visible spectroscopy and e.s.r. The fact

---

[170] Y. Matsuda, A. Shono, C. Iwakura, Y. Ohshiro, T. Agawa, and H. Tamura, *Bull. Chem. Soc. Japan*, 1971, **44**, 2960.
[171] G. Cauquis and J.-L. Cros, *Bull. Soc. chim. France*, 1971, 3760.
[172] G. Cauquis, J.-L. Cros, and M. Geniès, *Bull. Soc. chim. France*, 1971, 3765.
[173] G. Cauquis, *Pure Appl. Chem.*, 1971, **25**, 365.
[174] G. Cauquis, J. Cognard, and D. Serve, *Tetrahedron Letters*, 1971, 4645.

that it continued to be formed after the electrolysis was stopped (at the stage of 0.5 F mol^{-1}) was explained in terms of coupled disproportionation reactions:

$$DPB^{2+} + 4DPA \rightleftharpoons 2DPB + 2DPAH^+$$

$$DPB^{2+} + DPB \rightleftharpoons 2DPB^+$$

On increasing the basicity of the solvent the predominant product was tetraphenylhydrazine. The overall reaction path proposed is shown in Scheme 54.

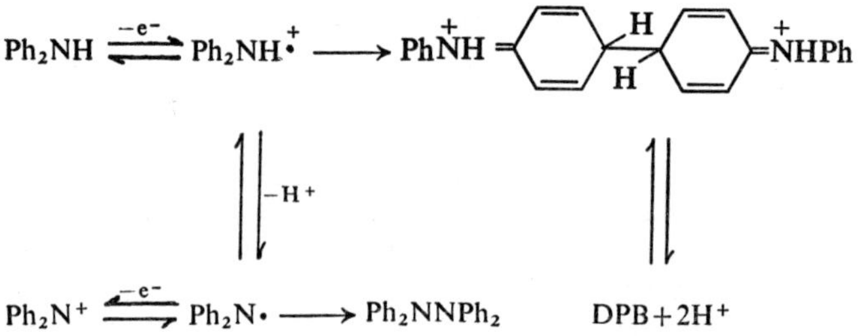

$$Ph_2NH \xrightleftharpoons{-e^-} Ph_2\overset{+}{N}H\cdot \longrightarrow Ph\overset{+}{N}H = \underset{H}{\overset{H}{\diagup\!\!\diagdown}} = \overset{+}{N}HPh$$

$$\Bigg\updownarrow_{-H^+} \qquad\qquad\qquad \Bigg\updownarrow$$

$$Ph_2N^+ \xrightleftharpoons{-e^-} Ph_2N\cdot \longrightarrow Ph_2NNPh_2 \qquad DPB+2H^+$$

Scheme 54

Hand et al.[175] found that the oxidation pathway of substituted dimethylanilines was dependent on the nature of the substituent, and, interestingly, on the concentration of the depolarizer. For example, electrolysis of NN-dimethylaniline (DMA) in acetonitrile gave the corresponding benzidine at a DMA concentration of 10^{-3} mol l^{-1}, 4,4'-methylenebis(dimethylaniline) at a DMA concentration of 10^{-2} mol l^{-1}, and Crystal Violet dye when the DMA concentration was 10^{-1} mol l^{-1}. The least stable radicals were obtained by electrolysis of p-halogeno-substituted DMAs, all of which underwent eliminations (Scheme 54a).

Barbey, Delahaye, and Caullet obtained similar results for the anodic oxidation of NN-diphenyl-p-toluidine in nitromethane. The main product

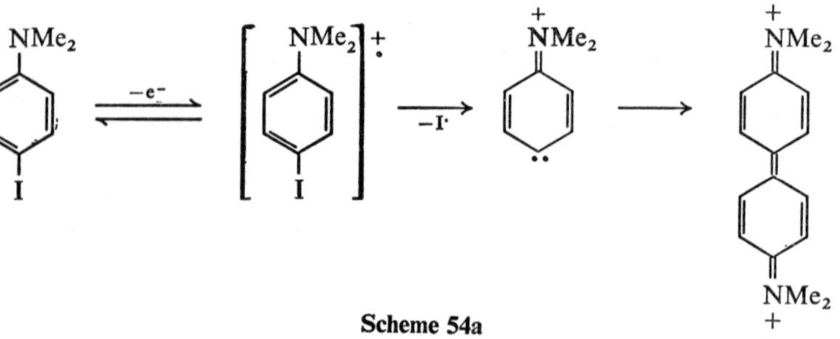

Scheme 54a

[175] R. Hand, M. Melicharek, D. I. Scoggin, R. Stotz, A. K. Carpenter, and R. F. Nelson, *Coll. Czech. Chem. Comm.*, 1971, **36**, 842.

Organic Electrochemistry – Synthetic Aspects 175

was the corresponding benzidine.[176] Electrochemical oxidation has also been used to initiate intramolecular cyclization of substituted anilines. For instance, Jenkins, Pedler, and Tatlow[177] reported that octafluoroacridone could be synthesized by the anodic oxidation of 2-aminononafluorobenzophenone at a platinum electrode (Scheme 54b).

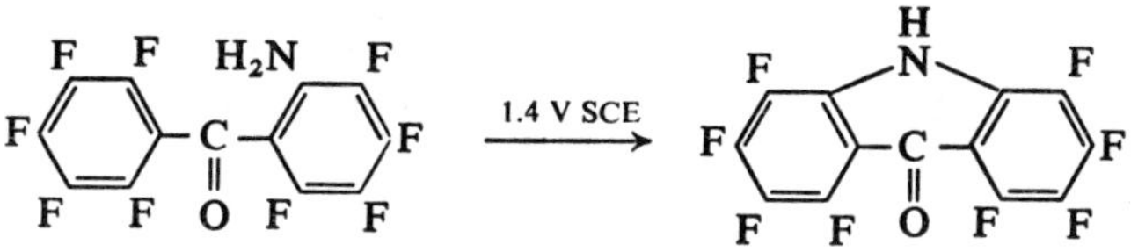

$$\text{F}_4\text{C}_6\text{-C(=O)-C}_6\text{F}_4\text{-NH}_2 \xrightarrow{1.4 \text{ V SCE}} \text{octafluoroacridone}$$

Scheme 54b

Other Nitrogen-containing Compounds.—Masui and Sayo investigated the oxidative dealkylation of tertiary dimethylamines at a glassy-carbon anode.[178] All the compounds studied, in addition to giving secondary amines, were dealkylated further to primary amines. Better yields (in fact almost quantitative) of the secondary amines were obtained when only a partial conversion of the starting compound was attempted. In contrast to the results of Smith and Mann, trimethylamine and N-substituted dimethylamines did undergo demethylation during oxidation. Cauquis and Geniès extended their own work on the oxidation of substituted hydrazines.[179] In acetonitrile containing lithium perchlorate, triphenylhydrazine was found not to give a well-defined oxidation wave. However, the addition of pyridine resolved one at 0.275 V (Ag|Ag$^+$ 10^{-2} mol l^{-1}). It was found to correspond to a diffusion-controlled two-electron transfer. Electrolysis at 0.5 V gave a quantitative yield of the triphenyldiazenium cation. Unfortunately, this was found to be less reactive towards olefins than the diphenyldiazenium cation. Reaction was only possible with olefins containing electron-donating substituents or by using very long reaction times. The electrochemical oxidation of piperidine at mercury and platinum anodes was studied by Barradas, Giordano, and Sheffield.[180] Polarography, chronopotentiometry, coulometry, rotating ring disc studies, and product analysis were used to elucidate the mechanism. It was thought that the reaction involved the transfer of two electrons and two hydroxyl ions to give piperidine N-oxide and not pyridine as proposed previously. Libert, Caullet, and Longchamp[181] investigated the anodic oxidation of 2,3,4,5-tetraphenylpyrrole in acetonitrile and alcohols. In acetonitrile it gave two oxidation waves at 0.61 and 1.23 V (Ag|Ag$^+$ 10^{-2} mol l^{-1}). A 60% yield of

[176] G. Barbey, D. Delahaye, and C. Caullet, *Bull. Soc. chim. France*, 1971, 3377.
[177] C. M. Jenkins, A. E. Pedler, and J. C. Tatlow, *Tetrahedron*, 1971, **27**, 2557.
[178] M. Masui and H. Sayo, *J. Chem. Soc. (B)*, 1971, 1593.
[179] G. Cauquis and M. Geniès, *Tetrahedron Letters*, 1971, 4677.
[180] R. G. Barradas, M. C. Giordano, and W. H. Sheffield, *Electrochim. Acta*, 1971, **16**, 1235.
[181] M. Libert, C. Caullet, and S. Longchamp, *Bull. Soc. chim. France*, 1971, 2367.

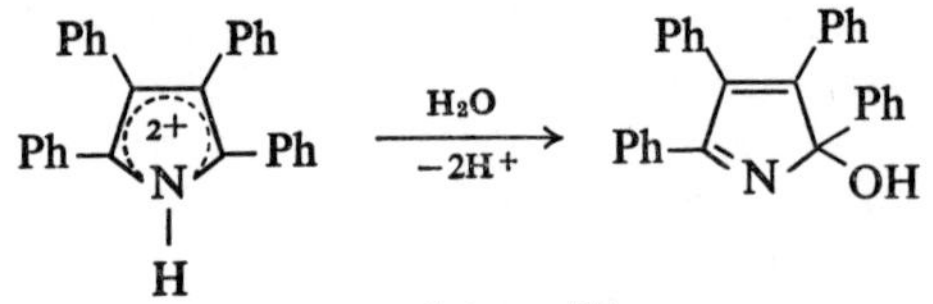

Scheme 55

2-hydroxy-2,3,4,5-tetraphenylpyrrolenine was obtained (Scheme 55). Its formation was assumed to occur *via* the formation of the dication and nucleophilic attack by the residual water. Oxidations in methanol and ethanol gave the corresponding methoxy- and ethoxy-compounds.

Dryhurst and co-workers published several papers on the oxidation of xanthines.[182—185] His previous investigation on the oxidation of uric acid was extended in order to obtain more evidence for the reaction intermediate

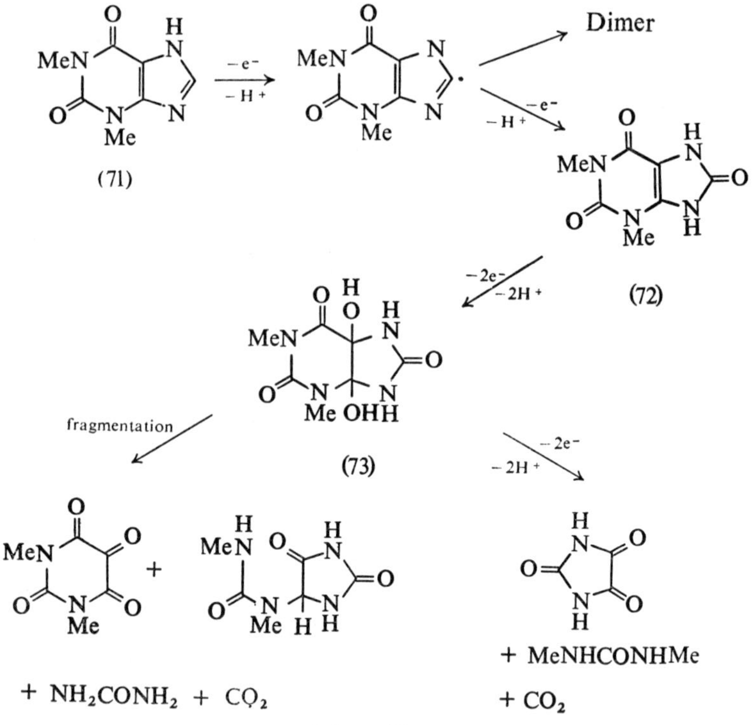

Scheme 55a

[182] B. H. Hansen and G. Dryhurst, *J. Electroanalyt. Chem. Interfacial Electrochem.*, 1971, **30**, 407.
[183] G. Dryhurst, *J. Electrochem. Soc.*, 1971, **118**, 699.
[184] B. H. Hansen and G. Dryhurst, *J. Electroanalyt. Chem. Interfacial Electrochem.*, 1971, **30**, 417.
[185] B. H. Hansen and G. Dryhurst, *J. Electroanalyt. Chem. Interfacial Electrochem.*, 1971, **32**, 405.

being uric acid 4,5-diol.[183] Oxidations in the presence of methanol were found to suppress the formation of the 4,5-diol and parabanic acid. This was attributed to the formation of the more stable 4,5-dimethoxyuric acid. The primary oxidation site of uric acid was therefore assumed to be at positions 4 and 5. Similarly it was considered that the oxidation paths of theobromine and caffeine also involved a uric acid 4,5-diol intermediate.[182] The effect of N-methylation of some xanthines was found to decrease the stability of the 4,5-diol. The destabilization resulting on methylation at N-3 and N-7 was so great that the diol could no longer be detected by fast-sweep cyclic voltammetry.[184] The investigation of the oxidation of theophylline (71) at a pyrolytic graphite electrode was also reported.[185] Voltammetry, c.p.e., and identification of the reaction products were used to elucidate the mechanism. The first step of the oxidation was thought to be the formation of a radical. Dimerization of the radical was in competition with oxidation to 1,3-dimethyluric acid (72), which was immediately oxidized further to 1,3-dimethyluric acid 4,5-diol (73). The latter compound was unstable and underwent either secondary oxidation or fragmentation to parabanic acid, dimethylurea, dimethylalloxan, urea, 6,8-dimethylallantoin, and carbon dioxide (Scheme 55a).

Oxygen-containing Compounds.—*Alcohols.* Sundholm reported the formation of acetals when anhydrous methanol and ethanol were electrochemically oxidized.[186] Using sodium perchlorate or tetrabutylammonium tetrafluoroborate the yields were high (58—92%). However, with sodium alkoxides as support electrolyte the corresponding aldehydes were formed by oxidation of the alkoxide ion itself. The oxidation potentials of a series of aliphatic alcohols in non-aqueous solvents was found to correlate with their ionization potentials.[187]

It is known that diacetone-2-keto-L-gulonic acid (DKGA) (an intermediate in vitamin C synthesis) can be generated in yields as high as 85% by the electro-oxidation of diacetone-L-sorbose. The reaction was normally carried out by an indirect electrochemical route via electrogenerated hypobromite oxidizing agent, in the presence of a nickelous or cobaltous ion catalyst. Fioshin *et al.*[188] have found that the basic electrode reaction is the bromide ion discharge leading to the formation of hypobromite, which transforms the nickelous ion into a nickel oxyhydroxide species. A similar catalytically oxidizing layer was thought to form on the surface of a nickel anode in alkaline solutions. With this system no other catalysts were needed and an 80% yield of DKGA was attainable.

Fleischmann, Korinek, and Pletcher have studied in some detail the oxidation of several classes of organic compounds at passivated nickel anodes in

[186] G. Sundholm, *J. Electroanalyt. Chem. Interfacial Electrochem.*, 1971, **31**, 265.
[187] G. Sundholm, *Acta Chem. Scand.*, 1971, **25**, 3188.
[188] M. Ya. Fioshin, I. A. Avrutskaya, A. I. Borisov, and L. A. Chupina, *Soviet Electrochem.*, 1971, **7**, 380.

strongly alkaline solutions.[189] Primary and secondary alcohols and amines were all found to oxidize at the potential where the nickel anode became oxidized. The cyclic voltammetric behaviour indicated that the compounds were oxidized by the mechanism:

$$Ni(OH)_2 \xrightarrow{fast} NiO(OH) + e^- + H^+$$

$$NiO(OH) + organic \xrightarrow{slow} Ni(OH)_2 + product$$

and not by a direct electron transfer from the organic compound to the anode. Controlled-potential electrolysis showed that the major products were derived by the routes:

(i) primary straight-chain amines, $RCH_2NH_2 \xrightarrow{-4e-} RCN$

(ii) primary branched-chain amines,

$$\begin{array}{c} R \\ \diagdown \\ CH-NH_2 \\ \diagup \\ R \end{array} \xrightarrow{-2e-} \begin{array}{c} R \\ \diagdown \\ C=NH \\ \diagup \\ R \end{array} \xrightarrow{H_2O} \begin{array}{c} R \\ \diagdown \\ C=O + NH_3 \\ \diagup \\ R \end{array}$$

(iii) secondary amines,

$$\begin{array}{c} RCH_2 \\ \diagdown \\ NH \\ \diagup \\ RCH_2 \end{array} \xrightarrow{-2e-} RCHO + RCH_2NH_2 \xrightarrow{-4e-} RCN$$

(iv) primary alcohols, $RCH_2OH \xrightarrow{-4e-} RCO_2H$

(v) secondary alcohols,

$$\begin{array}{c} R \\ \diagdown \\ CHOH \\ \diagup \\ R \end{array} \xrightarrow{-2e-} \begin{array}{c} R \\ \diagdown \\ C=O \\ \diagup \\ R \end{array}$$

and that very high yields were possible. The behaviour of organic compounds at oxidized nickel electrodes was quite different to that at carbon and platinum anodes. The results strongly suggested that the stereochemistry of the substrate played a vital role in determining the rate of the oxidations. The impor-

---

[189] M. Fleischmann, K. Korinek, and D. Pletcher, *J. Electroanalyt. Chem. Interfacial Electrochem.*, 1971, **31**, 39.

tance of surface oxide layers was also illustrated by Vijh for the oxidation of ethylene glycol at platinum anodes.[190]

Phenols. The electrochemical oxidation of phenols or naphthols is of considerable interest since several important quinones and naphthoquinones (*e.g.* intermediates for pharmaceuticals and dyestuffs) may be prepared in pure form. Wehrli and Caldwell have issued a patent on the electrosynthesis of some vitamin precursors. Details of the reaction conditions are given therein.[191] Parker and Ronlán have investigated the oxidation of 2,6-di-butyl-*p*-cresol (74) with a special interest in determining whether a quinone-methide (75) was an intermediate.[192] It was anticipated that in view of the difficulty in oxidizing (76), the proposed intermediate (75) should be suffi-

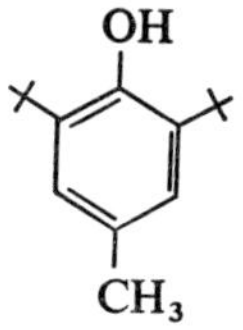

(74)

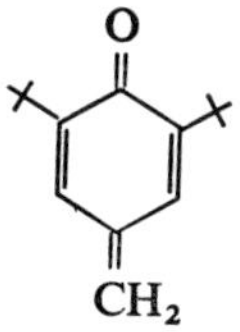

(75)

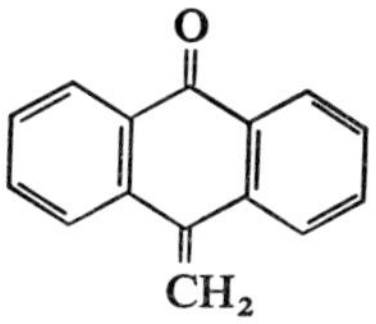

(76)

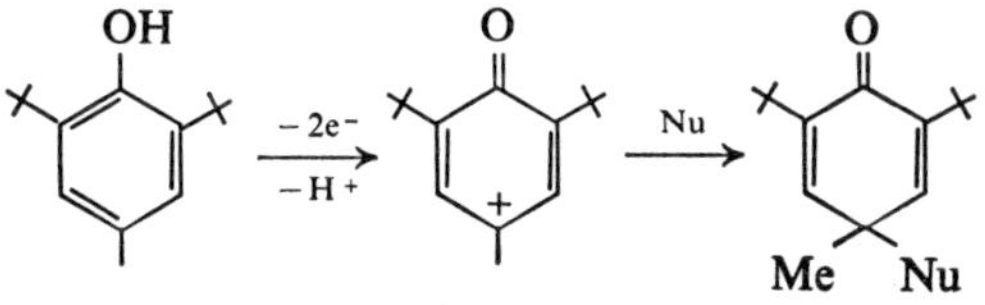

Scheme 56

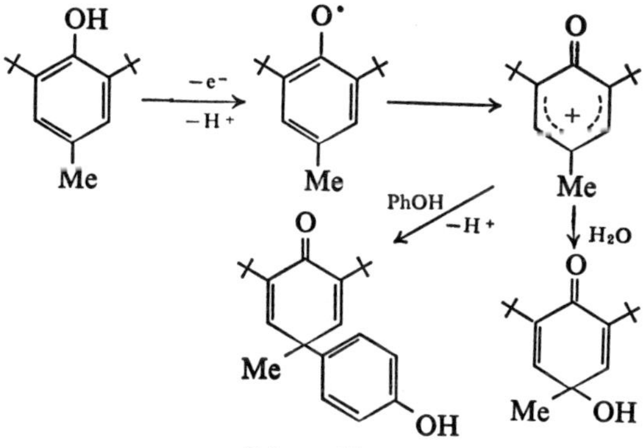

Scheme 56a

[190] A. K. Vijh, *Canad. J. Chem.*, 1971, **49**, 78.
[191] P. A. Wehrli, U.S.P. 3592748.
[192] V. D. Parker and A. Ronlán, *J. Electroanalyt. Chem. Interfacial Electrochem.*, 1971, **30**, 502.

ciently stable to detect. In fact, this was found not to be the case and the reaction products obtained were not those expected to derive from (75). A more likely reaction path was thought to be Scheme 56.

The anodic cross-coupling of phenols was studied by Ronlán.[193] In the presence of a second less easily oxidized phenol, oxidation of 2,6-di-butyl-*p*-cresol gave a good yield of the cross-coupled product. Addition of water suppressed the coupling, indicating that a cationic species (phenoxonium ion) was the reactive intermediate and not a radical The mechanism was likened to that of aromatic substitution (Scheme 56a).

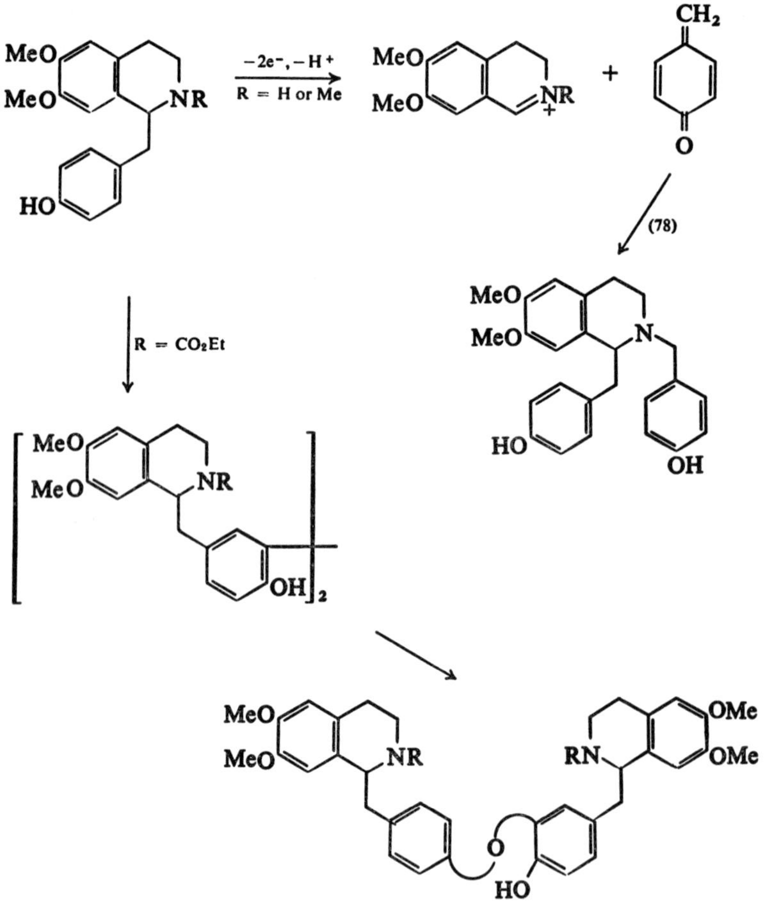

Scheme 57

[193] A. Ronlán, *Chem. Comm.*, 1971, 1643.

Bobbit, Noguchi, Yagi, and Weisgraber have found a stereoselective, stereospecific phenol-coupling reaction.[194] In contrast to chemically initiated processes, the anodic coupling of 1,2,3,4-tetrahydro-7-hydroxy-6-methoxy-1,2-dimethylisoquinoline gave a dimer in which only one of the enantiomeric pairs was obtained (yield 69%). The reaction was also stereoselective in that only molecules of identical configuration coupled. Bobbit and Hallacher[195] have also attempted the synthesis of the dauricine (77) skeleton by the anodic coupling of the phenolic armepavine (78). Only when the nitrogen was acylated was a dimer obtained and fragmentation of the molecule prevented. Although this dimer was not dauricine it was possible to induce a rearrangement to the required structure (Scheme 57).

Benzylic Compounds. Miller et al. have found that benzylic aldehydes and ketones[196] and ethers and esters[197] undergo an interesting oxidative cleavage reaction with high efficiency:

$$Ph-\underset{\underset{R^2}{|}}{CH}-OR^1 \xrightarrow{2e-} Ph-\overset{O}{\underset{\|}{C}}-R^2 + R^1OH$$

This is contrary to previous beliefs that even benzylic alcohol to aldehyde conversion was not in general possible. The cleavage of benzylic ethers by this method indicated an alternative technique for the protection and regeneration of alcohols. The success of the process was due in part to the application of a cathodic pulse, which prevented the accumulation of inhibiting species at the electrode surface. The mechanisms envisaged are shown in Scheme 58 for alcohols and ethers and Scheme 59 for the ketone benzoin.

Furans. The anodic oxidation of substituted furans has been discussed by Torii, Tanaku, Ogo, and Yamasita[198] and by Smirnov, Mil'man, and Krayanskii.[199] Torii et al, were interested in the relative ease of oxidation of the furan ring and the carboxylic acid group of 5-alkyl-2-furoic acids. The oxidation was found to proceed *via* the stable intermediate (79), since this compound gave the same reaction products as the furoic acid. Scheme 60 shows the reaction path proposed.

---

[194] J. M. Bobbitt, I. Noguchi, H. Yagi, and K. H. Weisgraber, *J. Amer. Chem. Soc.*, 1971, **93**, 3551.
[195] J. M. Bobbitt and R. C. Hallcher, *Chem. Comm.*, 1971, 543.
[196] L. L. Miller, V. R. Koch, M. E. Larscheid, and J. F. Wolf, *Tetrahedron Letters*, 1971, 1389.
[197] L. L. Miller, J. F. Wolf, and E. A. Mayeda, *J. Amer. Chem. Soc.*, 1971, **93**, 3306.
[198] S. Torii, H. Tanaka, H. Ogo, and S. Yamasita, *Bull. Chem. Soc. Japan*, 1971, **44**, 1079.
[199] V. A. Smirnov, V. I. Mil'man, and O. B. Krayanskii, *Soviet Electrochem.*, 1971, **7**, 799.

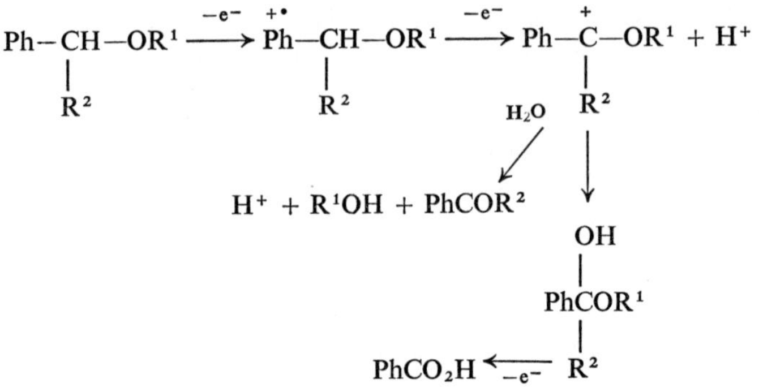

Scheme 58

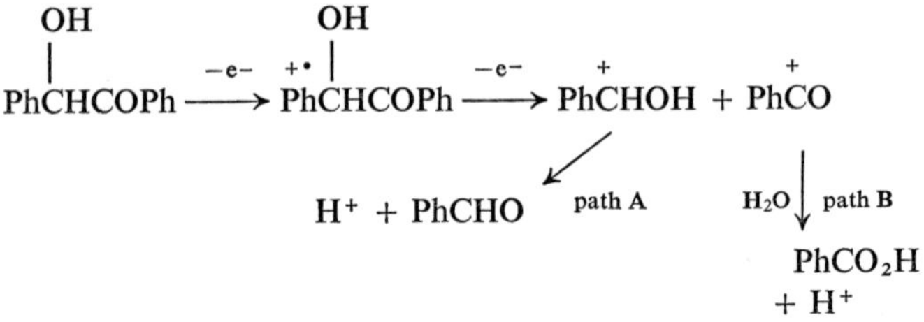

$$Ph-\underset{R^2}{\underset{|}{CH}}-OR^1 \xrightarrow{-e^-} \overset{+\bullet}{Ph-\underset{R^2}{\underset{|}{CH}}-OR^1} \xrightarrow{-e^-} \overset{+}{Ph-\underset{R^2}{\underset{|}{C}}-OR^1} + H^+$$

$$\xrightarrow{H_2O}$$

$$H^+ + R^1OH + PhCOR^2$$

$$\underset{R^2}{\underset{|}{PhCOR^1}}$$
$$\underset{OH}{|}$$

$$PhCO_2H \xleftarrow{-e^-} R^2$$

Scheme 59

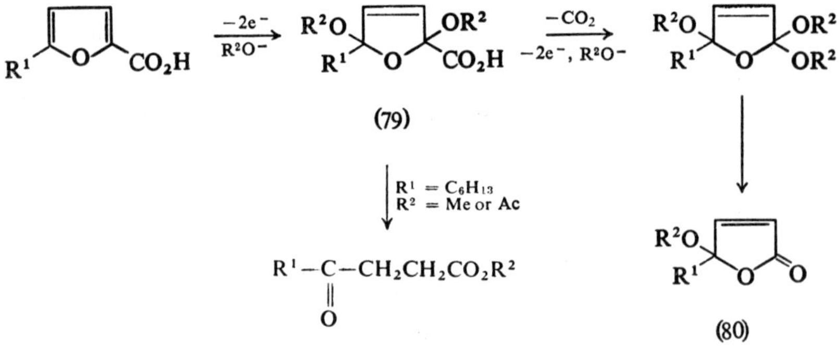

$$\underset{PhCHCOPh}{\underset{|}{OH}} \xrightarrow{-e^-} \overset{+\bullet}{\underset{PhCHCOPh}{\underset{|}{OH}}} \xrightarrow{-e^-} \overset{+}{PhCHOH} + \overset{+}{PhCO}$$

path A: $H^+ + PhCHO$

path B: $H_2O \downarrow$ then $PhCO_2H + H^+$

Scheme 60

$$R^1\underset{O}{\diagdown}CO_2H \xrightarrow[R^2O^-]{-2e^-} \underset{R^1 \quad O \quad CO_2H}{R^2O \quad OR^2} \xrightarrow[-2e^-, R^2O^-]{-CO_2} \underset{R^1 \quad O \quad OR^2}{R^2O \quad OR^2}$$

(79)

$$\begin{vmatrix} R^1 = C_6H_{13} \\ R^2 = Me \text{ or } Ac \end{vmatrix}$$

$$R^1-\underset{\underset{O}{\|}}{C}-CH_2CH_2CO_2R^2$$

$$\underset{R^1 \quad O \quad O}{R^2O}$$

(80)

Smirnov et al. obtained very similar behaviour with furfural. Oxidation at the first wave [1.30—1.35 V (S.C.E.)] gave β-formylacrylic acid in 85% yield. The mechanism probably follows path A in Scheme 59 with the additional step of oxidizing the aldehyde group. The hydroxy-lactone corresponding to (80) is probably so unstable that it decomposes immediately to give β-formylacrylic acid (81).

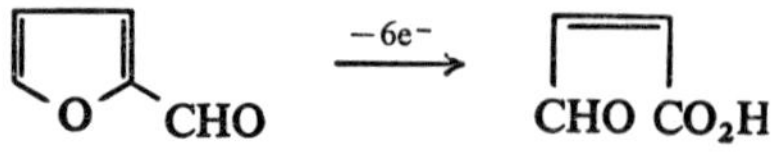

$$\underset{O}{\diagdown}CHO \xrightarrow{-6e^-} CHO \; CO_2H$$

Yoshida and Fueno[200] found that the cyanomethoxylation of 2,5-dimethylfuran was non-stereospecific. It was assumed that the mechanism of the formation of 2-cyano-5-methoxy-2,5-dimethyldihydrofuran was similar to anodic aromatic substitution; thiophen was found to behave similarly.[201] Markushina and Shulyakovkaya[202] studied the anodic methoxylation of carbonyl-containing furans. The ketones gave the normal 2,5-methoxylated products, but the aldehydes underwent an interesting intramolecular cyclization (Scheme 61). The efficiencies of these processes were high ($>70\%$).

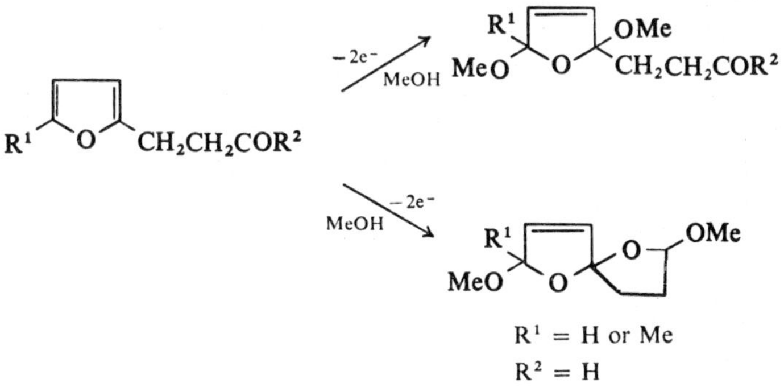

$$R^1 = H \text{ or } Me$$
$$R^2 = H$$

Scheme 61

Anodic Halogenation and Pseudohalogenation of Organic Compounds.—It is perhaps worth recalling that the electrochemical generation of chlorine is one of the most important single processes in the chemical industry. Most of the output is used for chlorinating organic compounds. It is therefore not surprising that direct methods of halogenating organic compounds should be sought. This is especially true in the case of fluorine, where it is both expensive and hazardous to handle it in elemental form. The industrial aspects of electrochemical fluorination have recently been reviewed by Rudge,[8] and several papers and patents have appeared.[203—208] The production of chlorofluorophosgene by the electrolysis of phosgene in anhydrous hydrogen fluoride was described by Leverkusen and Monheim.[203] The electrochemical fluorination of benzenesulphonyl chloride or fluoride was found to give perfluorocyclohexanesulphonyl fluoride as the main product.[204] The nuclear fluorination of

[200] K. Yoshida and T. Fueno, *J. Org. Chem.*, 1971, **36**, 1523.
[201] K. Yoshida, T. Saeki, and T. Fueno, *J. Org. Chem.*, 1971, **36**, 3673.
[202] I. A. Markushina and N. V. Shulyakovkaya, *Khim. geterotsikl. Soedinenii*, 1971, 1155.
[203] P. V. Leverkusen and H. N. Monheim, U.S.P. 3 595 763.
[204] G. Gambaretto, L. Marchesini, and R. Trevisan, *Ann. Chim. (Italy)*, 1971, **61**, 733.
[205] I. L. Knunyants, I. N. Rozhkov, and A. V. Bukhtiarov, *Izvest. Akad. Nauk S.S.S.R., Ser. khim.*, 1971, 1369.
[206] B. Chang, H. Yanase, N. Nakanishi, and W. Watanabe, *Electrochim. Acta*, 1971, **16**, 1179.
[207] W. V. Childs, U.S.P. 3 558 449.
[208] H. M. Fox, F. N. Ruehlen, and W. V. Childs, *J. Electrochem. Soc.*, 1971, **118**, 1246.

benzotrifluoride at high anodic potentials was reported by Knunyants, Rozhkov, and Bukhtiarov.[205] Chang et al.[206] investigated the anodic fluorination of n-propyldiethyl- and triethyl-amines at nickel anodes. Many products were formed, including a substantial percentage resulting from cleavage of C—C and C—N bonds. Bond cleavage was even more pronounced at low temperatures. It is obvious that the reaction conditions must be moderated in order to prevent bond scission if the process is to be at all interesting. Fox, Ruehlen, and Childs seem to have solved this problem, at least for gaseous feed-stocks.[207, 208] A feature of their technique was the use of a porous carbon anode that was not wetted by the KF,2HF electrolyte. The reactant to be fluorinated was circulated within the anode and the fluorinated product recovered continuously. The feedstock and product were not allowed to escape into the electrolyte. In this way conversions approaching 100% per pass with very high current efficiencies were realized, and virtually no carbon–carbon bond scission occurred. The electrochemical fluorination of ethane was thus investigated: at high feed rates a low conversion into partially fluorinated ethanes was obtained, whereas at lower rates the conversion and percentage of perfluorinated product were higher. In none of the runs did the yield of carbon tetrafluoride exceed 1%.

The electrolysis of a 20% sodium chloride solution saturated with an alkene at a graphite anode was reported by Mayell to give the monohalogenated alkene in good yield.[209] Cauquis and Pierre[210] described the thiocyanation and selenocyanation of aromatic compounds by the *in situ* generation of thiocyanogen or selenocyanogen. Solutions of sodium thiocyanate or selenocyanate in acetonitrile containing the aromatic compound were electrolysed at a potential just above the discharge potential of the pseudohalide ion (SCN^-, 0.32 V; $SeCN^-$, 0.06 V; $Ag|Ag^+ \ 10^{-2}$ mol l^{-1}). Aniline and *N*-alkyl substituted anilines, *o*- and *m*-toluidines, and phenol all gave *p*-substituted products with yields of *ca.* 70%.

Miscellaneous Oxidations.—Uneyama and Torii have investigated the electro-oxidation of phenyl sulphides.[211] The product distribution was found to be similar to that obtained by a bromine oxidation. The reaction involved two competing paths: one the oxidation to a sulphone and the other cleavage of the alkyl group and dimerization of the so-formed radical. The dimerization path seemed to be dependent on the property of R as a leaving group (Scheme 62). Other organosulphur compounds studied were benzenethiol[212] and β-mercaptopyruvic acid.[213]

Miller, Stermitz, and Falck[214] have found that the use of electro-oxidation

---
[209] J. S. Mayell, U.S.P. 3558453.
[210] G. Cauquis and G. Pierre, *Compt. rend.*, 1971, **272**, C, 609.
[211] K. Uneyama and S. Torii, *Tetrahedron Letters*, 1971, 329.
[212] F. Magno, G. Bontempelli, and G. Pilloni, *J. Electroanalyt. Chem. Interfacial Electrochem.*, 1971, **30**, 375.
[213] J. Tohier and M.-B. Fleury, *Bull. Soc. chim. France*, 1971, 3075.
[214] L. L. Miller, F. R. Stermitz, and J. R. Falck, *J. Amer. Chem. Soc.*, 1971, **93**, 5941.

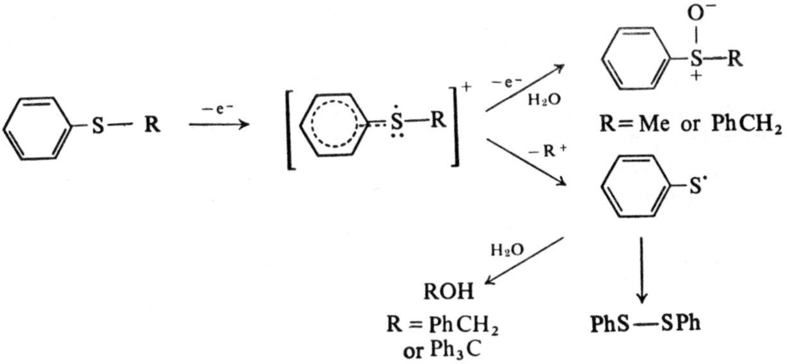

Scheme 62

in the synthesis of alkaloids can give striking improvements in yield. For example, the intermolecular coupling of laudanosine to a morphinandienone was achieved with a yield of 52%.

Laurent and Tardirel[215] showed that the carbonium ion formed by the oxidative cleavage of β-phenylethyl iodide (deuteriated on the α-carbon atom) was susceptible to nucleophilic attack at either the α- or β-positions. A carbonium ion intermediate having the structure shown in Scheme 63 was postulated.

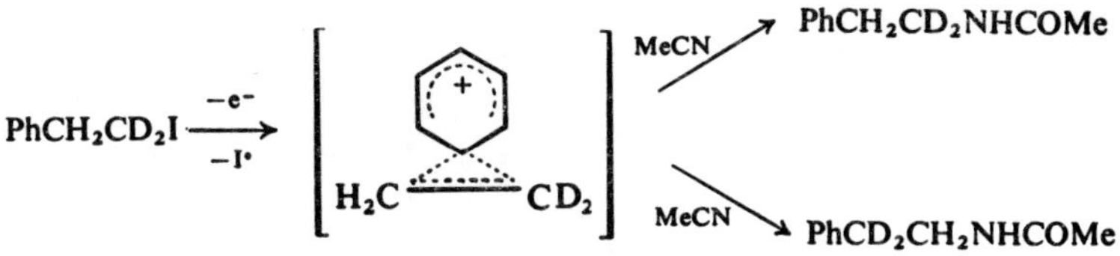

Scheme 63

With regard to the oxidation of mercury olefin complexes, Fleischmann, Pletcher, and Sundholm have found that the oxidative cleavage of a carbon-mercury bond was a reaction general to organomercury compounds.[216]

Indirect electrochemical oxidations using high-oxidation-state metal ions, which are regenerated electrochemically, have continued to attract attention. Rashid, Straka, and Kalvoda[217] have investigated the reaction of aliphatic alcohols with electrogenerated osmium(VIII). The rate of reaction was found to increase with the length of the carbon chain.

---

[215] A. Laurent and R. Tardivel, *Compt. rend.*, 1971, **272**, C, 8.
[216] M. Fleischmann, D. Pletcher, and G. Sundholm, *J. Electroanalyt. Chem. Interfacia Electrochem.*, 1971, **31**, 51.
[217] A. Rashid, P. Straka, and R. Kalvoda, *J. Eletroanalyt. Chem. Interfacial Electrochem.*, 1971, **29**, 383.

Fleischmann, Pletcher, and Rafinski[218] have studied the kinetics of the $Ag^I|Ag^{II}$ couple at a platinum electrode. The partial oxidation of several organic compounds by electro-generated Ag^{II} was reported to give good current efficiencies. A slight disadvantage with this system was the need to control the potential rather carefully because of the proximity to the potential for oxygen evolution and argentic oxide formation.

Tamat and Vecchi have found that it is feasible to generate Fentons reagent (Fe^{2+}–H_2O_2) electrochemically.[219] The method involved the electrolysis of a solution of a ferric salt saturated with oxygen at a potential of -0.35 V. At this potential the ferric ion and oxygen were reduced simultaneously:

$$Fe_3^+ + e^- \rightarrow Fe^{2+}$$

$$O_2 + 2H^+ + 2e^- \rightarrow H_2O_2$$

$$Fe^{2+} + H_2O_2 \rightarrow Fe_3^+ + OH\cdot + OH^-$$

The oxidative addition of the hydroxyl radical to benzene was investigated. Phenol was produced with a current efficiency of 60%. Similar 'cathodic oxidations' have already been described using cupric ions.

[218] M. Fleischmann, D. Pletcher, and A. Rafinski, *J. Appl. Electrochem.*, 1971, **1**, 1.
[219] R. Tomat and E. Vecchi, *J. Appl. Electrochem.*, 1971, **1**, 185.

6
Electrolyte Solutions

BY T. H. LILLEY

1 Introduction

Liquid water continues to be the most studied solvent system for ionic species and consequently most of the discussion in this Report will involve aqueous systems. Several books and reviews have appeared since the last Report[1] in this series. The sixth volume of Conway and Bockris' series has been published[2] and the chapter by Friedman on theoretical approaches to electrolyte solutions is essential reading for anyone interested in this field. A two-volume work on ionic solutions edited by Petrucci[3] has been published. The first volume[4] of a multi-volume series edited by Franks has also been published. This first book reviews experimental data of many types and theoretical models for water in some considerable detail. The concluding chapter by Frank is particularly interesting and stimulating. Further volumes in the series will consider aqueous solutions of non-electrolytes, electrolytes, disperse systems, and macromolecules. 'Water and Aqueous Solutions' is the title of a collection of articles edited by Horne[5] and contains some 19 chapters on aspects of water chemistry ranging from statistical mechanical models to sea-water and biofluids. There is necessarily some overlap between this book and that of Franks.[4] Other new books which have appeared include a volume of electrochemistry edited by Eyring,[6] two texts,[7,8] and a theoretically orientated

---

[1] A. K. Covington and T. H. Lilley, in 'Electrochemistry', ed. G. J. Hills (Specialist Periodical Reports), The Chemical Society, London, 1970, vol. 1, pp. 1, 31.
[2] 'Modern Aspects of Electrochemistry', ed. J. O'M. Bockris and B. E. Conway, Butterworths, London, 1971, no. 6.
[3] 'Ionic Interactions', ed. S. Petrucci, Academic Press, London, 1971, vols. I and II.
[4] 'Water, A Comprehensive Treatise', vol. 1, 'The Physics and Physical Chemistry of Water', ed. F. Franks, Plenum Press, New York, 1972.
[5] 'Water and Aqueous Solutions', ed. R. A. Horne, Wiley-Interscience, New York, 1972.
[6] 'Physical Chemistry—An Advanced Treatise', vol. IXA, 'Electrochemistry', ed. H. Eyring, Academic Press, London, 1970.
[7] J. O'M. Bockris and A. N. Reddy, 'Modern Electrochemistry', MacDonald, London, 1970.
[8] J. Koryta, J. Dvorák, and V. Boháčková, 'Electrochemistry', Methuen, London, 1970.

work by Falkenhagen, Ebeling, and Hertz.[9] Relevant reviews which have appeared include one on the molal volumes of electrolytes,[10] a summary[11] of investigations of, and theoretical approaches to, charge-transfer-to-solvent spectroscopy and a general broad-based review of electrolyte solutions by Blandamer.[12] Reviews on solvation by Hertz[13] and Hinton and Amis[14] illustrate two rather different approaches to a common problem. Some of the papers presented at the symposium in honour of T. F. Young have appeared.[15] Other collections of papers given at conferences include the proceedings of the 20th C.I.T.C.E. meeting[16] on ionic interactions and a series of papers on the general theme of molecular motions in liquids and solutions.[17] Wood and Reilly[18] have reviewed the thermodynamic properties of single and mixed electrolyte solutions up to 1969 and the recent report by Rosseinsky[19] is essentially complementary to this chapter. Reviews on non-aqueous electrolytic solutions have also appeared.[20,21] Two interesting and important reviews [22,23] discuss the medium-effect activity coefficients of ionic systems. A new journal, the *Journal of Solution Chemistry*, was published in 1972. Unfortunately the Reporter was unable to obtain copies of this, undoubtedly important, new contribution to the physical chemistry literature and consequently was unable to include papers from it in this article.

2 Liquid Water

Experimental data for the vapour pressure of (ordinary) water in the temperature range -2.5 to $+20$ °C have been obtained.[24] These and earlier data have been re-assessed by Ambrose and Lawrenson.[25] Accurate data on the effect of isotopic substitution (O and H) have been reported[26] over a range of temperatures. An estimate of the vapour pressure of HOD as a function of temperature and concentration in H_2O–D_2O mixtures has been made.[27] The pVT behaviour of liquid water at negative pressures is discussed by Winnick

---

[9] H. Falkenhagen, W. Ebeling, and H. G. Hertz, 'Theorie der Elektrolyte', Hirzel, Leipzig, 1970.
[10] F. J. Millero, *Chem. Rev.*, 1971, **71**, 147.
[11] M. J. Blandamer and M. F. Fox, *Chem. Rev.*, 1970, **70**, 59.
[12] M. J. Blandamer, *Quart. Rev.*, 1970, **24**, 169.
[13] H. G. Hertz, *Angew. Chem. Internat. Edn.*, 1970, **9**, 124.
[14] J. F. Hinton and E. S. Amis, *Chem. Rev.*, 1971, **71**, 627.
[15] Papers in *J. Phys. Chem.*, 1970, **74**, 3677–3822.
[16] Papers in *Electrochim. Acta*, 1971, **16**, 667–738.
[17] Papers in *Ber. Bunsengesellschaft phys. Chem.*, 1971, **75**, 183–402.
[18] R. H. Wood and P. J. Reilly, *Ann. Rev. Phys. Chem.*, 1970, **21**, 387.
[19] D. R. Rosseinsky, *Ann Reports (A)*, 1971, **68**, 81.
[20] Papers in *J. Electroanalyt. Chem. Interfacial Electrochem.*, 1971, **19**, 1–209.
[21] Papers in *Pure Appl. Chem.*, 1971, **25**, 305.
[22] O. Popovych, *Crit. Rev. Analyt. Chem.*, 1970, **1**, 73.
[23] A. J. Parker, *Chem. Rev.*, 1969, **69**, 1.
[24] D. R. Doustin, *J. Chem. Thermodynamics*, 1971, **3**, 187.
[25] D. Ambrose and I. J. Lawrenson, *J. Chem. Thermodynamics*, 1972, **4**, 755.
[26] J. Pupezin, G. Jackli, G. Jansco, and W. A. Van Hook, *J. Phys. Chem.*, 1972, **76**, 743.
[27] W. A. Van Hook, *J. Phys. Chem.*, 1972, **76**, 3040.

Electrolyte Solutions 189

and Cho.[28] The density of liquid water has been re-measured[29] from 5 to 80 °C and earlier data re-evaluated. A table is given of densities at 0.1 °C intervals in the range 0—80 °C. Liquid D_2O densities and viscosities have been obtained[30] between 5 and 70 °C. Refractive indices for H_2O from 1 to 60 °C and from 1 to 1406 kg cm^{-2} have been measured[31] and the refractive index data of water have been analysed[32] using a two-state model approach for the liquid. The self-diffusion coefficient of super-cooled liquid water has been measured[33] using the n.m.r. technique down to -31 °C. Lack of agreement with other recently reported work[34] was noted. The use of an activation-energy approach shows that the apparent energy of activation for diffusional motion changes from about 11 kcal at the lowest temperature studies to approximately 4.5 kcal at 25 °C. An interesting correlation is presented between the temperature variation of the diffusion coefficient and the fraction of broken hydrogen bonds as determined by Walrafen[35] and Luck and Ditter[36] using two-state models for spectroscopic results. The viscosity of $H_2^{18}O$ between 15 and 35 °C has been reported.[37] Further measurements on the Kerr constant and its temperature variation have been described[38] for liquid H_2O and D_2O. The spectroscopic properties of water and aqueous solutions have been reviewed elsewhere[4, 5, 39] and further discussion is unnecessary. Similarly theoretical treatments and approaches to liquid water have been or soon will be published[4, 5, 39] and mention will only be made of very recent papers. The most important papers of all those published recently come from Rahman and Stillinger.[40, 41] They describe molecular dynamic studies on liquid water using an earlier[42] expression for the pairwise interaction potential. No evidence is found for two classes of water molecules but the pair-potential distribution function changes with temperature in such a way as to suggest a basic hydrogen-bond rupture mechanism with an energy of about 2.5 kcal mol^{-1}. This is just the result which has been obtained by Walrafen[35] from spectroscopic studies.

[28] J. Winnick and S. J. Cho, *J. Chem. Phys.*, 1971, **55**, 2092.
[29] W. A. Gildseth, A. Habenschuss, and F. H. Spedding, *J. Chem. and Eng. Data*, 1972, **17**, 402.
[30] F. J. Millero, R. Dexter, and F. Hoff, *J. Chem. and Eng. Data*, 1971, **16**, 85.
[31] E. M. Stanley, *J. Chem. Eng. Data*, 1971, **16**, 454.
[32] S. K. Mitra, N. Dass and N. C. Varshneya, *J. Chem. Phys.*, 1972, **57**, 1798.
[33] K. T. Gillen, D. C. Douglass, and M. J. R. Hoch, *J. Chem. Phys.*, 1972, **57**, 5117.
[34] H. R. Pruppacher, *J. Chem. Phys.*, 1972, **56**, 101.
[35] G. E. Walrafen, Chapter 5 in ref. 4.
[36] W. A. P. Luck and W. Ditter, *J. Phys. Chem.*, 1970, **74**, 3687.
[37] A. I. Kudish, D. Wolf, and F. Steckel, *J. C. S. Faraday II*, 1972, **68**, 2041.
[38] Y. Chen and W. H. Orttung, *J. Phys. Chem.*, 1972, **76**, 216 (see also Y. Chen and W. H. Orttung, *J. Phys. Chem.*, 1968, **72**, 3069).
[39] 'Water, A Comprehensive Treatise', vols. II and III, ed. F. Franks, Plenum Press, New York, in the press.
[40] A. Rahman and F. H. Stillinger, *J. Chem. Phys.*, 1971, **55**, 3336.
[41] F. H. Stillinger and A. Rahman, *J. Chem. Phys.*, 1972, **57**, 1281.
[42] A. Ben-Naim and F. H. Stillinger, Chapter in ref. 5.

3 Thermodynamics of Single Electrolytes

An extremely important and valuable contribution has been made by Card and Valleau.[43] These authors have performed Monte Carlo (MC) calculations using the restricted primitive model (hard-sphere anions and cations of the same diameter, a, with the relative permittivity, ε, being fixed at that of the 'solvent') for a 1:1 electrolyte in a medium with $\varepsilon = 78.5$ and with $a = 4.25$ Å at 25 °C. Calculations such as these are very expensive to carry out, but are of great importance and serve as the 'experimental' check on results obtained from approximate statistical mechanical approaches. It has been shown[43] that one such approximation (the hypernetted chain approach used by Friedman and co-workers, see ref. 2) compares very favourably with the MC results. The earlier MC results of Vorontsov-Velmayinov and El'yashevich[44,45] were at variance with those obtained by Card and Valleau[43] but have been re-assessed[46] and the differences reconciled.

There has been considerable interest in the interpretation of the thermodynamic data of 2:2 electrolytes and in particular the bivalent sulphates. Gardner and Glueckauf[47] have treated the osmotic coefficient data of Brown and Prue[48] by assuming that in solution, as well as having free ions and ion pairs present, further ion association into triplets and quartets occurs. The electrostatic contributions to the activity coefficients of the various species were expressed using the equations advocated by Glueckauf[49] and three additional parameters were introduced which were assumed to account for hydration of the species. These assumptions lead to a reasonable fit to the experimental osmotic coefficients for the six sulphates studied. Estimation of the activity coefficients by use for the Gibbs–Duhem equation in conjunction with the observed[50] solubility of $CaSO_4$ in water gave a value of 2.49×10^{-5} mol^2 kg^{-2} for the solubility product of gypsum at 25 °C. This value is considerably smaller than that estimated by the original authors by use of an extended Debye–Hückel expression. In a second paper[51] restricted to $CaSO_4$ solutions, the same general treatment as that described above was used but the temperature dependences of the activity coefficients and the ion-pair association constant were considered. The solubility products were obtained in the temperature range 0—200 °C, again utilizing earlier data,[50] and this modified procedure gave a value of 2.26×10^{-5} mol^2 kg^{-2} for the solubility product of $CaSO_4,2H_2O$ at 25 °C.

[43] D. N. Card and J. P. Valleau, *J. Chem. Phys.*, 1970, **52**, 6232.
[44] P. N. Vorontsov-Velmaginov, A. M. El'gashevich, and A. K. Kron, *Elektrokhimiya*, 1966, **2**, 708.
[45] P. N. Vorontsov-Velmaginov and A. M. El'gashevich, *Elektrokhimiya*, 1968, **4**, 1430.
[46] P. N. Vorontsov-Velmaginov, A. M. El'gashevich, J. C. Rasaiah, and H. L. Friedman, *J. Chem. Phys.*, 1970, **52**, 1013.
[47] A. W. Gardner and E. Gluekauf, *Proc. Roy. Soc.*, 1969, **A313**, 131.
[48] P. G. M. Brown and J. E. Prue, *Proc. Roy. Soc.*, 1955, **A232**, 320.
[49] E. Glueckauf, *Proc. Roy. Soc.*, 1969, **A310**, 449.
[50] W. L. Marshall and R. Slusher, *J. Phys. Chem.*, 1966, **70**, 415.
[51] A. W. Gardner and E. Glueckauf, *Trans Faraday Soc.*, 1970, **66**, 1081.

A third paper by the same authors[52] dispensed with the ion-association concept and activity coefficients were derived using a numerical integration of the Poisson–Boltzmann equation. Extensive calculations for 2:2 electrolytes based upon Müller's[53] approach were performed (*cf.* with Gronwall, LaMer, and Sandved[54]). In calculations of this sort the only unknown is the distance of closest approach of the ions and by utilizing the low-concentration osmotic-coefficient data[48] to estimate this, a value of 2.23×10^{-5} mol^2 kg^{-2} was obtained for the solubility product of gypsum at 25 °C. The agreement between this value and that obtained[51] using the Bjerrum ion-pairing approach is excellent. However, one cannot help but conclude that it is probably fortuitous since it has been shown[55] that the numerical results obtained for activity coefficients using numerical integration approaches of this type depend upon whether the Debye or Guntleberg charging process is used. The discrepancy between the results obtained from the two processes points to the inadequacy of the Poisson–Boltzmann equation in describing systems of the 2:2 charge type. The bivalent sulphates have also been investigated by Pitzer[56] using experimental data from several sources, including heats of dilution. An extended Debye–Hückel expression was used for activity coefficients and association into ion triplets was considered. The possibility that triplet ionic species might exist in solution gains support from the work of Shamin and Spiro.[57] The anion-constituent transference numbers at 25 °C of picric and D-tartaric acid have been analysed as a function of concentration. In both instances, to obtain satisfactory agreement between the experimental results and the (assumed correct) equation of Robinson and Stokes[58] for the concentration dependence of the transference numbers, it was necessary to postulate the presence of aggregated ionic species. With picric acid a dimeric picrate ion was suggested and for tartaric acid a triple ion consisting of associated ionized and un-ionized tartrate species. The latter is of the same type as those suggested by Gluekauf[47] and Pitzer[56] and was also postulated to be present in iodic[59] and phosphoric acids.[60] Whether there are such species present as real chemical entities or whether all one is seeing is a deficiency in the theoretical expressions for the activity coefficient or the transference number is a matter of conjecture, but in view of the known limitations of Debye–Hückel expressions at finite concentrations one would guess that the equilibrium constants for, say, triple-ion formation have only a formal value.

Osmotic and Activity Coefficients.—A study similar to the classical work of Brown and Prue[48] on the freezing points of aqueous solutions of the bivalent

[52] A. W. Gardner and E. Glueckauf, *Proc. Roy. Soc.*, 1971, **A321**, 515.
[53] H. Muller, *Phys. Ztg.*, 1927, **28**, 324.
[54] T. H. Gronwall, V. K. La Mer, and K. Sandved. *Phys. Ztg.*, 1928, **29**, 358.
[55] D. M. Burley, V. C. L. Hutson, and C. W. Outhwaite. *Mol. Phys.*, 1972, **23**, 867.
[56] K. S. Pitzer, *J. C. S. Faraday II*, 1972, **68**, 101.
[57] M. Shamin and M. Spiro, *Trans Faraday Soc.*, 1970, **66**, 2863.
[58] R. A. Robinson and R. H. Stokes, 'Electrolyte Solutions', Butterworths, London, 1959.
[59] A. D. Pethybridge and J. E. Prue, *Trans. Faraday Soc.*, 1967, **63**, 2019.
[60] M. Selvaratnam and M. Spiro, *Trans. Faraday Soc.*, 1965, **61**, 360.

sulphates has been performed[61] for Mg, Ca, Ba, and Mn dithionates. Data are also given for potassium persulphate. A re-examination[62] of the vapour pressures of aqueous sodium chloride by direct measurement shows good agreement with the compilation of Robinson and Stokes.[58] The osmotic coefficients of ammonium bromide solutions at 25 °C have been determined.[63] A series of measurements [64, 65] on the osmotic properties of polystyrene sulphonate salts have been made. The osmotic coefficients[66] of benzenesulphonic acid up to 16 molal at 25 °C are reported. A considerable number of determinations of the osmotic coefficients of aqueous solutions at elevated temperatures have been carried out. Results for NaBr, NaI, KF, and $CaCl_2$ between 0 and 90 °C have been described,[67] and the concentration dependence fitted using the Scatchard–Lietzke and Stoughton equations.[68] Lindsay and Lin[69] have determined osmotic coefficients for approximately 1 molal LiCl and CsCl and for NaCl solutions[70] solutions in the temperature range 125—300 °C at 25° intervals using apparatus described elsewhere.[71] The osmotic coefficients of some salts at 80 °C are reported.[72] An extensive series of measurements on the osmotic coefficients and derivable properties of electrolytes in N-methylacetamide have been made.[73, 74] The results were obtained from careful freezing-point measurements and data are given for many alkali-metal halides and nitrates as well as some tetra-alkylammonium iodides and some alkali-metal carboxylates. Bonner has obtained[75] osmotic and activity coefficients of some solutes in D_2O.

Mussini and Pagella[76] have used calcium amalgam electrodes and obtained good agreement with Robinson and Stokes[58] for the activity coefficients of $CaCl_2$ solutions. The standard electrode potential of the amalgam electrodes was also determined as a function of temperature. Mussini has also investigated[77] the activity coefficients of perchloric acid using a perchlorate-responsive membrane cell. Reasonable, although not good, agreement is found with earlier experimental work. The solubility product of silver chromate in water over the range 5—80 °C has been measured.[78] Results are given[79] for

[61] M. R. Christoffersen and J. E. Prue, *Trans. Faraday Soc.*, 1970, **66**, 2878.
[62] C. N. Pepela and P. J. Dunlop, *J. Chem. Thermodynamics*, 1972, **4**, 255.
[63] A. K. Covington and D. E. Irish, *J. Chem. and Eng. Data*, 1972, **17**, 175.
[64] M. Reddy and J. A. Marinsky, *J. Phys. Chem.*, 1970, **74**, 3884.
[65] M. Reddy, J. A. Marinsky, and A. Sarkar, *J. Phys. Chem.*, 1971, **74**, 3891.
[66] A. E. Marinkowsky, *Canad. J. Chem.*, 1970, **48**, 2128.
[67] G. Jakli and W. A. Van Hook, *J. Chem. and Eng. Data*, 1972, **17**, 348.
[68] M. H. Lietzke and R. W. Stoughton, *J. Phys. Chem.*, 1962, **66**, 508.
[69] W. T. Lindsay and C. Liu, *J. Phys. Chem.*, 1971, **75**, 3723.
[70] C. Liu and W. T. Lindsay, *J. Phys. Chem.*, 1970, **74**, 341.
[71] W. T. Lindsay and T. S. Bulischeck, *Rev. Sci. Instr.*, 1970, **41**, 149.
[72] J. T. Moore, W. T. Humphries, and C. S. Patterson, *J. Chem. and Eng. Data*, 1972, **17**, 180.
[73] R. H. Wood, R. K. Wicker, and R. W. Kreis, *J. Phys. Chem.*, 1971, **75**, 2313.
[74] R. W. Kreis and R. H. Wood, *J. Phys. Chem.*, 1971, **75**, 2319.
[75] O. D. Bonner, *J. Chem. Thermodynamics*, 1971, **3**, 837.
[76] T. Mussini and A. Pagella, *J. Chem. and Eng. Data*, 1971, **16**, 49.
[77] T. Mussini, R. Galli, and E. Dubini-Paglia, *J. C. S. Faraday I*, 1972, **68**, 1322.
[78] A. L. Jones, H. G. Linge and I. R. Wilson, *Austral. J. Chem.*, 1971, **24**, 2005.
[79] K. P. Anderson, E. A. Butler, and E. M. Woolley, *J. Phys. Chem.*, 1971, **75**, 93.

Electrolyte Solutions 193

the solubility of silver chloride in methanol–water, acetone–water, and dioxan–water mixtures.

Enthalpies.—Vanderzee and Swanson,[80] using a straightforward approach, have determined the enthalpies of dilution of aqueous barium perchlorate solutions at 25 °C. An analogous study has been performed[81] on aqueous solutions of some sodium carboxylates. Enthalpies of dilution and of solution are described for lithium chlorate[82] and sodium chlorate[83] in water and dimethyl sulphoxide–water mixtures. An unusual approach has been used by Bailey and Larson[84] in which enthalpies of dilution of uranyl and vanadyl sulphate are treated to give the association constants for ion-pairing. Measurements on the enthalpies of solution (in water) for sodium benzene disulphonate,[85] some hydrated electrolytes[86] (also in water), and some 2:1 electrolytes[87] in formamide, N-methylformamide, and dimethylformamide are described.

An unusual approach to the heats of solution and dilution has been described by Bahe,[88] and Bockris and Saluja[89] have discussed the enthalpies and entropies of hydration from the viewpoint of several models. A study[90] of the partial molar heat capacities of sodium perchlorate in water and in methanol shows that, in both solvents, a marked temperature dependence is obtained.

Volumes.—The polemic between Conway and co-workers and Panckhurst on the use of various possible extrapolation procedures for the determination of ionic contributions to electrolyte partial molar volumes has continued.[91–93] A further paper by Yeager and Zana[94] on ionic vibration potentials is presented. Curthoys and Mathieson[95] have used an empirical equation of the form

$$V_i = Ar^3 - BE^2/r$$

to fit the available experimental data on partial molar ionic volumes. The quality of fit is reasonable and the resulting parameters may be of use when data are not available for a particular salt at a particular temperature. Apparent

---

[80] C. E. Vanderzee and J. A. Swanson, *J. Chem. and Eng. Data*, 1972, **17**, 488.
[81] S. Lindenbaum, *J. Chem. Thermodynamics*, 1971, **3**, 625.
[82] A. N. Campbell and O. N. Bhatnagar, *Canad. J. Chem.*, 1972, **50**, 1627.
[83] A. N. Campbell and O. N. Bhatnagar, *Canad. J. Chem.*, 1971, **49**, 217.
[84] A. R. Bailey and J. W. Larson, *J. Phys. Chem.*, 1971, **75**, 2368.
[85] C. F. Boudreau and C. A. Wulff, *J. Chem. Thermodynamics*, 1970, **2**, 125.
[86] H. Nakayama, *Bull. Chem. Soc. Japan*, 1971, **44**, 1709.
[87] A. Finch, P. J. Gardner, and C. J. Steadman, *J. Phys. Chem.*, 1971, **75**, 2325.
[88] L. W. Bahe, *J. Phys. Chem.*, 1972, **76**, 1608.
[89] J. O'M. Bockris and P. P. S. Saluja, *J. Phys. Chem.*, 1972, **76**, 2298.
[90] Mastroianni and C. M. Criss, *J. Chem. and Eng. Data*, 1972, **17**, 222.
[91] M. H. Panckhurst, *Rev. Pure Appl. Chem.*, 1969, **19**, 45.
[92] B. E. Conway, J. E. Desnoyers, and R. E. Verrall, *J. Phys. Chem.*, 1971, **75**, 3031.
[93] M. H. Panckhurst, *J. Phys. Chem.*, 1971, **75**, 3035.
[94] E. Yeager and R. Zana, *J. Phys. Chem.*, 1972, **76**, 1086.
[95] G. Curthoys and J. G. Mathieson, *Trans Faraday Soc.*, 1970, **66**, 43.

molar values of sodium chloride over a range of temperatures,[96,97] and for magnesium chloride[97] and some multiply charged electrolytes,[98] have been determined in water. A study has been made[99] of the apparent molar values of potassium chloride in formamide at 25 °C.

4 Thermodynamics of Mixed Electrolytes

The resurgence of interest in this aspect of the subject referred to[1] in the last Report in this series has continued. Scatchard, Rush, and Johnson[100] have treated the experimental results on systems containing Na^+, Mg^{2+}, Cl^-, and SO_4^{2-} ions using an 'ions as components' procedure. The quality of fit and predictive use of this approach is superior to earlier approaches.[101,102] The 'ions as components' procedure has also been applied[103] to some direct vapour-pressure measurements of synthetic sea-water in the temperature range 25—100 °C. Leyendekkers has shown[104,105] that there exists a linear correlation between the ionic entropy and the thermodynamic mixing functions and has used a volume fraction statistics approach to predict activity coefficients in mixtures of 1:1 electrolytes. The significance of Young's rules has been discussed by Wu,[106] and Reilly, Wood, and Robinson[107] have considered the prediction of osmotic and activity coefficients in mixed electrolyte solutions. A theoretical approach to excess mixing functions for ionic systems has been discussed by Friedman and Ramanathan[108] and applied to lithium chloride–caesium chloride mixtures. Other discussions on theoretical approaches to mixed electrolytes are given by Lakshmanan and Rangarajan[109] and Rosseinsky and Hill.[110] An important paper by Robinson *et al.*[111] treats short-range interactions between ionic species in mixed solutions in terms of equilibrium constants and longer-range interactions in terms of the Debye–Hückel expression and an expression for the higher-order limiting law (predicted by Friedman[112]) results. Some experimental data are given. The reason for the agreement between this[111] 'chemical' approach and the

[96] F. J. Millero, *J. Phys. Chem.*, 1970, **74**, 356.
[97] H. E. Wirth and F. K. Bangert, *J. Phys. Chem.*, 1972, **76**, 3488.
[98] A. Indelli and R. Zamboni, *J. C. S. Faraday I*, 1972, **68**, 1831.
[99] L. A. Dunn, *Trans Faraday Soc.*, 1971, **67**, 2525.
[100] G. Scatchard, R. M. Rush, and J. S. Johnson, *J. Phys. Chem.*, 1970, **74**, 3786.
[101] G. Scatchard, *J. Amer. Chem. Soc.*, 1968, **90**, 3124.
[102] G. Scatchard, *J. Amer. Chem. Soc.*, 1969, **91**, 2410.
[103] H. F. Gibbard and G. Scatchard, *J. Chem. and Eng. Data*, 1972, **17**, 498.
[104] J. V. Leyendekkers, *J. Phys. Chem.*, 1970, **74**, 2225.
[105] J. V. Leyendekkers, *J. Phys. Chem.*, 1971, **75**, 946.
[106] Y. C. Wu, *J. Phys. Chem.*, 1970, **74**, 3781.
[107] P. J. Reilly, R. H. Wood, and R. A. Robinson, *J. Phys. Chem.*, 1971, **75**, 1305.
[108] H. L. Friedman and P. S. Ramanathan, *J. Phys. Chem.*, 1970, **74**, 3756.
[109] S. Lakshmanan and S. K. Rangarajan, *J. Electroanalyt. Chem. Interfacial Electrochem.*, 1970, **27**, 170.
[110] D. R. Rosseinsky and R. J. Hill, *J. Electroanalyt. Chem. Interfacial Electrochem.*, 1971, **30**, App. 7.
[111] R. A. Robinson, R. H. Wood, and P. J. Reilly, *J. Chem. Thermodynamics*, 1971, **3**, 461.
[112] H. L. Friedman, 'Ionic Solution Theory', Interscience, New York, 1962.

Electrolyte Solutions

cluster-expansion approach of Friedman obviously lies in the similarity of the expressions for the equilibrium constant and the 'virial' coefficient for ion–ion interactions.

Isopiestic investigations on mixtures of electrolytes in water at 25 °C include studies on $HClO_4$–$UO_2(ClO_4)_2$ and $NaClO_4$–$UO_2(ClO_4)_2$,[113] $CaCl_2$–$Ca(ClO_4)_2$,[114] $NaBr$–$ZnBr_2$,[115] some alkali-metal chlorides and nitrates,[116] some alkali-metal halides with some tetra-alkylammonium halides,[117] $LiCl$–$BaCl_2$ and $CsCl$–$BaCl_2$,[118] and mixtures containing the ions Mg^{2+}, Ca^{2+}, Cl^-, and NO_3^-.[119] Measurements using the isopiestic technique are reported[120] for $LiCl$, $LiBr$, $LiNO_3$, $Ca(NO_3)_2$, and some binary mixtures of those salts between 100 and 150 °C. Childs and Platford[121] have investigated the systems $NaCl$–Na_2SO_4 and $NaCl$–$MgSO_4$ in water at 0 and 15 °C.

Measurements of activity coefficients in $CaCl_2$–$NaCl$ mixtures using a calcium ion-exchange membrane electrode have been described by Leyendekkers and Whitfield.[122] The results are considered in the light of the approaches discussed earlier.[104, 105] The activity coefficients of $CdCl_2$ in $CdCl_2$–$NaCl$ mixtures have been determined[123] using an amalgam electrode and the results discussed in terms of the stability constants of cadmium complexes. The activity coefficients of HCl at 25 °C in HCl–$MnCl_2$ mixtures have been obtained[124] using hydrogen and silver–silver chloride electrodes. A series of constant ionic strength mixtures was used and it was found that Harned's rule represents the system rather well. An application of Friedman's approach[112] to the excess Gibbs function of unsymmetrical mixtures would be an interesting exercise. An ion-pairing treatment has been used[125] for the systems $NaCl$–$NaHCO_3$ and $NaCl$–Na_2CO_3. The activity coefficients of HCl in various mixed electrolytes have been measured.[126, 127] Potassium-responsive glass electrodes have been used[128] to measure the activity coefficients of KCl in $MgCl_2$, $CaCl_2$, $BaCl_2$, $MgSO_4$, and K_2SO_4 at a total ionic strength of unity. There is good agreement with earlier[129, 130] isopiestic results.

[113] R. M. Rush and J. S. Johnson, *J. Chem. Thermodynamics*, 1971, **3**, 779.
[114] R. A. Robinson and C. K. Lin, *J. Chem. and Eng. Data*, 1971, **16**, 203.
[115] G. E. Boyd, S. Lindenbaum, and R. A. Robinson, *J. Phys. Chem.*, 1971, **75**, 3153.
[116] C. P. Bezborouah, A. K. Covington, and R. A. Robinson, *J. Chem. Thermodynamics*, 1970, **2**, 431.
[117] W.-Y. Wen, K. Mijajima, and A. Otsuka, *J. Phys. Chem.*, 1971, **75**, 2148.
[118] S. Lindenbaum, R. M. Rush, and R. A. Robinson, *J. Chem. Thermodynamics*, 1972, **4**, 381.
[119] R. F. Platford, *J. Chem. Thermodynamics*, 1971, **3**, 319.
[120] H. Braunstein and J. Braunstein, *J. Chem. Thermodynamics*, 1971, **3**, 419.
[121] C. W. Childs and R. F. Platford, *Austral. J. Chem.*, 1971, **24**, 2487.
[122] J. V. Leyendekkers and M. Whitfield, *J. Phys. Chem.*, 1971, **75**, 957.
[123] P. J. Reilly and R. H. Stokes, *Austral. J. Chem.*, 1970, **23**, 1397.
[124] C. J. Downes, *J. C. S. Faraday I*, 1972, **68**, 1964.
[125] J. N. Butler and R. Huston, *J. Phys. Chem.*, 1970, **74**, 2976.
[126] H. Ohtaki and G. Biedermann, *Bull. Chem. Soc. Japan*, 1971, **44**, 1515.
[127] C. J. Downes, *J. Phys. Chem.*, 1970, **74**, 2153.
[128] P. G. Christenson and J. M. Gieskes, *J. Chem. and Eng. Data*, 1971, **16**, 398.
[129] R. A. Robinson and A. K. Covington, *J. Res. Nat. Bur. Stand. Sect. A*, 1968, **72**, 239.
[130] R. A. Robinson and V. E. Bower, *J. Res. Nat. Bur. Stand., Sect. A*, 1965, **69**, 439.

An interesting and critical approach to ionic interactions in mixed electrolytes is presented by Prue and co-workers,[131] and Steigman and Monica[132] discuss interactions in solutions containing $CuSO_4$ and tetra-n-butylammonium ions. Lietzke and Danford[133] have made further measurements on the system $HCl–LaCl_3$, up to ionic strengths of 5 in the range 25—75 °C. This supplements the earlier work of Lietzke and Stoughton.[134] The solubility of gypsum in Na_2SO_4–NaCl, Li_2SO_4–$LiNO_3$, and LiCl–Li_2SO_4 solutions has been determined[135] and, not too surprisingly, difficulties were experienced in correlating the behaviour in the different systems.

The precise work of Wood (see previous Report[1] for references) has been extended to an investigation[136] of the heats of mixing of solutions containing the ions Mg^{2+}, Na^+, Cl^-, and Br^-. Anderson and co-workers have investigated the temperature dependence of the heats of mixing for a series of salts.[137, 138] The thermodynamics of tetra-n-butylammonium bromide in aqueous NaCl solutions have been discussed.[139] The volume change on mixing $MgCl_2$ and NaCl solutions have been discussed[140] by Wirth and Bangert.

5 Solutions containing Tetra-alkylammonium Ions

In view of the large number of papers describing the physico-chemical properties of tetra-alkylammonium and similar salts, it was decided that this category of species deserved a separate section.

Several discussions[141—143] on the effect of large ions with alkyl and aryl side-chains have been published and many discussions are given in the papers mentioned below. It is fairly apparent that, for systems such as these, we are still at the qualitative argument stage, but one can say that when the correct theory does appear there will be no shortage of experimental data with which to verify it.

In the Reporter's opinion the most refreshing and interesting paper to appear on alkylammonium ions is a novel approach by Blandamer and Waddington[144] in which a comparison is made between the thermodynamic properties of some solutions and their ultrasonic absorption characteristics. A study has also been made[145] of the ultrasonic relaxation of some tetra-

[131] J. E. Prue, A. J. Read, and G. Romero, *Trans. Faraday Soc.*, 1971, **67**, 420.
[132] J. Steigman and M. D. Monica, *J. Phys. Chem.*, 1970, **74**, 516.
[133] M. H. Lietzke and M. D. Danford, *J. Chem. and Eng. Data*, 1972, **17**, 459.
[134] M. H. Lietzke and R. W. Stoughton, *J. Phys. Chem.*, 1967, **71**, 662.
[135] L. B. Yeatts and W. L. Marshall, *J. Chem. and Eng. Data.*, 1972, **17**, 163.
[136] P. J. Reilly and R. H. Wood, *J. Phys. Chem.*, 1972, **76**, 3474.
[137] H. L. Anderson and L. A. Petree, *J. Phys. Chem.*, 1970, **74**, 1455.
[138] H. L. Anderson, R. D. Wilson, and D. E. Smith, *J. Phys. Chem.*, 1971, **75**, 1125.
[139] B. Chawla and J. C. Ahluwalia, *J. Phys. Chem.*, 1972, **76**, 2582.
[140] H. E. Wirth and F. K. Bangert, *J. Phys. Chem.*, 1972, **76**, 3491.
[141] H. Snell and J. Greyson, *J. Phys. Chem.*, 1970, **74**, 2148.
[142] S. Lindenbaum, *J. Phys. Chem.*, 1970, **74**, 3027.
[143] R. B. Hermann, *J. Phys. Chem.*, 1971, **75**, 363.
[144] M. J. Blandamer and D. Waddington, *J. Chem. Phys.*, 1970, **52**, 6247.
[145] G. S. Darbari and S. Petrucci, *J. Phys. Chem.*, 1970, **74**, 268.

Electrolyte Solutions

alkylammonium salts in acetone. Several n.m.r. studies have been described. The solvent (water) protons chemical shift have been investigated[146—150] at several temperatures for several alkylammonium ions. The study of Davies et al.[147] includes an investigation on sodium carboxylates. At room temperature, marked differences were obtained between the observed solvent chemical shifts and those obtained earlier.[151] It was found that at low temperatures the longer-chain alkylammonium salts gave rise to a downfield shift which was ascribed to a 'structure-making' effect of the cation. Increase in temperature reverses the sign of the shift. It was suggested that the solvent shifts of the carboxylate solution are principally controlled by the basicity of the carboxylate group, although a slight structural effect was evident. Non-protonic n.m.r. investigations include a study[152] by Forsén and co-workers using ^{79}Br as a probe for mono-, di-, and tri-alkylammonium ions in water, two papers by Larsen on ^{14}N relaxation[153] in solutions containing tetra-alkylammonium ions, and ^{75}As relaxation studies[154] on the corresponding arsonium ions. ^{15}N N.m.r. chemical shifts and coupling constants have been made[155] for methylamine hydrochlorides in water.

The near-i.r. spectra (see ref. 156 for a far-i.r. investigation) of water in solutions containing tetra-alkyl- and -arylammonium, -phosphonium, and -arsonium ions and tetrafluoroborate anion show[157] that the Bu_4N^+ ion changes the water spectrum in a similar way to a decrease in temperature, whereas for some of the other ions the changes are apparently not related to a simple modification of the solvent structure. Stable nitroxide spin labels have been used[158, 159] as a possible probe into hydrophobic bonding.

The viscosity behaviour of electrolyte solutions has been extensively used in the past to give information on the nature of solvent adjacent to ions. The earlier measurements[160] on dilute tetra-alkylammonium salts have been extended by Eagland and Pilling[161] to high concentrations. (See also Breslau and Miller[162]). A semi-empirical treatment using equations which are usually applied to polymer solutions enables extrapolations to be made from quite high concentrations (up to 4 mol l^{-1}) to infinite dilution. The viscosities (and

---

[146] J. S. Davies, S. Ormonroyd, and M. C. R. Symons, *Chem. Comm.*, 1970, 1426.
[147] J. S. Davies, S. Ormonroyd, and M. C. R. Symons, *J. C. S. Faraday II*, 1972, **68**, 686.
[148] J. F. Coetzee and W. R. Sharpe, *J. Phys. Chem.*, 1971, **75**, 3141.
[149] M.-M. Marciaq-Rousselot, A. de Trobriand, and M. Lucas, *J. Phys. Chem.*, 1972, **76**, 1455.
[150] A. LoSurdo and H. E. Wirth, *J. Phys. Chem.*, 1972, **76**, 130.
[151] H. G. Hertz and W. Spalthoff, *Z. Electrochem.*, 1963, **63**, 1096.
[152] B. Lindman, H. Wennerström, and S. Forsén, *J. Phys. Chem.*, 1970, **74**, 754.
[153] D. W. Larsen, *J. Phys. Chem.*, 1970, **74**, 3380.
[154] D. W. Larsen, *J. Phys. Chem.*, 1971, **75**, 3880.
[155] M. Alei, A. E. Florin, and W. M. Litchman, *J. Phys. Chem.*, 1971, **75**, 1758.
[156] J. R. Kludt, G. Y. W. Kwong, and R. L. McDonald, *J. Phys. Chem.*, 1972, **76**, 339.
[157] C. Jolicoeur, N. D. Thé, and A. Cabana, *Canad. J. Chem.*, 1971, **49**, 2008.
[158] C. Jolicoeur and H. L. Friedman, *J. Phys. Chem.*, 1971, **75**, 165.
[159] C. Jolicoeur and H. L. Friedman, *Ber. Bunsengesellschaft phys. Chem.*, 1971, **75**, 248.
[160] R. L. Kay, T. Vituccio, C. Zawoyski, and D. F. Evans, *J. Phys. Chem.*, 1966, **70**, 2336.
[161] D. Eagland and G. Pilling, *J. Phys. Chem.*, 1972, **76**, 1902.
[162] B. R. Brezlau and I. F. Miller, *J. Phys. Chem.*, 1970, **74**, 1056.

densities) of some unsymmetrical quaternary ammonium halides have been measured[163] in dilute solutions. Viscosity B coefficients are reported for tetra-alkylammonium salts in N-methylformamide[164] and carbon tetrachloride.[165] Lowe and Rendall have continued their studies on unsymmetrical quaternary ammonium ions by investigating the conductance[166] behaviour of a large number of unsymmetrical quaternary ammonium iodies. A conductance study has also been carried out[167] in hydrogen-bonding solvents. The self-diffusion coefficients of a large number of electrolytes, including some large substituted ions, have been reported.[168] A considerable number of measurements are reported on the partial molar heat capacities of tetra-alkylammonium and similar salts. Mastroianni and Criss[169] have shown that for Me_4NBr and Bu^n_4NBr the heat-capacity change for the transition from solid salt to infinitely dilute solution shows significant variations with temperature. The effects in methanol and water are rather different for the methylammonium salt. The temperature variation observed in water is at variance with that obtained by other[170] workers. It is suggested that the effects may be qualitatively interpreted by assuming that the Me_4N^+ ion disrupts water structure and the $Bu^n_4N^+$ ion promotes water structure, but to an extent which diminishes as the temperature is raised. A series of investigations into the temperature variation of the partial molar heat capacities of salts with large organic ions have been performed by Ahlawalia and co-workers.[171—177] The data have been obtained using an integral heat of solution procedure. The behaviour observed is certainly very complex (*e.g.* in the temperature range 10—80 °C the change in the partial molar heat capacity of $NaBPh_4$ shows maxima at about 40 and 70 °C and a minimum at about 50 °C). The authors infer[173] that the interpretation of effects such as these is difficult.

Further measurements on the apparent molar volume of tetra-alkylammonium salts have been reported. Perron and Desnoyers,[178] in an extensive study of aqueous solutions at 25 °C, found good agreement with earlier work (see ref. 1 for earlier studies). The temperature dependence of the apparent and partial molar values of some tetra-alkylammonium salts have been des-

[163] B. M. Lowe and H. M. Rendall, *Trans. Faraday Soc.*, 1971, **67**, 2318.
[164] P. P. Rastogi, *Bull. Chem. Soc. Japan*, 1970, **43**, 2442.
[165] K. F. Denning and J. A. Plambeck, *Canad. J. Chem.*, 1972, **50**, 1600.
[166] B. M. Lowe and H. M. Rendall, *J. C. S. Faraday I*, 1972, **68**, 2191.
[167] M. Godffredi and R. Triolo, *J. C. S. Faraday I*, 1972, **68**, 2324.
[168] A. Weiss and K. H. Nothnagel, *Ber. Bunsengesellschaft phys. Chem.*, 1971, **75**, 216.
[169] M. J. Mastroianni and C. M. Criss, *J. Chem. Thermodynamics*, 1972, **4**, 321.
[170] E. M. Arnett and J. J. Campion, *J. Amer. Chem. Soc.*, 1970, **92**, 7097.
[171] T. S. Sarma, R. K. Mohanty, and J. C. Ahluwalia, *Trans. Faraday Soc.*, 1969, **65**, 2333.
[172] T. S. Sarma and J. C. Ahluwalia, *J. Phys. Chem.*, 1970, **74**, 3547.
[173] R. K. Mohanty, S. Sunder, and J. C. Ahluwalia, *J. Phys. Chem.*, 1972, **76**, 2577.
[174] S. Subramanian and J. C. Ahluwalia, *J. Phys. Chem.*, 1968, **72**, 2525.
[175] R. K. Mohanty, T. S. Sarma, and J. C. Ahluwalia, *Trans. Faraday Soc.*, 1971, **67**, 305.
[176] R. K. Mohanty and J. C. Ahluwalia, *J. Chem. Thermodynamics*, 1972, **4**, 53.
[177] T. S. Sarma and J. C. Ahluwalia, *Trans. Faraday Soc.*, 1971, **67**, 2528.
[178] G. Perron and J. E. Desnoyers, *J. Chem. and Eng. Data*, 1972, **17**, 136.

cribed.[179] A report[180] on the partial molar volumes of some salts and a comparison with the scaled particle theory for salting out coefficients has been made. Density measurements[181] on tetraphenylarsonium solutions at 0, 25, and 50 °C have been used by Millero[182] in a discussion of ionic contributions to electrolyte partial molar volumes. An extensive investigation by Desnoyers and co-workers[183] on salts with large cations has been carried out. Heat capacities and viscosities were measured, as were density measurements in methanol and water at 25 °C. The results obtained were compared with an earlier [184] spectroscopic study. Conway and Laliberté[185] have continued their earlier[186—188] studies on the partial molar volumes of ions by making measurements of some alkali-metal halides and tetra-alkylammonium ions in D_2O. Measurements are reported at temperatures other than 25 °C. Partial molar ionic volumes were estimated using the same procedure as before[187] (see also refs. 91—93). The small but experimentally significant differences between the ionic partial molar volumes in H_2O and D_2O are discussed from an electrostatic and structural viewpoint. Studies of the partial molar volumes of some tetra-alkylammonium chlorides in aqueous acetone and aqueous dimethyl sulphoxide are described by MacDonald and Hyne.[189] A discussion[190] of the isothermal apparent molar compressibilities (calculated from other experimental thermodynamic properties) for tetra-alkylammonium ions has been given. Conway has extended his interest in the volumetric properties of solutions with a study of polyimine salts.[191]

Lindenbaum and co-workers have investigated the osmotic coefficients of aqueous tetra-alkylammonium halides[192] as a function of temperature and the apparent molal enthalpies and osmotic coefficients of tetrabutylammonium carboxylates[193] at 25 °C.

6 Interactions between Electrolytes and Non-electrolytes

Somewhat similar problems arise in the interpretation of the thermodynamic behaviour of mixed electrolyte solutions, where the evidence suggests that the excess functions are determined by short-range interactions between ions, and

---

[179] A LoSurdo and H. E. Wirth, J. Phys. Chem., 1972, 76, 1333.
[180] M. Lucas and A. de Trobriand, J. Phys. Chem., 1971, 75, 1803.
[181] F. J. Millero, J. Chem. and Eng. Data., 1971, 16, 229.
[182] F. J. Millero, J. Phys. Chem., 1971, 75, 280.
[183] C. Jolicoeur, P. R. Philip, G. Perron, R. A. Leduc and J. E. Desnoyers, Canad. J. Chem., 1972, 50, 3167.
[184] C. Jolicoeur, N. D. Thé, and A. Cabana, Canad. J. Chem., 1971, 49, 2008.
[185] B. E. Conway and L. H. Laliberté, Trans. Faraday Soc., 1970, 66, 3032.
[186] L. H. Laliberté and B. E. Conway, J. Phys. Chem., 1970, 74, 4116.
[187] B. E. Conway, R. E. Verrall, and J. E. Desnoyers, Trans. Faraday Soc., 1966, 62, 2738.
[188] B. E. Conway and L. H. Laliberté, J. Phys. Chem., 1968, 72, 4317.
[189] D. D. MacDonald and J. B. Hyne, Canad. J. Chem., 1970, 48, 2416.
[190] J. E. Desnoyers and P. R. Philip, Canad. J. Chem., 1972, 50, 1094.
[191] J. Lawrence and B. E. Conway, J. Phys. Chem., 1971, 75, 2353, 2362.
[192] S. Lindenbaum, L. Leifer, G. E. Boyd, and J. W. Chase, J. Phys. Chem., 1970, 74, 761.
[193] S. Lindenbaum, J. Phys. Chem., 1971, 75, 3733.

the behaviour of non-electrolytes in ionic solutions in which the thermodynamics presumably are determined by short-range non-electrolyte–ion interactions. There are several ways of formalizing the thermodynamics of non-electrolyte–salt–solvent systems and in the past the most used method has been through the Setschenow coefficient. Masterton and Lee[194] have discussed the use of scaled particle theory (see original paper for references to this theory) in the calculation of Setschenow coefficients and have attempted a calculation[195] of ionic radii from the experimentally determined coefficients. Application of the scaled particle theory was also used in the thorough study by Wilcox and Schrier[196] on the activity coefficient of NaCl in dilute alcoholic solvent systems. A combination of sodium-responsive glass electrodes and conventional Ag|AgCl electrodes were used in solutions containing methanol, ethanol, propanol, and ethylene glycol. A similar study has been carried out[197] in alkali-metal halide–acetone–water systems. The use of an approach such as the scaled particle theory is interesting since it is a 'non-structural' approach to aqueous systems, which at the present time is unusual.

Wen and Hung[198] have made a comprehensive study of the thermodynamics of hydrocarbon gases in tetra-alkylammonium and other salt solutions. Measurements were made over a range of temperatures and the corresponding enthalpy and entropy terms were evaluated. A striking feature of the results is that for all of the systems studied at all temperatures investigated, a linear correlation exists[199] between the enthalpy and entropy. This is an example of the compensation behaviour between enthalpy and entropy terms which is so common in aqueous systems. A study is described[200] of the solubility and partial molar volumes of nitrogen and methane in water and aqueous sodium chloride solutions from 50 to 125 °C and from 100 to 600 atm. Tiepel and Gubbins[201] have investigated the partial molar volumes of gases dissolved in electrolytes.

Stern[202] has continued his investigations of mixtures of electrolytes and non-electrolytes with a paper on the thermodynamic behaviour of nitromethane in water and in 1 molal KCl solutions. The activity coefficients of p-nitroanilinium chloride and bromide have been measured[203] in aqueous concentrated salt solutions. The solubility behaviour of thiocyanic acid in some electrolytic solutions has been discussed.[204] Uedaira and Uedaira have

---

[194] W. L. Masterton and T. P. Lee, *J. Phys. Chem.*, 1970, **74**, 1766 (see also S. K. Shoor and K. E. Gubbins, *J. Phys. Chem.*, 1969, **73**, 498).
[195] W. L. Masterton, D. Bolocofsky, and T. P. Lee, *J. Phys. Chem.*, 1971, **75**, 2809.
[196] F. Wilcox and E. E. Schrier, *J. Phys. Chem.*, 1971, **75**, 3757.
[197] M. Y. Spink and E. E. Schrier, *J. Chem. Thermodynamics*, 1970, **2**, 821.
[198] W.-Y. Wen and J. H. Hung, *J. Phys. Chem.*, 1970, **74**, 170.
[199] T. H. Lilley, personal communication to W.-Y. Wen, 1970.
[200] T. D. O'Sullivan and N. O. Smith, *J. Phys. Chem.*, 1970, **74**, 1460.
[201] E. W. Tiepel and K. E. Gubbins, *J. Phys. Chem.*, 1972, **76**, 3044.
[202] J. H. Stern and J. T. Swearingen, *J. Phys. Chem.*, 1970, **74**, 167.
[203] M. Lucas and J. Steigman, *J. Phys. Chem.*, 1970, **74**, 2699.
[204] K. S. de Haas, *J. Chem. and Eng. Data*, 1971, **16**, 457.

described[205] isopiestic measurements on the systems sodium benzenesulphonate–xylose and sodium benzenesulphonate–urea in water and analogous systems[206] with the toluenesulphonate salt replacing the benzenesulphonate. Isoiestic studies have also been made on aqueous systems of sucrose–NaCl[207] and β-alanine–n-propylammonium bromide.[208] Measurements on the enthalpy changes which occur when solutes are transferred from one medium to another are described by Stern and Bottenberg[209] (the transfer of NaCl from water to aqueous hydrogen peroxide solutions), by Chawla et al. (the transfer of NaBPh$_4$ from water to aqueous urea)[210], Sarma and Ahluwalia[211] and Cassel and Wen[212] (the transfer of tetra-alkylammonium salts from water to aqueous urea solutions), and by Lucas and Feillolay[213] (the transfer of CHCl$_3$, butanol, and isopropanol from water to some tetra-alkylammonium salt solutions). The same workers have measured[214] the solubility of helium and methane in tetra-n-butylammonium bromide solutions at 25 and 35 °C.

Solvation.—Although the basic problems of long range ion–ion interactions and their effect on thermodynamic properties are not yet solved, sufficient progress has been made (see the very readable discussion by Friedman[2]) to enable many electrolyte solution chemists to turn their attention from ion–ion interactions to ion–solvent interactions. The approaches to problems such as these are many and varied, but a noticeable and welcome change of emphasis from simple electrostatic (Born-type) approaches is evident in the recent literature. A much-tilled area of research in the past few years has been the investigation of systems of the type salt–non-electrolyte A–non-electrolyte B. (In most of the studies one of the non-electrolytes has been water). In certain cases complete miscibility of the two non-electrolytes occurs and many measurements have been reported on thermodynamic properties of salts in such systems. Most interest has centred on processes in which the salt at infinite dilution in one mixture (often a pure solvent) is transferred (remaining at infinite dilution) to another mixture (often the other pure solvent). Changes in the thermodynamic properties for processes such as these represent modifications in ion–solvent (and solvent–ion) interactions. It is usually assumed either implicitly or explicitly that what might be called 'internal' changes in the ions can be neglected and the observed changes reflect only changes in

[205] H. Uedaira and H. Uedaira, *J. Phys. Chem.*, 1970, **74**, 1931.
[206] H. Uedaira, *J. Chem. Eng. Data*, 1972, **17**, 241.
[207] R. A. Robinson, R. H. Stokes, and K. N. Marsh, *J. Chem. Thermodynamics*, 1970, **2**, 745.
[208] C. N. Pepela and P. J. Dunlop, *J. Chem. Thermodynamics*, 1972, **4**, 115.
[209] J. H. Stern and W. A. Bottenberg, *J. Phys. Chem.*, 1971, **75**, 2229.
[210] B. Chawla, S. Subramanian, and J. C. Ahluwalia, *J. Chem. Thermodynamics*, 1972, **4**, 575.
[211] T. S. Sarma and J. C. Ahluwalia, *J. Phys. Chem.*, 1972, **76**, 1366.
[212] R. B. Cassel and W.-Y. Wen, *J. Phys. Chem.*, 1972, **76**, 1369.
[213] M. Lucas and A. Feillolay, *J. Phys. Chem.*, 1971, **75**, 2330.
[214] A. Feillolay and M. Lucas, *J. Phys. Chem.*, 1972, **76**, 3068.

the solvent adjacent to the ions. Often it is found that one of the non-electrolytes is only sparingly soluble in the other non-electrolyte and it is then usual to investigate the thermodynamics of the process:

(Non-electrolyte A) ⟶ (Non-electrolyte A)
in pure solvent B in solution containing salt + solvent B.

Systems of this type in which one of the non-electrolytes (solvent B) is water have, of course, been extensively studied over many years and others are discussed below. As may be inferred from the above discussion, most of the studies in this area have been of a thermodynamic nature but several investigations have been performed (and are at present being carried out) using spectroscopic techniques. A great advance was made in this area by Fratiello (see ref. 1 for a discussion of this work and also references given later) when he observed separate nuclear magnetic resonance signals for bound and unbound solvent adjacent to ions. The availability of spectrometers which enable ionic nuclei to be studied has great potential and recent work[215, 216] indicates how thermodynamic and spectroscopic data can be combined in a self-consistent manner. Other approaches are discussed below but this Reporter confidently predicts that the further application of spectroscopic methods to ion–solvent interactions will be an area of great fruitfulness in the future.

A series of quantum mechanical calculations which use the CNDO approximation have been described on ions with associated water molecules.[217-220] The nature of the approximations involved necessarily means that comparisons with thermodynamic parameters in liquid water at ambient temperatures must be treated with considerable circumspection. Other studies in a theoretical vein include those of dePaz, Ehrenson, and Friedman[221] and Clementi and Popkie.[222] The former workers[221] show that chain structures of the hydrates of H^+ and OH^- appear to be favoured, which is consistent with the accepted mechanism of proton transfer in water. The energy surface of a water molecule adjacent to a lithium ion forms the subject of the most comprehensive and detailed calculations[222] which have to the present been performed.

Kebarle and co-workers have continued their studies on hydration of ions in the gas phase using mass spectroscopic techniques. Recent work includes the hydration of alkali-metal ions,[223] negative ions,[224] and OH^- and O_2^-.[225]

[215] A. K. Covington, T. H. Lilley, K. E. Newman, and G. A. Porthouse, *J. C. S. Faraday I*, 1973, **69**, 963.
[216] A. K. Covington, T. H. Lilley, and K. E. Newman, submitted to *J. C. S. Faraday I*, 1973, **69**, 973.
[217] R. E. Burton and J. E. Daly, *Trans. Faraday Soc.*, 1970, **66**, 1281.
[218] J. Daly and R. E. Burton, *Trans. Faraday Soc.*, 1970, **66**, 2408.
[219] R. E. Burton and J. E. Daly, *Trans. Faraday Soc.*, 1971, **67**, 1219.
[220] D. W. Clack and M. S. Farrimond, *J. Chem. Soc. (A)*, 1971, 299.
[221] M. de Paz, S. Ehrenson, and L. Friedman, *J. Chem. Phys.*, 1970, **52**, 3362.
[222] E. Clementi and H. Popkie, *J. Chem. Phys.*, 1972, **57**, 1077.
[223] I. Dzidic and P. Kebarle, *J. Phys. Chem.*, 1970, **74**, 1466.
[224] M. Arshadi, R. Yamadagni, and P. Kebarle, *J. Phys. Chem.*, 1970, **74**, 1475.
[225] M. Arshadi and P. Kebarle, *J. Phys. Chem.*, 1970, **74**, 1483.

Electrolyte Solutions

The relation and importance of work of this type to solvation problems in solution is obvious and one would imagine that the results obtained would find considerable use as the best available basis for comparison with theoretical studies such as those discussed earlier.

The enthalpies of solution and where possible the enthalpies of solvation of ions in several solvents have been determined. In an extensive series of investigations, Krishnan and Friedman have obtained results on various ions in H_2O and D_2O,[226] alkylammonium ions in H_2O, D_2O, propylene carbonate, and dimethyl sulphoxide,[227] a large number of solutes (including some salts in CH_3OH and CH_3OD),[228] and dimethylformamide.[229] Complementary studies on hydrocarbons and n-alcohols in some polar solvents have also been carried out.[230] The results have been discussed in terms of the changes in the solvation shells (co-spheres) on transferring ions from one solvent medium to another. The enthalpies changes for the transfer of several salts from water to D_2O have been measured.[231] Good agreement was obtained compared with earlier[226, 232] studies. The role of aqueous solution structure to transfer thermodynamics from H_2O to D_2O has been discussed.[233] de Visser, Weeda, and Somsen have investigated the enthalpies of solution of some tetra-alkylammonium bromides in formamide[234, 235] and N-methylformamide and N-methylacetamide.[236] There is a marked difference for the three solvents and it is suggested that the enthalpies of solvations of the tetra-alkylammonium ions in water, methanol, and, perhaps, formamide include an effect which could be called 'solvophobic'. This effect is not present with N-methylacetamide and N-methylformamide as solvents. An electrostatic approach has been presented for the solvation of ions in water and propylene carbonate by Salomon.[237]

The standard electrode potentials of the Ag|AgCl electrode in various mixed solvents and at various temperatures have been determined. The solvents used were t-butanol–water,[238, 239] isopropanol–water,[240—242] and

[226] C. V. Krishnan and H. L. Friedman, *J. Phys. Chem.*, 1970, **74**, 2356.
[227] C. V. Krishnan and H. L. Friedman, *J. Phys. Chem.*, 1970, **74**, 3900.
[228] C. V. Krishnan and H. L. Friedman, *J. Phys. Chem.*, 1971, **75**, 388.
[229] C. V. Krishnan and H. L. Friedman, *J. Phys. Chem.*, 1971, **75**, 3606.
[230] C. V. Krishnan and H. L. Friedman, *J. Phys. Chem.*, 1971, **75**, 3598.
[231] J. Greyson and H. Snell, *J. Chem. and Eng. Data*, 1971, **16**, 73.
[232] E. M. Arnett and D. R. McKelvey, in 'Solute-solvent Interactions' ed. J. F. Coetzee and C. D. Ritchie, Marcel Dekker, New York, 1969, p. 371.
[233] D. B. Dahlberg, *J. Phys. Chem.*, 1972, **76**, 2045.
[234] C. de Visser and G. Somsen, *Rec. Trav. chim.*, 1971, **90**, 1129.
[235] G. Somsen and L. Weeda, *Rec. Trav. chim.*, 1971, **90**, 81.
[236] C. de Visser and G. Somsen, *J. Chem. Thermodynamics*, 1972, **4**, 313.
[237] M. Salomon, *J. Phys. Chem.*, 1970, **74**, 2519.
[238] R. N. Roy, W. Vernon, A. Bothwell and J. Gibbons, *J. Chem. and Eng. Data*, 1972, **17**, 79.
[239] R. N. Roy, W. Vernon, and A. L. M. Bothwell, *J. Chem. Soc. (A)*, 1971, 1242.
[240] R. N. Roy, W. Vernon, and A. L. M. Bothwell, *J. Chem. Thermodynamics*, 1971, **3**, 769.
[241] R. N. Roy and A. Bothwell, *J. Chem. and Eng. Data*, 1970, **15**, 548.
[242] R. N. Roy, W. Vernon, and A. L. M. Bothwell, *J. Chem. Thermodynamics*, 1971, **3**, 6.

glycerol–water,[243, 244] and methanol–water[245] has been used for an analogous study for the Ag|AgBr electrode.[246, 247] The standard electrode potentials of the Ag|AgBr electrode in propylene glycol and the Ag|AgBr electrode in propylene glycol and the Ag|AgI electrode in ethylene and propylene glycols have been determined[248] as a function of temperature, and the thermodynamics of HCl in n-propanol[249] and isopropanol[250, 251] have been investigated over a temperature range. Widmer describes measurements[252] on HCl in the water–methyl isobutyl ketone system. The usefulness of studies such as these is well illustrated by the recent work[253] of Feakins and Voice on the free energy of transfer of alkali-metal chlorides from water to methanol–water mixtures. Similar studies have also been described.[254—256] Discussions on the primary medium effect have been given by Hall[257] and Kolthoff and Chantooni[258] and Coetzee and Sharp[259] have discussed the use of large cations and anions of the same size to disentangle separate ion contributions to thermodynamic properties.

The n.m.r. technique developed by Fratiello (see ref. 1 for earlier references) for the determination of solvation numbers has been exploited in a very elegant manner by Green and Sheppard.[260] Measurements were carried out on magnesium perchlorate in acetone–water mixtures at temperatures below 200 K. Separate ^{1}H n.m.r. peaks were found for six hydrated forms, the monohydrate at lowest field and the hexahydrate at highest field. The results indicate that, at low concentrations of water, one perchlorate ion enters the solvation sphere of Mg^{2+}. Fratiello and co-workers have used the same solvent system to investigate solvation of uranyl salts[261] and tin salts.[262] In the latter investigation, ^{110}Sn resonance measurements were performed to complement the ^{1}H measurements. Haas and Navon[263] have investigated europium(III) in water–acetonitrile mixtures, and studies of $AlCl_3$ and

[243] K. H. Khoo, *J. C. S. Faraday I*, 1972, **68**, 554.
[244] R. N. Roy, W. Vernon, and A. L. M. Bothwell, *J. Electrochem. Soc.*, 1971, **118**, 1302.
[245] K. H. Khoo, *J. Chem. and Eng. Data*, 1972, **17**, 82.
[246] K. H. Khoo, *J. Chem. Soc. (A)*, 1971, 1177.
[247] D. Feakins and K. H. Khoo, *J. Chem. Soc. (A)*, 1970, 361.
[248] K. K. Kundu, D. Jana, and M. N. Das, *J. Phys. Chem.*, 1970, **74**, 2625.
[249] R. N. Roy, A. Bothwell, J. Gibbons, and W. Vernon, *J. C. S. Dalton*, 1972, 530.
[250] R. N. Roy, W. Vernon, J. J. Gibbons, and A. L. M. Bothwell, *J. Chem. Thermodynamics*, 1971, **3**, 883.
[251] R. N. Roy, W. Vernon, and A. L. M. Bothwell, *J. C. S. Faraday I*, 1972, **68**, 2047.
[252] H. M. Widmer, *J. Phys. Chem.*, 1970, **74**, 3251.
[253] D. Feakins and P. J. Voice, *J. C. S. Faraday I*, 1972, **68**, 1390.
[254] D. Feakins and A. R. Willmott, *J. Chem. Soc. (A)*, 1970, 1321.
[255] D. Feakins, K. G. Lawrence, P. J. Voice, and A. R. Willmott, *J. Chem. Soc. (A)*, 1970, 837.
[256] K. H. Khoo, *J. Chem. Soc. (A)*, 1971, 2932.
[257] D. G. Hall, *J. C. S. Faraday II*, 1972, **68**, 25.
[258] I. M. Kolthoff and M. K. Chantooni, *J. Amer. Chem. Soc.*, 1971, **93**, 7104.
[259] J. F. Coetzee and W. R. Sharpe, *J. Phys. Chem.*, 1971, **75**, 3141.
[260] R. D. Green and N. Sheppard, *J. C. S. Faraday II*, 1972, **68**, 821.
[261] A. Fratiello, V. Kubo, R. E. Lee, and R. E. Schuster, *J. Phys. Chem.*, 1970, **74**, 3726.
[262] A. Fratiello, S. Peak, R. E. Schuster, and D. D. Davis, *J. Phys. Chem.*, 1970, **74**, 3730.
[263] Y. Haas and G. Navon, *J. Phys. Chem.*, 1972, **76**, 1449.

Electrolyte Solutions 205

$Al(ClO_4)_3$ in acetonitrile[264] have also been described. Popov and co-workers have applied spectroscopic techniques to the study of ion solvation in pyridine[265] and other[266, 267] non-aqueous solvents and mixtures of non-aqueous solvents. Infrared studies of the solvation of Na^+ in tetrahydrofuran (THF) and THF–cyclohexane mixtures have been performed by Olander and Day.[268] The solvation of Li^+ in dimethylformamide[269] is the subject of a paper by Lassigne and Baine. The link between thermodynamic data and n.m.r. spectroscopic parameters is the subject of a discussion by Frankel and co-workers[270] (see also refs. 215 and 216). Preferential solvation in binary solvent systems is discussed by Arnett et al.[271] and by Grunwald and co-workers.[272] These latter workers investigated the ^1H n.m.r. shifts of water at low concentrations in propylene carbonate salt solutions and estimated equilibrium constants for the association of water molecules with ions.

7 Protolytic Equilibria

The study of Brönsted acids and bases in solution continues unabated. Three investigations[273—275] describe the thermodynamics of the auto-ionization of water, and the ionic product of water in some aqueous–organic mixtures has been determined.[276] Two very elegant papers from the Grunwald school discuss the temperature dependence of the standard heat-capacity change for the self-ionization of water[277] and methanol.[278] Corresponding data are also given for the benzoic acid ionization in both solvents and for acetic acid in water. Glass-electrode measurements on the lyonium ion concentration in H_2O–D_2O mixtures are reported.[279]

The thermodynamics of auto-ionization in methanol–water mixtures[280] and of ethylene and propylene glycols[281] have been discussed. Irish and co-workers, with customary thoroughness, have re-investigated the sulphuric acid–water system[282] using the Raman technique. The constitution, equili-

[264] J. F. O'Brien and M. Alei, *J. Phys. Chem.*, 1970, **74**, 743.
[265] W. J. McKinney and A. I. Popov, *J. Phys. Chem.*, 1970, **74**, 535.
[266] R. H. Erlich and A. I. Popov, *J. Amer. Chem. Soc.*, 1971, **93**, 5620.
[267] M. K. Wong, W. J. McKinney, and A. I. Popov, *J. Phys. Chem.*, 1971, **75**, 56.
[268] J. A. Olander and M. C. Day, *J. Amer. Chem. Soc.*, 1971, **93**, 3384.
[269] C. Lassigne and P. Baine, *J. Phys. Chem.*, 1971, **75**, 3188.
[270] L. S. Frankel, C. H. Langford, and T. R. Stengle, *J. Phys. Chem.*, 1970, **74**, 1736.
[271] E. M. Arnett, II. C. Ko, and R. J. Minasz, *J. Phys. Chem.*, 1972, **76**, 2474.
[272] D. R. Cogley, J. N. Butler, and E. Grunwald, *J. Phys. Chem.*, 1971, **75**, 1477.
[273] R. E. Mesmer, C. F. Baes, and F. H. Sweeton, *J. Phys. Chem.*, 1970, **74**, 1937.
[274] J. W. Larson, *J. Phys. Chem.*, 1970, **74**, 685.
[275] G. J. Bignold, A. D. Brewer, and B. Hearn, *Trans. Faraday Soc.*, 1971, **67**, 2419.
[276] E. M. Woolley, D. G. Hurkot, and L. G. Hepler, *J. Phys. Chem.*, 1970, **74**, 3908.
[277] C. S. Leung and E. Grunwald, *J. Phys. Chem.*, 1970, **74**, 687.
[278] C. S. Leung and E. Grunwald, *J. Phys. Chem.*, 1970, **74**, 696.
[279] H. Kakihana, M. Maeda, and T. Amaya, *Bull. Chem. Soc. Japan*, 1970, **43**, 1377.
[280] G. H. Parsons and C. H. Rochester, *J. C. S. Faraday I*, 1972, **68**, 523.
[281] K. K. Kundu, P. K. Chattopadhay, D. Jana, and M. N. Das, *J. Phys. Chem.*, 1970, **74**, 2633.
[282] H. Chen and D. E. Irish, *J. Phys. Chem.*, 1971, **75**, 2672 (for a discussion of earlier work see T. F. Young, L. F. Maranville and H. M. Smith, in 'Structure of Electrolytic Solutions', ed. W. J. Hamer, Wiley, New York, 1959, p. 35).

bria, and kinetics of proton transfer in aqueous H_2SO_4, $NaOH-H_2SO_4-H_2O$, and $HCl-H_2SO_4-H_2O$ mixtures are discussed. This work is a corollary to other work on the $HSO_4^--H_2O$ system.[283—285] Turner[286] has carried out a similar but less extensive study on HSO_4^- in water and interprets his results in terms of two hydrated forms of the HSO_4^- ion.

The n.m.r. and Raman techniques have been used in an investigation[287] into the ionization and dissociation of trifluoroacetic and trichloroacetic acids. Comparison of the results with those obtained from other methods of measurement leads the authors to suggest that a distinction should be made between ionization and dissociation. The Raman technique has also been used in another[288] investigation into trichloroacetic acid. The experimental results agree quite well with the work above,[287] but the interpretation of the results is conventional. The n.m.r. technique has been used to investigate equilibria of $HClO_4$, fluorosulphuric acid, disulphuric acid, and $HSbF_6$ in sulpholane.[289]

The thermodynamics of the ionization of phenols in methanol have been discussed.[290] The enthalpy of ionization of the ammonium ion in water is the subject of two papers.[291, 292] The latter paper also discusses the enthalpies of formation of mono- and di-ammonium succinates. Studies of ionization of acids and bases using more conventional methods include several anilinium ions in aqueous dimethyl sulphoxide,[293] some aliphatic diamines in methanol–water mixtures,[294] protonated 2,2-bis(hydroxymethyl)-2,2′,2″-nitrilotriethanol in water from 0 to 50 °C,[295] D_3PO_4 in D_2O from 5 to 50 °C,[296] tris(hydroxymethyl)acetic acid and some related acids[297] in water and aqueous methanol as a function of temperature, some amides in water,[298] the imadozolium ion, 6-uruciicarboxylic acid (pK_1 and pK_2), and hypoxanthine (pK_1, pK_2, and pK_3) at 25 °C in water[299] (the standard enthalpy changes were also determined calorimetrically), some chloroanilinium ions[300] as a function of temperature, some naphthyl azoxy-acids[301] in water, squaric (1,2-dihydroxycyclobutenedione) acid in water,[302] m- and p-hydroxybenzo-

---

[283] D. E. Irish and H. Chen, *J. Phys. Chem.*, 1970, **74**, 3796.
[284] H. Chen and D. E. Irish, *J. Phys. Chem.*, 1971, **75**, 2681.
[285] D. E. Irish and R. C. Meatherall, *J. Phys. Chem.*, 1971, **75**, 2684.
[286] D. J. Turner, *J. C. S. Faraday II*, 1972, **68**, 643.
[287] A. K. Covington, J. G. Freeman, and T. H. Lilley, *J. Phys. Chem.*, 1970, **74**, 3773.
[288] O. D. Bonner, H. B. Flora, and H. W. Aitken, *J. Phys. Chem.*, 1971, **75**, 2492.
[289] R. C. Benoit, C. Buisson, and G. Choux, *Canad. J. Chem.*, 1970, **48**, 2353.
[290] P. D. Bolton, C. H. Rochester, and B. Rossall, *Trans. Faraday Soc.*, 1970, **66**, 1348.
[291] C. E. Vanderzee, D. L. King, and I. Wadsö, *J. Chem. Thermodynamics*, 1972, **4**, 685.
[292] C. E. Vanderzee, M. Månsson, I. Wadsö, and S. Sunner, *J. Chem. Thermodynamics*, 1972, **4**, 541.
[293] K. Yates and G. Welch, *Canad. J. Chem.*, 1972, **50**, 474.
[294] H. Ohtaki and N. Tanaka, *J. Phys. Chem.*, 1971, **75**, 90.
[295] M. Paabo and R. G. Bates, *J. Phys. Chem.*, 1970, **74**, 702.
[296] M. Paabo and R. G. Bates, *J. Phys. Chem.*, 1970, **74**, 706.
[297] S. Goldman, P. Ságner and R. G. Bates, *J. Phys. Chem.*, 1971, **75**, 826.
[298] G. Wada and T. Tanenaka, *Bull Chem. Soc. Japan*, 1971, **44**, 2877.
[299] E. M. Woolley, R. W. Witton, and L. G. Hepler, *Canad. J. Chem.*, 1970, **48**, 3249.
[300] A. H. Gandhi and S. R. Patel, *Bull Chem. Soc. Japan*, 1971, **44**, 455.
[301] A. Dolenko, K. Mahendran, and E. Buncel, *Canad. J. Chem.*, 1970, **48**, 1736.
[302] L. M. Schwartz and L. O. Howard, *J. Phys. Chem.*, 1970, **74**, 4374.

trifluoride,[303] tris(hydroxymethyl)aminomethane,[304] acetic acid and some chloroacetic acids[305] in aqueous dimethyl sulphoxide, and the bisulphate ion in water at 25 °C.[306]

Lown and Thirsk have discussed[307, 308] the ionization of acetic acid over a range of temperatures and pressures and the effect of the solvent vapour pressure on the temperature dependence is emphasized.[308] The effect of pressure on the conductance of orthophosphoric acid solution has also been studied[309] by the same workers.

The use and applicability of cryoscopic and ebullioscopic measurements in the determination of dissociation constants is the subject of a paper[310] by Rogers and Mukherjee. It has been shown[311] that a good correlation exists between the pK values in water and the chemical shifts of the hydroxy-groups on some 55 aromatic hydroxy-compounds. Katz and Miller have discussed[312] medium effects on the protonation of carboxylate groups.

The use of acidity functions is admirably discussed in a book by Rochester.[313] This book will undoubtedly serve as the standard text in this field for many years. Leng and Lenzi[314] have discussed the use of acidity functions in hydration problems along the lines of earlier work.[315, 316] Several other determinations and uses of acidity functions have also been described.[317—321]

8 Ion Association

The description of short-range interactions between ions in solution using mass-action models continues to be popular. In this field as in others the use of Raman and n.m.r. experimental techniques is of increasing popularity. The application of the Raman technique to problems of ion association has been reviewed elsewhere[322] and the reader is referred to this. Recent papers using

---

[303] C. L. Liotta, D. F. Smith, H. P. Hopkins, and K. A. Rhodes, *J. Phys. Chem.*, 1972, **76**, 1909.
[304] G. Olofsson, *J. Chem. Thermodynamics*, 1971, **3**, 217.
[305] N. M. Ballash, E. B. Robertson, and M. D. Sokolowski, *Trans. Faraday Soc.*, 1970, **66**, 2622.
[306] C. E. Evans and C. B. Monk, *Trans. Faraday Soc.*, 1971, **67**, 2652.
[307] D. A. Lown and H. R. Thirsk, *Trans. Faraday Soc.*, 1970, **66**, 51.
[308] D. A. Lown and H. R. Thirsk, *J. C. S. Faraday I*, 1972, **68**, 1982.
[309] D. A. Lown and H. R. Thirsk, *Trans. Faraday Soc.*, 1971, **67**, 149.
[310] R. S. Rogers and L. M. Mukherjee, *J. Chem. Phys.*, 1970, **52**, 4550.
[311] G. Socrates, *Trans. Faraday Soc.*, 1970, **66**, 1052.
[312] S. Katz and J. E. Miller, *J. Phys. Chem.*, 1972, **76**, 2778.
[313] C. H. Rochester, 'Acidity Functions', Academic Press, London and New York, 1970.
[314] T. T. Leng and F. Lenzi, *Canad. J. Chem.*, 1972, **50**, 3283.
[315] K. N. Bascombe and R. P. Bell, *Discuss. Faraday Soc.*, 1957, no. 24, p. 158.
[316] P. A. H. Wyatt, *Discuss. Faraday Soc.*, 1957, no. 24, p. 162.
[317] K. Yates and S. A. Shapiro, *Canad. J. Chem.*, 1972, **50**, 581.
[318] E. Buncel, E. A. Symons, D. Dolman, and R. Stewart, *Canad. J. Chem.*, 1970, **48**, 3354.
[319] D. G. Lee, *Canad. J. Chem.*, 1970, **48**, 1919.
[320] T. R. Essig and J. A. Marinsky, *Canad. J. Chem.*, 1972, **50**, 2254.
[321] K. Yates and G. Welch, *Canad. J. Chem.*, 1972, **50**, 1513.
[322] T. H. Lilley, in 'Water, A Comprehensive Treatise', vol. III, 'Aqueous Solutions of Simple Electrolytes', ed. F. Franks, Plenum Press, New York, in the press.

Raman spectroscopy to investigate ionic equilibria include studies of ion pairing in aqueous $Zn(NO_3)_2$ solutions[323] and $Mg(NO_3)_2$ solutions.[324] Irish and co-workers have investigated[325] ion association in aqueous lithium nitrate solutions and have used a lattice model to estimate the equilibrium constant for the process

$$Li^+(H_2O)_3 + NO_3^- \rightleftharpoons Li^+(H_2O)_2 NO_3^- + H_2O.$$

This approach has been extended to a study[326] of the sodium nitrate–D_2O system. The Raman measurements were supplemented by i.r., conductance, viscosity, and partial molar volume data. The spectroscopic studies indicated ion pairs and solvated nitrate ions. Degrees of association were calculated and the results discussed in terms of the other measured properties. Conductance, densities, and viscosities of $NaNO_3$ in H_2O and in dioxan–H_2O mixtures have also been reported.[327]

An interesting paper by Covington and co-workers[328] describes the association between trimethylsulphonium and iodide ions in water and acetonitrile. Degrees of association were determined from the chemical shifts of the methyl protons and reasonable agreement was found with earlier[329] conductance studies. Symons has continued his investigations on solvation spectra with an n.m.r. study[330] at low temperatures in methanol–water mixtures containing some salts of bivalent metals. The experimental results suggest that acetate ions can enter the solvation shell of Mg^{2+}.

Interactions between 2:2 electrolytes were discussed earlier in connection with the bivalent sulphates. Other[331, 332] studies have also been described. The paper[332] on ion pairing in complex-ion sulphates provoked a comment by Prue et al.[333] It is concluded from a conductance study[334] that rubidium chloride is associated to a significant extent in water at 25 °C. The raw conductance data were analysed using the Fuoss–Hsia[335] and Guggenheim[336] approaches with substantial agreement between the association constants. A similar study by D'Aprano is described[337] for alkali-metal perchlorates in water. Other ion-association investigations were sodium carbonate and

---

[323] A. T. G. Lemley and R. A. Plane, *J. Chem. Phys.*, 1972, **57**, 1648.
[324] M. Peleg, *J. Phys. Chem.*, 1972, **76**, 1019.
[325] D. E. Irish, D. L. Nelson, and M. H. Brooker, *J. Chem. Phys.*, 1971, **54**, 654.
[326] J. D. Riddell, D. J. Lockwood, and D. E. Irish, *Canad. J. Chem.*, 1972, **50**, 2951.
[327] E. M. Kartzmark, *Canad. J. Chem.*, 1972, **50**, 2845.
[328] A. K. Covington, M. L. Hassall, and I. R. Lantzke, *J. C. S. Faraday II*, 1972, **68**, 1352.
[329] D. F. Evans and T. L. Broadwater, *J. Phys. Chem.*, 1968, **72**, 1037.
[330] R. N. Butler, J. Davies, and M. C. R. Symons, *Trans. Faraday Soc.*, 1970, **66**, 2426.
[331] J. W. Larson, *J. Phys. Chem.*, 1970, **74**, 3392.
[332] W. L. Masterton and T. Bierly, *J. Phys. Chem.*, 1970, **74**, 139.
[333] E. M. Hanna, A. D. Pethybridge, and J. E. Prue, *J. Phys. Chem.*, 1971, **75**, 291.
[334] H. S. Dunsmore, S. K. Jalota, and R. Paterson, *J. C. S. Faraday I*, 1972, **68**, 1583, [see also R. Paterson, S. K. Jalota, and H. S. Dunsmore, *J. Chem. Soc. (A)*, 1971, 2116].
[335] R. M. Fuoss and K. L. Hsai, *Proc. Nat. Acad. Sci. U.S.A.*, 1967, **57**, 1550.
[336] E. A. Guggenheim, *Trans. Faraday Soc.*, 1969, **65**, 2474.
[337] A. D'Aprano, *J. Phys. Chem.*, 1971, **75**, 3290.

Electrolyte Solutions

bicarbonate in water,[338] alkali-metal cations with picrate,[339, 340] Zn^{2+}–F^-,[341] and bivalent ions with carboxylate anions.[342, 343] The association between Cs^+ and tetraethylammonium ions and ClO_4^- in ethanol–acetone mixtures has been studied.[344] The interaction between long-chain substituted ammoninm and phosphonium salts in ethanol has been investigated[345] by vapour-phase osmometry.

The ionization behaviour of NaCl in dioxan–water mixtures up to 100 °C and 4000 bar has been discussed,[346] as has the general approach[347, 348] to ion association at high temperatures and pressures.

9 Dielectric Studies and Ultrasonic Absorption

The already large number of investigations on relaxation phenomena in 2:2 electrolytes in water has been supplemented by three further studies.[349—351] The experimental results obtained are slightly different but the general interpretation is similar to former studies[352—355] although details do differ. The system $MgSO_4$–dioxan–water has been investigated.[356] 'Non-associating' electrolytes have been the subject of several[357—361] investigations, the most comprehensive of which is that of Pottel and co-workers[357] which describes the permittivity and dielectric and proton magnetic relaxation in aqueous solutions of alkali-metal halides.

The relative permittivities (and dielectric relaxation times) of several salts have been determined.[362] The interpretation of the results is similar to that used earlier.[363—365]

[338] F. S. Nakayama. *J. Phys. Chem.*, 1970, **74**, 2726.
[339] K. Sano, M. Sakuma, S. Motomizu, T. Iwachido, and K. Tôei, *Bull. Chem. Soc. Japan*, 1970, **43**, 2457.
[340] M. Yamane, T. Iwachido, and K. Tôei, *Bull. Chem. Soc. Japan*, 1971, **44**, 745.
[341] R. O. Cook, A. Davies, and L. A. K. Staveley, *J. Chem. Thermodynamics*, 1971, **3**, 907.
[342] J. W. Bunting and K. M. Thong, *Canad. J. Chem.*, 1970, **48**, 1655.
[343] C. E. Evans and C. B. Monk, *Trans. Faraday Soc.*, 1970, **66**, 1491.
[344] G. Pistoia and G. Pecci, *J. Phys. Chem.*, 1970, **74**, 1450.
[345] I. Paligoric and I. J. Gal, *J. C. S. Faraday I*, 1972, **68**, 1093.
[346] L. B. Yeatts, L. A. Dunn, and W. L. Marshall, *J. Phys. Chem.*, 1971, **75**, 1099.
[347] W. L. Marshall, *J. Phys. Chem.*, 1970, **74**, 346.
[348] W. R. Gilkerson, *J. Phys. Chem.*, 1970, **74**, 746.
[349] K. Fritsch, C. J. Montrose, J. I. Hunter, and J. F. Dill, *J. Chem. Phys.*, 1970, **52**, 2242.
[350] A. Bochtler, K. G. Breitschwerdt, and K. Tamm, *J. Chem. Phys.*, 1970, **52**, 2975.
[351] L. G. Jackopin and E. Yeager, *J. Phys. Chem.*, 1970, **74**, 3766.
[352] H. Diebler and M. Eigen, *Z. phys. Chem. (Frankfurt)*, 1959, **20**, 299.
[353] M. Eigen and K. Tamm, *Z. Electrochem.*, 1962, **66**, 93, 107.
[354] G. Atkinson and S. Petrucci, *J. Phys. Chem.*, 1966, **70**, 3122.
[355] G. Atkinson and S. K. Kor, *J. Phys. Chem.*, 1967, **71**, 673.
[356] F. H. Fisher, *J. Phys. Chem.*, 1972, **76**, 1571.
[357] K. Giese, U. Kaatze, and R. Pottel, *J. Phys. Chem.*, 1970, **74**, 3718.
[358] N. Purdie and A. J. Barlow, *J. C. S. Faraday II*, 1972, **68**, 33.
[359] J. Barthel, H. Behret, and F. Schmithals, *Ber. Bunsengesellschaft phys. Chem.*, 1971, **75**, 305.
[360] C. T. Moynihan, R. D. Bressel, and C. A. Angell, *J. Chem. Phys.*, 1971, **55**, 4414.
[361] D. P. Fay and N. Purdie, *J. Phys. Chem.*, 1970, **74**, 1160.
[362] P. S. K. M. Rao and D. Premaswarup, *Trans. Faraday Soc.*, 1970, **66**, 1974.
[363] J. B. Hasted, D. M. Ritson, and C. H. Collie, *J. Chem. Phys.*, 1948, **16**, 1.
[364] G. H. Haggis, J. B. Hasted, and T. S. Buchanan, *J. Chem. Phys.*, 1952, **20**, 1452.
[365] E. Gluekauf, *Trans. Faraday Soc.*, 1964, **60**, 1637.

7
Electroanalytical Chemistry: Voltammetry

BY B. FLEET AND R. D. JEE

1 Introduction

In this introductory review of the field of electroanalytical chemistry the aim has not been to present a comprehensive coverage, but rather to outline the developments in some of the more promising areas, including classical polarography together with its more sophisticated variants and related methods such as coulometry.

The period under review encompasses the fiftieth anniversary of the invention of the polarographic technique[1] and it is encouraging to note that we appear to be entering a period of increased awareness of the capabilities of the technique. Advances in instrumentation[2] have demonstrated the applicability to trace analysis which is of particular relevance to current problems of monitoring environmental pollution.

The past few years have seen a growing interest in more sophisticated instrumentation, though the practical applications of polarography are still mainly limited to the classical d.c. method. This state of affairs is presumably due to the lack of suitable commercial instrumentation coupled with the lack of information concerning the relative merits of the wide range of techniques now available. The situation has been further hindered by the confusion in nomenclature of the various techniques; for example radio-frequency, high-frequency, and detector polarography all refer to the same technique. Literal translations of Russian terminology cause added confusion.

The trend towards the use of multifunctional instrumentation will undoubtedly assist the problem of assessment of the relative merits of different polarographic techniques. Bond and Canterford,[3] for example, have compared the determination of copper in 1M aqueous $NaNO_3$ by conventional d.c., rapid d.c., fast d.c., derivative fast d.c., conventional and rapid a.c. (both phase-sensitive and non-phase-sensitive), fast a.c., pulse, derivative pulse, differential pulse, and anodic stripping d.c., a.c., and pulse polarography, from which they concluded that rapid phase-sensitive a.c. polarography was

[1] J. Heyrovsky, *Chem. listy*, 1922, **16**, 256.
[2] J. B. Flato, *Analyt. Chem.*, 1972, **44**, 75A.
[3] A. M. Bond and J. R. Canterford, *Analyt. Chem.*, 1972, **44**, 721.

generally best. Unfortunately this comparative study did not include examples of irreversible or adsorption systems or it would have provided an even more useful guideline for the choice of technique for a particular problem. General polarographic reviews on organic polarography by Okuda et al.[4] and Kitagawa et al.[5] have appeared. The applications of modern electroanalytical techniques to pharmaceutical chemistry have been reviewed by Adams.[6] Van Kerchove[7] and Grofcsik[8] have also reviewed similar applications. The use of polarography in the study of clinical problems[9] and the analysis of biological substances[10] has been reviewed. A further excellent review on a.c. polarography by Smith has appeared.[11] He reviews in great detail the theoretical developments in electrochemical reaction kinetics, instrumentation, and practical analytical applications of a.c. polarography, together with a practical guide to computerized electroanalytical instrumentation.

2 Polarography

D.C. Polarography.—In the context of this review, d.c. polarography encompasses not only classical polarography but any technique which involves the application of a linear ramp voltage to the polarographic working electrode. Thus, in addition to fast-sweep voltammetry, the technique of cyclic voltammetry is included in this section.

Although the dropping-mercury electrode remains understandably the most widely used electrode for polarography, other electrode systems are still being investigated. Fleet and Alder[12] have evaluated some thirty materials, ranging from metals to semiconductors, as possible electrodes for anodic voltammetry. Their conclusion was that, with the possible exception of chromium and tungsten carbide, vitreous carbon is the best electrode material. Independently, Jennings et al.[13] have determined the reproducibility that can be achieved using a vitreous carbon electrode.

Hanging mercury drop electrodes have been described by Roffia and Vianello[14] and Piccardi and Guidelli.[15] The electrode used by the former group could be refilled from an external reservoir without removing the electrode from the solution, whereas Guidelli's electrode used a motor-driven piston system to renew the mercury drop.

[4] J. Okuda, G. Okuda, and I. Miwa, *Chem. and Pharm. Bull. (Japan)*, 1970, **18**, 1945.
[5] T. Kitagawa, T. Wasa, and M. Takagi, *Japan Analyst*, 1969, 57R.
[6] R. N. Adams, *J. Pharm. Sci.*, 1969, **58**, 1171.
[7] C. Van Kerchove, *J. Pharm. Belg.*, 1970, **25**, 215.
[8] M. N. Grofcsik, *Gyogyszereszet*, 1969, **13**, 167.
[9] J. Homolka, *Methods Biochem. Analyt.*, 1971, **19**, 435.
[10] J. H. Assikis, *Ellenike Kteniatriki*, 1970, 198.
[11] D. E. Smith, *CRC Critical Reviews in Analytical Chemistry*, 1971, 247.
[12] J. F. Alder and B. Fleet, *J. Electroanalyt. Chem. Interfacial Electrochem.*, 1971, **30**, 427.
[13] V. J. Jennings, T. E. Forster, and J. Williams, *Analyst*, 1970, **95**, 718.
[14] S. Roffia and E. Vianello, *J. Electroanalyt. Chem. Interfacial Electrochem.*, 1969, App. 9—11.
[15] G. Piccardi and R. Guidelli, *Analyt. Letters*, 1968, **1**, 771.

In many electrochemical experiments using the D.M.E. it is convenient to be able to control the drop time independently of the mass flow rate in the capillary. There would also appear to be an increasing trend to use current sampling and mechanically controlled drop times in polarography. This trend can be expected to continue owing to the increased ease with which accurate measurements of polarographic currents may be made. Fisher et al.[16] have described a precision mechanically operated drop-time controller, and Neeb and Williams[17] have described an arrangement for the synchronization of a switching program with the drop time of the D.M.E. using the natural signals of the polarographic cell. The influence on the drop time on a polarizing voltage for a time interval during the growth was also investigated.[18] Mark and Means[19] have developed a simple instrument for D.M.E. drop-time and current-sampling control, and a peak-follower circuit has been described by Borello et al.[20]

The use of a short, mechanically controlled drop time can give a considerable improvement in the speed of analysis;[3] however, it must be remembered that the detection limit of d.c. polarography will generally be degraded owing to the larger residual current. Bond[3] has compared d.c. polarography with short drop times with other techniques. Vibrating dropping-mercury electrodes have been the subject of detailed study by Cover and Connery.[21-23] Studies with the vibrating D.M.E. at different vibrational frequencies demonstrated that it is significantly superior to the ordinary D.M.E. for most analytical purposes. At a frequency of 210 Hz maxima of the first and second kinds are suppressed without the addition of surfactants. Catalytic and kinetic currents could also be minimized or eliminated. In addition, the vibrating D.M.E. permitted the analysis of agitated solutions. Advantages in using the vibrating D.M.E. for analytical purposes where adsorption phenomena normally inhibit electrode response were also found. A fundamental limitation, however, is the extremely large charging current which would tend to limit the use of the vibrating D.M.E. to fairly concentrated solutions.

Fisher et al.[24] have described a solid-state controlled-potential d.c. polarograph with cyclic scanning and calibrated first and second derivative facilities. The instrument uses a D.M.E. with a precisely controlled drop time, so allowing the use of a fairly sophisticated filter circuit to remove the current oscillations due to drop growth. An evaluation of the instrument for cadmium

[16] D. J. Fisher, W. L. Belew, H. C. Jones, and M. T. Kelley, *Analyt. Chem.*, 1969, **41**, 779.
[17] G. Willems and R. Neeb, *J. Electroanalyt. Chem. Interfacial Electrochem.*, 1969, **21**, 69.
[18] G. Willems and R. Neeb, *J. Electroanalyt. Chem. Interfacial Electrochem.*, 1970, **25**, 61.
[19] H. B. Mark, jun. and D. K. Means, *Analyt. Letters*, 1971, **4**, 23.
[20] A. Borello, M. De Carolis and G. R. Guidotti, *J. Electroanalyt. Chem. Interfacial Electrochem.*, 1971, **30**, 231.
[21] R. E. Cover and J. G. Connery, *Analyt. Chem.*, 1969, **41**, 918.
[22] R. E. Cover and J. G. Connery, *Analyt. Chem.*, 1969, **41**, 1191.
[23] R. E. Cover and J. G. Connery, *Analyt. Chem.*, 1969, **41**, 1797.
[24] D. J. Fisher, H. C. Jones, W. L. Belew, R. D. Stelzner, and T. R. Mueller, *Analyt. Chem.*, 1969, **41**, 772.

is given. Cahan and Haynes[25] have described a differential and derivative polarograph for continuous analysis in a flowing system. The details of an automated polarograph for use with a hanging mercury drop electrode has been given by Fano and Scalvini,[26] and a simple attachment for converting a conventional two-electrode polarograph into a three-electrode instrument is described by Tenji.[27]

Several designs for signal generators suitable for electroanalytical experiments have been described.[28-30] Lee and Shain[28] have described a unit which will produce ramp and hold signals, single cycle and multicycle triangular waves with independently selectable initial potential, anodic and cathodic limit potentials, and hold potential. The scan rates available range from 0.2 mV s^{-1} to 1000 V s^{-1}. The details of a somewhat similar instrument are given by Fsuji and Takahashi,[29] and a computer-controlled sweep generator has been described by Ramaley.[31] Some of the most interesting work carried out over the past few years concerns the application of computers to polarography. Of considerable interest, because of its implications, is the demonstration by Perone et al.[32, 33] of using a computer on a real-time basis. A small digital computer was interfaced with a fast-sweep derivative polarograph and used to aid the analysis of multi-component samples. To minimize the interference usually caused by the reduction current of a more positively reduced species on that of a more negatively reduced one an interrupted voltage sweep technique was used. The computer was used to detect the presence of a reduction peak and then interrupt the voltage sweep for a time period proportional to the size of the peak; in this way the faradaic current was allowed to decay to a small value before continuing the voltage scan to potentials at which the next component would be reduced. In another paper Perone and Sybrandt[34] have described a computerized learning machine approach for qualitative analysis of mixtures. The limiting effects of concentration ratios, degree of peak overlap, and peak-potential variation were investigated. Their results indicated that resolution was limited by the precision of the experimental data and the concentration ratios. Ideally, overlapping polarograms could be identified with 2 mV peak separation over a twenty-fold range of concentration ratios. For data obtained under ordinary experimental conditions, resolution was limited to peak separation of 35—40 mV for about a ten-fold range of concentration ratios. Peak identification could be performed reliably under conditions where peak overlap precluded

[25] B. Cahan and R. Haynes, *J. Electroanalyt. Chem. Interfacial Electrochem.*, 1969, **22**, 339.
[26] V. Fano and M. Scalvini, *J. Microchem.*, 1970, **15**, 97.
[27] K. Tsuji, *Analyt. Chim. Acta*, 1969, **47**, 386.
[28] R. L. Myers and I. Shain, *Chem. Instr.*, 1969, **2**, 203.
[29] K. Tsuji and K. Takahashi, *Analyt. Chim. Acta*, 1971, **57**, 473.
[30] R. H. Bull and G. C. Bull, *Analyt. Chem.*, 1971, **43**, 1342.
[31] L. Ramaley, *Chem. Instr.*, 1970, **2**, 415.
[32] S. P. Perone, D. O. Jones, and W. F. Gutknecht, *Analyt. Chem.*, 1969, **41**, 1154.
[33] S. P. Perone and D. O. Jones, *Analyt. Chem.*, 1970, **42**, 1151.
[34] S. P. Perone and L. B. Sybrandt, *Analyt. Chem.*, 1971, **43**, 382.

visual resolution. The use of an on-line computer for data processing has been carried out by Perone et al.,[35-37] who devised a procedure for the numerical deconvolution of overlapping polarographic curves.[37] An empirical equation was developed to describe the shape of stationary electrode polarograms, and a standard set of constants was obtained for a range of ions by measurements on pure solutions. When a mixture was analysed, these standard constants were used to regenerate the polarograms, which were then fitted to the signal obtained from the unknown. During the course of fitting the regenerated curves to the mixture signal, corrections were automatically made for slight potential shifts. This approach enabled the quantitative resolution of overlapping peaks of similar magnitude with peak separations of less than 40 mV.

Ogle et al.[38] have used a computer to process data for d.c. polarography with a rapid dropping-mercury electrode ($t = 0.25$ s). The data are fed directly into the computer and this procedure permits the automatic calculation of the slope of the log plot, half-wave potential, and limiting current constant. Two papers on the subject of digital polarography[39, 40] have appeared. A general discussion on the subject is presented, followed by a brief report on efforts made to digitize polarographic currents for computer input. Smith et al.[41] have evaluated a computerized sampling technique for digital data acquisition of cyclic voltammograms.

A digital electrochemical system in which all the data were stored in a pulse-height analyser has been described by Clem and Goldsworthy.[42] The instrument allowed the selection of the current-sampling period and its location during the life of a drop to be optimized for maximum sensitivity. The results of different voltage scans could be held in different memories so reducing subtractive polarography to one of transferring data between memory locations. This technique may be used to subtract background currents and increased the polarographic sensitivity from 10^{-7} mol l^{-1} to 5×10^{-9} mol l^{-1}. The same authors[43] have given details of a digital potentiostat in which digital current feedback is used. This obviates the need for elaborate phase-correcting networks required with analog potentiostats. An instrument which gives direct conversion of charge into a digital number has also been described.[44]

A number of examples concerned with integration procedures applied to d.c. polarography have appeared. Lingane[45] has described a simple time-

[35] S. P. Perone, J. E. Harrar, F. B. Stephens, and R. E. Anderson, *Analyt. Chem.*, 1968, **40**, 899.
[36] S. P. Perone, J. W. Frazer, and A. Kray, *Analyt. Chem.*, 1971, **43**, 1485.
[37] S. P. Perone and W. F. Gutknecht, *Analyt. Chem.*, 1970, **42**, 906.
[38] J. Ogle, P. Tomsons, I. Tutani, and J. Stradins, *Zavod. Lab.*, 1970, **36**, 1180.
[39] H. Shirai, *Kagaku Kogyi*, 1970, **21**, 1127.
[40] E. Niki and H. Shirai, *Bunsiki Kagaku*, 1971, **20**, 1022.
[41] D. E. Smith, C. Creason, and R. J. Loyd, *Analyt. Chem.*, 1972, **44**, 1159.
[42] R. G. Clem and W. W. Goldsworthy, *Analyt. Chem.*, 1971, **43**, 918.
[43] R. G. Clem and W. W. Goldsworthy, *Analyt. Chem.*, 1971, **43**, 1718.
[44] R. G. Clem and W. W. Goldsworthy, *Analyt. Chem.*, 1972, **44**, 1360.
[45] J. J. Lingane, *Analyt. Chim. Acta*, 1969, **44**, 411.

integration technique for improving the precision. The electrode potential was held constant at some value on the diffusion plateau and the current integrated for a fixed time period. A precision of ±0.1% for ideal systems could be obtained. Long-term integration of polarographic currents has been shown by Matsuda and Tamamushi[46] to be a useful method of compensating for residual currents. These authors used a two-cell differential system and integrated the difference current over a period of 10 min at a fixed potential on the plateau of the reduction process. Reasonable accuracy ($< 2\%$ s.d.), even at concentrations as low as 10^{-6} mol l^{-1}, was obtained.

D.C. Polarography: Applications.—Classical polarography remains the most popular technique for applications, a state of affairs which will surely change with the current trends in instrumentation.

Humphrey and Laird[47] have developed methods for the determination of chloride, cyanide, fluoride, sulphate, and sulphite ions based upon their reactions with metal chloranilates. A method for fluoride determination based on the shift of the U^V–U^{III} half-wave potential has been described by Bond and O'Donnell.[48] The method was suitable for fluoride in the concentration range 0.05—20 p.p.m. A procedure for the determination of perbromates alone or in the presence of small amounts of bromate is given by Jaselskis and Huston.[49]

The determination of thorium[50] and silicon[51] in plutonium metal by linear-sweep oscillographic polarography has been described by Plock and Vasquez. Plock[52] has also developed an anodic voltammetric determination procedure for plutonium by the oxidation of plutonium(III) to plutonium(IV) at a glassy carbon electrode in mineral-acid media.

Pottkamp and Umland[53] have described the determination of traces of phosphorus and silicon in mixtures of organic solvents. The phosphorus and silicon were extracted as dodecamolybdo-phosphoric and -silicic acids into butyl acetate. This extract was then polarographically examined in a lithium chloride ethanolic supporting electrolyte. An indirect polarographic method for the determination of aluminium based on the determination of cadmium(II) displaced from its edta complex allowed 0.1 µg ml^{-1} of aluminium to be measured.[54] The method was used for aluminium in semiconductor-grade materials.

McMaster[55] has described the determination of acid-soluble and acid-insoluble tin in tough petit copper after separation from other interfering ele-

---

[46] K. Matsuda and R. Tamamushi, *Bull. Chem. Soc. Japan*, 1969, **42**, 439.
[47] R. E. Humphrey and C. E. Laird, *Analyt. Chem.*, 1971, **43**, 1895.
[48] A. M. Bond and T. A. O'Donnell, *Analyt. Chem.*, 1968, **40**, 1405.
[49] B. Jaselskis and J. L. Huston, *Analyt. Chem.*, 1971, **43**, 581.
[50] C. E. Plock and J. Vasquez, *Analyt. Chim. Acta*, 1971, **57**, 113.
[51] C. E. Plock and J. Vasquez, *Analyt. Chim. Acta*, 1970, **50**, 338.
[52] C. E. Plock, *Analyt. Chim. Acta*, 1970, **49**, 83.
[53] F. Pottkamp and F. Umland, *Z. analyt. Chem.*, 1971, **255**, 367.
[54] K. Schoene, *Talanta*, 1971, **18**, 339.
[55] C. H. McMaster, *Analyt. Chem.*, 1969, **41**, 1489.

ments. A comparative survey of spectrography and polarography for the determination of impurities in high-purity cadmium has been made by Rajie and Markovic.[56] Bond has made a comparative study of a number of polarographic techniques including classical d.c., sampled d.c., and rapid d.c. for the determination of tin(IV),[57] and has described methods for the determination of tin in geological samples.[58] A polarographic method for cobalt and nickel in iron meteorites has been described by Tackett and Ong.[59]

Hydrofluoric acid, although a valuable solvent for the dissolution of refractory samples, usually causes serious interference to the subsequent analytical procedure. Bond et al.[60-62] have made a number of polarographic studies using this medium. Conventional polarographic procedures cannot be used because of etching of the glass capillary of the D.M.E., although rapid d.c. polarography (and other rapid techniques) were found suitable for use in HF up to 30% because of the increased speed of analysis compared with that when using a conventional D.M.E. When concentrated HF solutions are employed it is necessary to use either a mercury-pool electrode[61] or a teflon D.M.E.[62] Rason[63] describes the determination of uranium in 1M-HF using rapid polarography with a vertical orifice PTFE D.M.E.

A method for determining selenium(IV) in the presence of selenium(VI) has been described.[64] Fleet et al.[65] have investigated some continuous voltammetric sensors for the determination of calcium and magnesium. The method was to determine the total calcium and magnesium content from the decrease in height of the anodic wave of edta on complexation with the metal ions after which the calcium alone was estimated in a similar manner using egta.

Single-sweep polarographic methods for zinc in uranium,[66] antimony in water,[67] and arsenic in drinking water[68] have been described. Polarographic sensors for dissolved oxygen in municipal and industrial waste water have been reviewed by Stickley and Blakeley.[69] A number of procedures for the determination of nitrilotriacetic acid in natural waters have also been described.[70-72] All the methods are based upon measurement of a metal–nitrilotriacetic acid complex.

[56] S. R. Rajie and S. V. Markovic, *Analyt. Chim. Acta*, 1970, **50**, 169.
[57] A. M. Bond, *Analyt. Chem.*, 1970, **42**, 1165.
[58] A. M. Bond, T. A. O'Donnell, and A. B. Waugh, *Analyt. Chem.*, 1970, **42**, 1168.
[59] S. L. Tackett and P. T. Ong, *Analyt. Letters*, 1970, **3**, 169.
[60] A. M. Bond and T. A. O'Donnell, *Analyt. Chem.*, 1969, **41**, 1801.
[61] A. M. Bond, T. A. O'Donnell, and R. J. Taylor, *Analyt. Chem.*, 1972, **44**, 464.
[62] A. M. Bond and T. A. O'Donnell, *Analyt. Chem.*, 1972, **44**, 590.
[63] H. P. Rason, *Analyt. Chim. Acta*, 1969, **48**, 427.
[64] L. R. Williams and P. R. Haskett, *Analyt. Chem.*, 1969, **41**, 1138.
[65] M. D. Booth, B. Fleet, So. Win, and T. S. West, *Analyt. Chim. Acta*, 1969, **48**, 329.
[66] C. E. Plock and J. Vasquez, *Analyt. Chim. Acta*, 1971, **55**, 278.
[67] P. E. Toren, *Analyt. Chem.*, 1968, **40**, 1152.
[68] G. C. Whitnack and R. G. Brophy, *Analyt. Chim. Acta*, 1969, **48**, 123.
[69] J. D. Stickley and C. P. Blakeley, *Analyt. Instr.*, 1969, **7**, 285.
[70] B. K. Afghan and P. D. Goulden, *Environ. Sci. Technol.*, 1971, **5**, 601.
[71] J. Asplund and E. Wanninen, *Analyt. Letters*, 1971, **4**, 267.
[72] J. P. Haberman, *Analyt. Chem.*, 1971, **43**, 63.

It has been encouraging to see an increasing use of polarography in the pharmaceutical and related fields. Fidelus et al.[73] have developed a method for the measurement of 4,6-dinitro-2-isopropylphenol in clotted blood, dried blood, and blood from cadavers, without prior extraction. To remove the blood oxygen, which would otherwise interfere, a reagent containing metol, 4-hydroquinone, and sodium thiosulphate was added. The determination of diazepam in cadaver blood without prior extraction has also been described by Fidelus et al.[74] A rapid method for the measurement of D-glucose using β-D-glucose oxidase in 5 µl blood samples capable of giving a result within 5 min has been developed by Okuda et al.[75] Nitrazepam in the concentration range 0.5—80 µg ml^{-1} in serum has been determined by Halvorsen and Jacobsen,[76] and Jacobsen and Jacobsen[77] have developed a procedure for the direct determination of chlordiazepoxide in serum without separation. A two-hour assay for penicillins and cephalosporins in serum has been described by Benner: three cephalosporins and seven penicillin antibiotics were examined.[78]

A number of assay procedures for pharmaceuticals in various formulations have been described. A rapid, specific, and accurate method for niacinamide has been developed and compared with the Konig colorimetric assay method.[79] The polarographic procedure compares favourably and eliminates the use of cyanogen bromide. Pyridoxal hydrochloride in pure or multivitamin preparations after oxidation to pyridoxal has been measured with a relative error of ≤ 3%.[80] Schertel and Sheppard[81] have examined the determination of riboflavin, thiamine hydrochloride, and niacinamide in multivitamin preparations. The polarographic results were compared with the official analytical procedures, indicating that the polarographic method has potential for rapid screening of multivitamin formulations. The determination of the Δ^4-3-ketosteroid flurandrenolone in pharmaceutical formulations using fast-scan polarography has been described by Heasman and Wood.[82] For the determination of this compound in ointments a preliminary extraction procedure was required.

Tolbutamide and chlorpropamide in tablets have been determined via their nitro-derivatives.[83] Jacobsen and Jacobsen[77] have developed a procedure for chlordiazepoxide in tablets containing propantheline bromide, an instance where the normal u.v. assay method fails. Pharmaceutical preparations containing 1,4-benzodiazepine derivatives (chlordiazepoxide, diazepam, and oxazepam) were analysed polarographically in a methanolic medium.[84]

[73] J. Fidelus, A. Mikolajek, and M. Zietek, Pharmazie, 1971, 26, 507.
[74] J. Fidelus, M. Zietek, A. Mikolajek, and Z. Grochowska, Mikrochim. Acta, 1972, 84.
[75] J. Okuda, G. Okuda, and I. Miwa, Chem. and Pharm. Bull. (Japan), 1970, 18, 1945.
[76] S. Halvorsen and E. Jacobsen, Analyt. Chim. Acta, 1972, 59, 127.
[77] E. Jacobsen and T. V. Jacobsen, Analyt. Chim. Acta, 1971, 55, 293.
[78] E. J. Benner, Antimicrob. Agents Chemother., 1970, 201.
[79] J. M. Moore, J. Pharm. Sci., 1969, 58, 1117.
[80] I. Kruse, Tartu Rukliku Uli-Kooli Toim., 1971, 270, 157.
[81] M. E. Schertel and A. J. Sheppard, J. Pharm. Sci., 1971, 60, 1070.
[82] M. J. Heasman and A. J. Wood, J. Pharm. Pharmacol., 1971, 23 (suppl.), 176S.
[83] S. Tammilehto, Farm. Aikak., 1971, 80, 367.
[84] G. Caille, J. Braun, and J. A. Mockle, Canad. J. Pharm. Sci., 1970, 5, 78.

Accurate assays, however, could not be made in the presence of other active ingredients. The determination of bismuth in pharmaceutical dosage forms such as tablets, emulsions, suspensions, injectable, and bulk powders has been described by Plank.[85] A collaborative study has shown the method to be superior in accuracy and precision to the present official colorimetric method.

Cohen[86] has shown that ethacrynic acid gives well-defined polarographic steps which could be used to assay ethacrynic acid in the presence of its dimer. Dimerization occurs under a variety of conditions and is the principal degradative pathway for the compound in pharmaceutical dosage forms.

A method for the simultaneous determination of p-nitro-2-acetamido-3-hydroxypropiophenone and 1-p-nitrophenyl-2-aminopropane-1,3-diol has been proposed.[87] These compounds are used in the synthesis of chloramphenicol by the Parke–Davis process and could be determined polarographically in the reaction mixture. Fossdal and Jacobsen[88] have investigated the determination of chloramphenicol by polarographic means. They point out that polarographic methods are more rapid and less susceptible to interference than the titrimetric, spectrophotometric, and chromatographic procedures and should be advantageous for routine analysis of this extremely valuable antibiotic. The determination of chloramphenicol cinnamate in complex syrup has been described by Corsi and Mecarelli.[89]

A micro method for the estimation of mono- and di-iodotyrosine, liothyronine, and thyronine based on the polarography of the nitro-derivatives, so permitting determinations in the range 1—10 µg ml^{-1}, have been described.[90] Kerchove[91] has made a polarographic study of iodinated contrast agents. The analytical utilization of polarographic and voltammetric behaviour of some sulphur-containing purines has been studied by Dryhurst.[92] These compounds are of special interest because of their use in the treatment of various types of cancer. The polarographic behaviour of 6-thiopurine, purine-6-sulphinic acid, purine-6-sulphonic acid, purine-6-sulphonamide, and bis-(6-purinyl) disulphide is discussed and procedures are described for the determination of the single compounds or mixtures of several compounds.

The determination of aliphatic ketones by polarography of their Girard T derivatives has been described by Booth and Fleet.[93] The derivative was formed *in situ* in a 75% ethanolic acetate buffer solution which was 0.1 mol l^{-1} with respect to Girard T reagent.

Two areas which receive very little attention, yet would seem to hold great promise, are the use of non-aqueous solvent systems and anodic oxidation

---

[85] W. M. Plank, *J. Assoc. Offic. Analyt. Chemists*, 1972, **55**, 155.
[86] E. M. Cohen, *J. Pharm. Sci.*, 1971, **60**, 1702.
[87] D. Dismanovic, J. Volki, and R. Jovanovic, *J. Ass. Offic. Analyt. Chem.*, 1971, **54**, 884.
[88] K. Fossdal and E. Jacobsen, *Analyt. Chim. Acta*, 1971, **56**, 105.
[89] R. Corsi and E. Mecarelli, *Bull. Chim. Farm.*, 1971, **110**, 147.
[90] E. Wachholz and S. Pfeifer, *Pharmazie*, 1972, **27**, 43.
[91] C. Van Kerchove, *J. Pharm. Belg.*, 1971, **26**, 304.
[92] G. Dryhurst, *Analyt. Chim. Acta*, 1969, **47**, 275.
[93] M. D. Booth and B. Fleet, *Analyst*, 1970, **95**, 649.

reactions. A proposed analytical procedure encompassing both areas has been described by Shiozaki et al.[94] These authors have determined α-tocopheryl acetate in commercially available vitamin E products. α-Tocopheryl acetate gives a single-electron anodic oxidation wave in acetonitrile using a glassy carbon electrode. α-Tocopherol also gives an anodic oxidation wave, but sufficiently separated to allow the simultaneous determination of these compounds. An investigation of the polarographic responses of 24 pharmaceutically important compounds in acetonitrile–tetrabutylammonium perchlorate media has been undertaken by Woodson and Smith,[95] showing that many of the compounds produce ideal one-electron reversible waves suitable for analytical purposes.

Lund and Iversen[96] have proposed analytical procedures for some hydroxylamines, hydrazines, and hydrazides based on the anodic waves obtained in an aqueous alkaline medium containing sulphite ions. A polarographic method of determining acids has been described by Takamura and Hayakawa,[97] based on the fact that methyl-*p*-benzoquinone gives a single polarographic wave in methyl cellosolve medium containing perchlorate as supporting electrolyte, whereas a pre-wave appears when acid is added. The half-wave potential of the pre-wave depends upon the pK_a value of the acid so that, in favourable cases, mixtures may be determined.

Electrochemical detectors for liquid chromatography have been investigated by several workers. Kemula[98] has reviewed the determination of organic impurities in organic substances by chromatopolarography. Joynes and Maggs[99] have compared the D.M.E. and the carbon silicon rubber membrane electrode as polarographic detectors. The latter was superior in terms of sensitivity, giving a detection limit of 1.6×10^{-9} mol l^{-1} for 2,6-dinitrophenol. Takemori and Honda,[100] on the other hand, have used a thin-walled capillary D.M.E. for chromatopolarography in conjunction with a three-electrode potentiostat. A rapid method for the determination of methylparathion and parathion on crops by liquid chromatopolarography, giving a detection limit of about 0.03 p.p.m., has been described by Koen and Huber.[101] Radiolytic products in an aqueous nitrate–ethylene system have also been investigated by means of chromatopolarography.[102]

Pulse Polarography.—The technique of pulse polarography is well established and still remains one of the most sensitive polarographic techniques available. However, it is surprising how few applications of this technique are to be

[94] K. Shiozaki, K. Fukui, and T. Kitagawa, *Bunseki Kagaku*, 1971, **20**, 438.
[95] A. L. Woodson and D. E. Smith, *Analyt. Chem.*, 1970, **42**, 242.
[96] P. E. Iversen and H. Lund, *Analyt. Chem.*, 1969, **41**, 1322.
[97] K. Takamura and Y. Hayakawa, *Analyt. Chim. Acta*, 1968, **43**, 273.
[98] W. Kemula, *Pure Appl. Chem.*, 1971, **25**, 763.
[99] P. L. Joynes and R. J. Maggs, *J. Chromatogr. Sci.*, 1970, **8**, 427.
[100] Y. Takemori and M. Honda, *Rev. Polarog.*, 1970, **16**, 96.
[101] J. G. Koen and J. F. K. Huber, *Analyt. Chim. Acta*, 1970, **51**, 303.
[102] R. K. Broszkiewicz and Z. Przybylowicz, *Analyt. Chem.*, 1969, **41**, 1121.

found in the literature, although what examples there are generally make full use of its extremely high sensitivity and ability to cope with high concentration ratios.

Keller and Osteryoung[103] have described the application of a computer to pulse polarography at a hanging mercury drop electrode. Application to the analysis of extremely dilute solutions by ensemble averaging and digital smoothing is described and the effect on signal-to-noise ratio demonstrated. Verbeek et al.[104-110] have determined trace impurities in a large range of high-purity metals and compounds. Detection of traces of nickel in cobalt and its compounds, without separation and using a pyridine solvent system, is described.[104] A detection limit for nickel of 1 p.p.m. in metallic cobalt and 0.25 p.p.m. in cobalt compounds was obtained. Using a potassium thiocyanate supporting electrolyte these detection limits could be decreased to 0.25 and 0.06 p.p.m., respectively. For the determination of traces of zinc and manganese in cobalt, a separation step using an anion-exchange process was required;[105] the method, however, permitted as little as 0.15 p.p.m. Zn and 0.03 p.p.m. Mn to be determined in a 1 g sample. The determination of indium in cobalt also containing minor quantities of lead and cadmium was accomplished by using ferric hydroxide as a collector. The separation was quantitative and no co-precipitation of cadmium occurred, permitting as little as 0.01 p.p.m. of indium in a 10 g sample to be determined.[107] Thallium in cadmium samples could be detected down to 1 p.p.m., and down to 0.008 p.p.m. if a co-precipitation separation step using manganese dioxide was used.[106] Methods for the determination of antimony (0.004 p.p.m.), tin (0.006 p.p.m.), and arsenic (0.003 p.p.m.),[108] and manganese (0.003 p.p.m.)[109] in cadmium have also been described by these authors. Finally, they have described the determination of gallium in arsenic compounds.[110]

The simultaneous determination of six metals (Cu, Pb, Cd, Ni, Zn, and Co) in natural waters has been described by Abdullah and Royle.[111] A concentration step using an ion-exchange resin in its calcium form was used. The eluate from the resin was found to contain sufficient calcium to act as the supporting electrolyte for the subsequent polarographic determination. Parry and Oldham[112] have shown that palladium at the 50 ng level may be determined in an ammonia–pyridine medium with few interferences.

[103] H. E. Keller and R. A. Osteryoung, *Analyt. Chem.*, 1971, **43**, 342.
[104] A. Lagrou and F. Verbeek, *J. Electroanalyt. Chem. Interfacial Electrochem.*, 1968, **19**, 125.
[105] A. Lagrou and F. Verbeek, *J. Electroanalyt. Chem. Interfacial Electrochem.*, 1968, **19**, 413.
[106] E. Temmerman and F. Verbeek, *J. Electroanalyt. Chem. Interfacial Electrochem.*, 1968, **19**, 423.
[107] A. Lagrou and F. Verbeek, *J. Electroanalyt. Chem. Interfacial Electrochem.*, 1969, **20**, 443.
[108] E. Temmerman and F. Verbeek, *Analyt. Chim. Acta*, 1968, **43**, 263.
[109] E. Temmerman and F. Verbeek, *Analyt. Chim. Acta*, 1970, **50**, 505.
[110] W. Demerie, E. Temmerman, and F. Verbeek, *Analyt. Letters*, 1971, **4**, 247.
[111] M. I. Abdullah and L. G. Royle, *Analyt. Chim. Acta*, 1972, **58**, 283.
[112] E. P. Parry and K. B. Oldham, *Analyt. Chem.*, 1968, **40**, 1031.

Vorlickova et al.[113] have pointed out that pulse polarography can be used to increase greatly the sensitivity of nucleic acid analysis. Double-helical polynucleotide solutions produce, in neutral media, a smaller current which also occurs at more positive potentials than that for the corresponding single-stranded polynucleotides. This difference in behaviour allows the estimation of single- and double-stranded polynucleotides in mixtures; it is possible to detect about 1% denatured DNA in a natural DNA sample. The authors also noted that pulse polarography may be used to study the conformational changes which precede the 'melting' of the double helix and to follow inter-actions between low molecular weight polarographically active substances with polynucleotides.

A mechanism for the electrode process of bis(diethylthiocarbamoyl) disulphide (disulfiram) at the D.M.E. and its determination in tablets has been given by Prue et al.[114]

A.C. Polarography.—The field of a.c. polarography is becoming increasingly complex owing to the large variety of techniques which it now encompasses. The a.c. signals may be applied to the cell as either a voltage or a current and may be of sinusoidal, square-wave, triangular-wave, or amplitude-modulated form. The measured signal can be at the same frequency as the applied signal, or can be a harmonic or intermodulation harmonic. Ultimately, the purpose of all of the variations when used for analytical purposes is simply to discriminate between the faradaic and non-faradaic currents.

Brocke[115] has described a form of a.c. polarography based on demodulation of a sinusoidal signal by a reversible electrode process. The sinusoidal polarization signal (100 kHz) was amplitude-modulated by a low-frequency (37 Hz) sinusoidal signal produced by faradaic rectification. The instrument was examined for cadmium and a detection limit of 5×10^{-7} mol l^{-1} quoted. The same author[116] has described a similar technique except that the high-frequency signal was triangularly amplitude-modulated, which apparently improved the discrimination against the non-faradaic current component.

High-frequency detector polarography, or radio-frequency polarography as it was originally termed by Barker,[117] has at long last received some attention from analytical chemists. Kambara,[118] who considers it to be one of the most elegant polarographic methods for analysis, has postulated two theoretical expressions and verified them experimentally for the dependence of the wave height on depolarizer concentration and the effect of cell resistance. Vasileva et al.[119, 120] have studied the effect of sweep rate on the shape

[113] M. Vorlickova, G. Jezkova, V. Brabec, Z. Pechan, and E. Palecek, *Stud. Biophys.*, 1970, **24–25**, 131.
[114] D. G. Prue, C. R. Warner, and B. T. Kho, *J. Pharm. Sci.*, 1972, **61**, 249.
[115] W. A. Brocke, *J. Electroanalyt. Chem. Interfacial Electrochem.*, 1971, **30**, 237.
[116] W. A. Brocke, *J. Electroanalyt. Chem. Interfacial Electrochem.*, 1971, **33**, Appl. 1.
[117] G. C. Barker, *Analyt. Chim. Acta*, 1958, **18**, 118.
[118] T. Kambara, S. Tanaka, and K. Hasebe, *J. Electroanalyt. Chem. Interfacial Electrochem.*, 1969, **21**, 49.
[119] L. N. Vasileva and N. V. Lukashenkova, *Zhur. analit. Khim.*, 1970, **25**, 412.
[120] L. N. Vasileva and N. B. Kogan, *Zhur. analit. Khim.*, 1971, **26**, 1932.

of the polarogram obtained at a stationary mercury electrode. A design for a transistorized high-frequency polarograph and its use for the determination of tin and lead has been described by Roughton et al.[121] This instrument uses twin cells in order to suppress the influence of capacitative currents. Bruk and Sternberg[122] and Salikhdzhanova and Bryksin[123] have also given details of high-frequency polarographs. Zheleztsov[124, 125] has described both the theory and the experimental side of an a.c. technique using an amplitude-modulated sinusoidal voltage giving a sensitivity of $< 5 \times 10^{-9}$ mol l^{-1} towards cadmium. The same author[126] has also described an instrument using a tone-modulated sinusoidal voltage.

Sluyters et al.[127, 128] have proposed a new method for eliminating the influence of the double-layer capacity by the use of a triangular voltage superimposed upon the d.c. sweep potential. In this instance the capacity current is a square wave which can be easily eliminated after full wave rectification and filtering. A sawtooth voltage wave form may be used instead of the triangular wave.

A.c. polarographs, like d.c. and pulse polarographs, have been interfaced with a computer. Smith and Glover[129] have described what they term 'alternating current polarography in the harmonic multiplex mode'. This refers to the simultaneous acquisition of direct current, fundamental harmonic, and second harmonic a.c. polarographic responses, a small on-line computer being used for the data acquisition, which included both amplitude and phase characteristics of the a.c. responses. Smith and Huebert[130] have described a computer system to measure simultaneously the cell response at each of a number of different frequency inputs. This technique was referred to as 'alternating current polarography in the non-coherent wave frequency multiplex mode'. Kojima and Fujiwara[131] have used an on-line computer coupled to a potentiostat to perform Fourier analysis of a.c. polarographic fundamental and second harmonic responses. These a.c. polarograph–computer systems have been primarily designed for studying the kinetics of electrochemical processes rather than for analytical applications. However, in the light of Perone's work[32–37] with computers on-line to fast-sweep voltammetry, the future appears very promising.

The measurement of higher harmonics in a.c. polarography offers a means of increasing the sensitivity of polarographic determinations by at least one

[121] C. L. Roughton, M. Harrison, and B. Surfleet, *Analyst*, 1970, **95**, 894.
[122] B. S. Bruk and B. M. Sternberg, *Zavod. Lab.*, 1970, **36**, 365.
[123] R. M. Salikhdzhanova and I. E. Bryksin, *Zavod. Lab.*, 1971, **37**, 765.
[124] A. V. Zheleztsov, *Zhur. analit. Khim.*, 1971, **26**, 644.
[125] A. V. Zheleztsov, *Zhur. analit. Khim.*, 1971, **26**, 650.
[126] A. V. Zheleztsov, *Zavod. Lab.*, 1971, **37**, 528.
[127] J. H. Sluyters, J. S. M. C. Breukel, and M. Sluyters-Rehbach, *J. Electroanalyt. Chem. Interfacial Electrochem.*, 1971, **31**, 201.
[128] J. H. Sluyters and M. Sluyters-Rehbach, *J. Electroanalyt. Chem. Interfacial Electrochem.*, 1972, **34**, 542.
[129] D. E. Glover and D. E. Smith, *Analyt. Chem.*, 1972, **44**, 1140.
[130] B. J. Huebert and D. E. Smith, *Analyt. Chem.*, 1972, **44**, 1179.
[131] H. Kojima and S. Fujiwara, *Bull. Chem. Soc. Japan*, 1971, **44**, 2158.

order of magnitude. The optimum conditions for the measurement of second and third harmonic polarograms has been described by Devay et al.[132] Bauer and Britz[133] have described the use of a commercial lock-in amplifier for phase-sensitive second harmonic a.c. polarography, and McAllister and Dryhurst[134] have used a commercial lock-in amplifier for the direct recording of phase-angle changes. The analytical utility of phase-angle measurements was also investigated and found to have no advantages over phase-sensitive a.c. polarography.

Krause and Ramaley[135, 136] have constructed a fast-sweep square-wave polarograph in which the square wave was superimposed upon a staircase ramp (with suitable synchronization). The instrument uses a hanging mercury drop electrode and provides a fairly rapid potential sweep (50 s), so permitting rapid analyses. The system was evaluated for a number of metals and gave detection limits of the order of 5×10^{-8} mol l^{-1}.

A.c. polarography employing short, controlled drop times and fast scan rates, or rapid a.c. polarography as it is usually termed, has been considered by Bond et al.[3, 137-139] The authors point out that the rapid method is superior to the use of a conventional D.M.E. not only because of the considerable saving in time but also because of the improvement in reproducibility. In a paper devoted to phase-sensitive rapid a.c. polarography, Bond and Canterford[138] have shown that theoretical relationships derived for natural-drop-time a.c. polarography can be used for the rapid technique as well. The excellent discrimination obtained against the charging current for the rapid technique was, if anything, better than that obtained with natural-drop-time phase-sensitive a.c. polarography.

Two integration procedures for improving the precision of a.c. polarography have been studied by Fleet and Jee.[140] Both methods are single-electrode techniques and involve the integration of the rectified a.c. current over the complete peak or at the peak potential, respectively. Integration at the peak potential resulted in poor reproducibility owing to small unavoidable potential shifts. Standard deviations were measured for a range of inorganic and organic species and typical values of $\pm 0.25\%$ at the concentration level 4×10^{-4} mol l^{-1} were obtained for the technique involving integration over the complete wave.

Theoretical data for the shapes of fundamental and second and third harmonic current polarograms for comparison with experimental data have

[132] J. Devay, T. Garai, L. Meszaros, and B. Palagyi-Fenyes, *Magyar Kém. Folyóirat*, 1969, **75**, 460.
[133] H. H. Bauer and D. Britz, *Chem. Instr.*, 1970, **2**, 361.
[134] D. L. McAllister and G. Dryhurst, *Analyt. Chim. Acta*, 1972, **58**, 373.
[135] L. Ramaley and M. S. Krause, jun., *Analyt. Chem.*, 1969, **41**, 1362.
[136] M. S. Krause, jun. and L. Ramaley, *Analyt. Chem.*, 1969, **41**, 1365.
[137] A. M. Bond, *J. Electrochem. Soc.*, 1971, **118**, 1588.
[138] A. M. Bond and J. H. Canterford, *Analyt. Chem.*, 1972, **44**, 1803.
[139] A. M. Bond and G. Hefter, *J. Electroanalyt. Chem. Interfacial Electrochem.*, 1972, **34**, 227.
[140] B. Fleet and R. D. Jee, *J. Appl. Electrochem.*, 1971, **1**, 269.

been given.[141] The theoretical data were obtained by differentiating the equation for a d.c. wave, which seems rather strange considering that theoretical equations for fundamental and second harmonic polarography are readily available. The question of resolving power of a.c. polarographic techniques has provoked a number of papers. Bond[142] has made a theoretical comparison of the resolution of fundamental, second, and third harmonic a.c. polarography. The resolution was found to decrease in the order fundamental, second, third harmonic. This surprising result is partly due to Bond's definition of resolution as percentage overlap. However, if one considers a single peak of the second or third harmonic responses then a small gain in resolution is obtained with higher derivative shaped responses. A theoretical and experimental evaluation of multi-element analysis by fundamental harmonic current polarography has been given by Bond and Canterford.[143] A method is given for the quantitative calculation of concentration ratios allowable for the direct simultaneous determination of two reversibly reduced species with similar summit potentials for a precision of 1%. Senkevich[144] and Zheleztsov[145] have both commented on the increase of resolution on going to higher harmonics.

A number of papers on the general field of analytical a.c. polarography have appeared,[146–150] emphasizing the advantages and disadvantages of the technique and including suggestions for the reporting of a.c. polarographic analytical data. A major drawback at the present time to the application of a.c. polarographic techniques to analytical problems is the general lack of data.

The use of adsorption–desorption effects (tensammetry) for analytical purposes is still rare. Booth and Fleet[151] have studied the adsorption–desorption phenomena of a range of methyl carbamate insecticides, and have derived the optimum experimental conditions for their analytical determination. Supporting-electrolyte effects in the tensammetry of n-amyl alcohol have been investigated by Bauer et al.,[152] and Sharma[153] has examined the effects of different cations on the tensammetric responses of organic compounds. Gundersen and Jacobsen[154] have examined a range of surfactants for maximum suppressors in a.c. polarography, concluding that sodium dodecyldiphenyletherdisulphonate is best. A.c. polarographic analysis of solutions

---

[141] A. M. Bond, *J. Electroanalyt. Chem. Interfacial Electrochem.*, 1972, **35**, 343.
[142] A. M. Bond, *J. Electroanalyt. Chem. Interfacial Electrochem.*, 1972, **36**, 235.
[143] A. M. Bond and J. H. Canterford, *Analyt. Chem.*, 1972, **44**, 732.
[144] V. V. Senkevich, *Zhur. analit. Khim.*, 1971, **26**, 461.
[145] A. V. Zheleztsov, *Zhur. analit. Khim.*, 1971, **26**, 869.
[146] A. M. Bond and J. H. Canterford, *Analyt. Chem.*, 1971, **43**, 228.
[147] A. M. Bond, *Analyt. Chem.*, 1972, **44**, 315.
[148] A. M. Bond and J. H. Canterford, *Analyt. Chem.*, 1971, **43**, 1658.
[149] A. M. Bond and J. H. Canterford, *Analyt. Chem.*, 1971, **43**, 228.
[150] A. M. Bond and J. H. Canterford, *Analyt. Chem.*, 1971, **43**, 393.
[151] M. D. Booth and B. Fleet, *Talanta*, 1970, **17**, 491.
[152] H. H. Bauer, H. R. Campbell, and A. K. Shallal, *J. Electroanalyt. Chem. Interfacial Electrochem.*, 1969, **21**, 45.
[153] S. K. Sharma, *Chem. Age India*, 1970, **21**, 721.
[154] N. Gundersen and E. Jacobsen, *Analyt. Chim. Acta*, 1969, **45**, 346.

containing adsorbable complexes has been studied by Kalvoda et al.[155] As would generally be expected, the adsorption of the depolarizer leads to an increase in the sensitivity. Lead, cadmium, copper, nickel, and thallium were examined in ammonium thiocyanate–thiourea and ammonium thiocyanate–urotropine mixtures. An interesting example of the application of surfactants in analytical chemistry has been given by Jacobsen and Tandberg.[156] A method for the determination of cadmium without any previous separation from large amounts of indium is described, based on the fact that the anionic surfactant Benax in a citrate medium serves as an electrochemical masking agent for indium. Cadmium could be determined in the presence of a 10^4-fold excess of indium by the addition of 0.03% Benax.

A.c. polarography in aqueous and anhydrous HF has been studied by Bond et al.[60-62, 157] A glass D.M.E. could be used for HF concentrations up to ~30% if rapid a.c. polarography were used. The use of a teflon end for a normal D.M.E. was examined[62] for cadmium, lead, tin(II), and thallium in 50% HF. The fabrication of a teflon D.M.E. by electrical discharge for use in anhydrous hydrogen fluoride has also been described.[157]

The determination of tin(IV) by conventional a.c. polarography, rapid a.c. polarography, and a.c. anodic stripping has been examined.[57] A.c. stripping was recommended for the concentration range 10^{-6}—10^{-8} mol l^{-1}, and conventional a.c. for the range 10^{-3}—10^{-6} mol l^{-1}; the results were applied to the determination of tin in geological samples.[58] Lyalikov et al.[158] have developed a procedure for the determination of zinc, germanium, and phosphorus in semiconductors. The germanium was reduced from its complex with gallic acid and the phosphorus was determined after conversion into phosphate ions by their effect on the peak height of UO^{2+}. Traces of phosphorus and silicon in organic solvents have been determined after extraction as dodecamolybdophosphoric and silicic acids.[53] Antimony(III) was found by Jacobsen and Rojahn[159] to give a nearly reversible a.c. wave in H_2SO_3–thiocyanate, the peak height being proportional to concentration over the range 2×10^{-5}—5×10^{-4} mol l^{-1}. The a.c. wave was not affected by the presence of a 1000-fold amount of arsenic or a 10-fold amount of iron, bismuth, or lead; only copper caused significant interference.

Odier and Plichon[160] were able to determine copper down to 5×10^{-8} mol l^{-1} in sea water and, from $E_\frac{1}{2}$ potential considerations, to deduce that copper in sea water is present as Cu^{2+}, $CuCl^+$, and $Cu(HCO_3)_2(OH)^-$. An a.c. polarographic procedure for the determination of nitrilotriacetic acid in surface water and sewage at levels down to 0.1 p.p.m. has been described by Wernet and Wahl.[161]

[155] R. Kalvoda, W. Anstine, and M. Heyrovsky, *Analyt. Chim. Acta*, 1970, **50**, 93.
[156] E. Jacobsen and G. Tandberg, *Analyt. Chim. Acta*, 1969, **47**, 285.
[157] A. M. Bond, T. A. O'Donnell, and A. B. Waugh, *J. Electroanalyt. Chem. Interfacial Electrochem.*, 1972, **39**, 137.
[158] Yu. S. Lyalikov, L. S. Kopanskaya, and N. S. Odobesku, *Zavod. Lab.*, 1971, **37**, 1158.
[159] E. Jacobsen and T. Rojahn, *Analyt. Chim. Acta*, 1971, **54**, 261.
[160] M. Odier and V. Plichon, *Analyt. Chim. Acta*, 1971, **55**, 209.
[161] J. Wernet and K. Wahl, *Zhur. analit. Chem.*, 1970, **251**, 373.

The a.c. polarography of a large number of trialkyltin derivatives has been studied by Shkorbatova et al.[162] The d.c. polarograms of these compounds show two waves, the first of which gives rise to an a.c. wave. Booth and Fleet[163] have examined the fungicide triphenyltin. The a.c. polarogram showed two peaks over the concentration range 10^{-3}—10^{-4} mol l^{-1}, the first of which was tensammetric in nature, the second being due to the reduction of the triphenyltin cation to a radical species. Fleet and Jee[164] have described the determination of a large range of olefinic compounds by their reaction with mercuric acetate in a methanolic medium. The organomercury derivative formed gives an a.c. wave suitable for analytical purposes.

Woodson and Smith[95] have surveyed the a.c. responses of some 24 pharmaceuticals in acetonitrile containing tetrabutylammonium perchlorate, and found that most yielded analytically useful fundamental and second harmonic a.c. waves. The use of non-aqueous systems can occasionally 'streamline' an assay procedure by eliminating certain steps such as back extraction. Sugii and Kabasawa utilized this principle for the determination of anethole trithione.[165] Lippold et al.[166] have studied the adsorption behaviour of nicotinic esters and poly(ethylene glycol) fatty alcohol ethers on mercury electrodes. The estimation of iodinated contrast agents, 3,5-di-iodo-4-pyridone and N-(2,3-dihydroxypropyl)-3,5-di-iodo-4-pyridone, has been described by Kerchove,[91] and rapid a.c. polarography was used by Bieder and Brunel[167] to study the urinary excretion of ethionamide and propionamide metabolites in man.

A number of applications of square-wave polarography have been proposed. A method for the determination of zinc traces in bismuth after preconcentration by solid–liquid extraction has been developed by Mizuiki and Kono.[168] Yamate et al.[169, 170] have described an automatic continuous analyser for lead in the atmosphere. Air containing <15 μg Pb m^{-3} is passed through nitric acid for a given time and then analysed for lead using square-wave polarography. Lead in river water down to 2 p.p.b. could be determined without a preconcentration step in a sodium perchlorate and sodium fluoride medium.[171] The sensitivity of the method was roughly 100 times that of atomic absorption. Guerther and Chu-Xuan-Anh[172] have measured U^{VI} over the concentration range 7×10^{-7}—4×10^{-5} mol l^{-1} by square-wave polaro-

[162] T. L. Shkorbatova, D. A. Kochkin, L. D. Sirak, and T. V. Khavalits, *Zhur. analit. Khim.*, 1971, **26**, 1521.
[163] M. D. Booth and B. Fleet, *Analyt. Chem.*, 1970, **42**, 825.
[164] B. Fleet and R. D. Jee, *Talanta*, 1969, **16**, 1561.
[165] A. Sugii and Y. Kabasawa, *Nippon Doigaku Yogugaku Kenkyu Hokoku*, 1969, **10**, 36.
[166] B. Lippold, E. Ullman, and K. Thoma, *Pharmazie*, 1971, **26**, 486.
[167] A. Bieder and P. Brunel, *Ann. pharm. franç*, 1971, **29**, 461.
[168] A. Mizuike and T. Kono, *Mikrochim. Acta*, 1970, 665.
[169] N. Yamate, Y. Matsumura, and M. Tonomura, *Eisei Shikenjo Hokoku*, 1969, **87**, 28.
[170] N. Yamate, *Kuki Seijo*, 1971, **9**, 55.
[171] E. B. Buchanan, jun., T. D. Schroeder, and B. Novosel, *Analyt. Chem.*, 1970, **42**, 370.
[172] O. Gürtler and Chu-Xuan-Anh, *Mikrochim. Acta*, 1970, 941.

Electroanalytical Chemistry: Voltammetry 227

graphy in the presence of many other metals, and Okochi and Sudo[173] have determined titanium down to 0.0002% in iron and steel.

High-frequency or radio-frequency polarography has been used for the determination of toluene-*p*-sulphonic acid in food[174] and for the simultaneous determination of zinc and cobalt or nickel.[175] Chernega *et al.*[176] have examined the behaviour of indium, cadmium, tin, and antimony in various supporting electrolytes by high-frequency polarography. A similar type of study has been made by Bruk and Sternberg.[122]

3 Stripping Voltammetry

The technique of stripping voltammetry (stripping polarography or inverse polarography) has received considerable attention owing to its extreme sensitivity coupled with the relatively simple instrumental requirements. In its simplest form the method involves two steps: electrolytic pre-concentration of the species of interest at a stationary electrode followed by the electro-dissolution, or stripping of the deposit.

The authoritative review by Barendrecht[177] has been supplemented by the bi-annual review in *Analytical Chemistry*.[178] The book by Neeb[179] (in German) is relatively little known but does provide a detailed account of experimental method and an extensive bibliography of applications. Several other reviews have appeared: by Stromberg and Zacharova,[180] Monien,[181] and Brainina,[182] the latter being confined to thin-film stripping at solid electrodes.

There have been relatively few contributions to the theory of stripping voltammetry during the period under review. Since the essential importance of the method is as a trace analysis technique this is perhaps not surprising. Exceptions to this rule have been the investigations of Stromberg and co-workers[183] on effects of temperature on the stripping process, and several groups[184–187] have studied medium effects particularly the effect of buffer components and complexing agents.

---

[173] H. Okochi and E. Sudo, *Trans. Nat. Res. Inst. Metals*, 1971, **13**, 105.
[174] Y. Osajima, K. Matsumoto, M. Nakashima, F. Hashinaga, and S. Furutani, *Bunseki Kagaku*, 1971, **20**, 1292.
[175] S. Furutani, *Japan Analyst*, 1967, **16**, 103.
[176] L. P. Chernega, V. I. Bodyu, and Yu. S. Syalikov, *Zhur. analit. Khim.*, 1971, **26**, 1686.
[177] E. Barendrecht, in 'Electroanalytical Chemistry', ed. A. J. Bard, Arnold, London, 1967, Vol. 2.
[178] R. S. Nicholson, *Analyt. Chem.*, 1972, **44**, 478R.
[179] R. Neeb, 'Inverse Polarographie und Voltammetrie', Weinheim Bergstr., Verlag Chemie, 1969.
[180] A. G. Stromberg and E. A. Zacharova, *Sovrem. Metody analit. Materialy*, 1969, 91.
[181] H. Monien, *Chem.-Ing.-Tech.*, 1970, **42**, 857.
[182] Kh.Z. Brainina, *Talanta*, 1971, **18**, 513.
[183] Y. A. Karbainov, A. G. Stromberg, and S. N. Karbainova, *Izvest. V. U. Z., Khim. khim. Technol.*, 1970, **13**, 345.
[184] Y. A. Karbainov, A. G. Stromberg, and S. N. Karbainova, *Zavod. Lab.*, 1970, **36**, 257.
[185] A. G. Stromberg and A. V. Kon' Kova, *Elektrokhimiya*, 1971, **7**, 603.
[186] E. N. Vinogradova, Y. V. Granovskii, and A. I. Kamenev, *Zhur. analit. Khim.*, 1971, **26**, 238.
[187] L. Zieglerova, K. Stulik, and J. Dolezal, *Talanta*, 1971, **18**, 603.

The choice of an optimum electrode system continues to be the all-important problem. The observations of early workers in the field[188] on deposit–electrode interactions, which have important practical implications in stripping analysis, continue to arouse interest. Robbins and Enke[189] have investigated the interaction of mercury at a platinum electrode and have concluded that a compound $PtHg_4$ is formed at the metal–mercury interface. Several papers have discussed the apparent anomalies in stripping yield obtained on certain metal surfaces.[190–192] Eisner and Mark[193] have investigated the stripping of silver on various graphite electrodes. Although multiple dissolution peaks were observed at a pyrolytic graphite electrode these consisted of one sharp main peak and one or two minor peaks. Reproducibility was adequate for analytical purposes.

A wide range of electrode configurations has been described. The hanging mercury drop electrode is still the most widely used although it suffers from poor sensitivity and resolution owing to back-diffusion of metal ions within the drop.[194] A mercury-pool electrode has been described by Tamba and Vantina[195] and using a medium-exchange technique has been applied to the determination of antimony and lead.[196] Although in principle only stationary electrodes of constant area are applicable, dropping-mercury electrodes have found limited use in stripping voltammetry. Velghe and Claeys[197] have described the use of a D.M.E. with a long drop time of about 18 min. A similar system has been described by Booth[198] for the semi-continuous monitoring of a flow stream.

The use of thin mercury-film electrodes offers improved sensitivity and resolution over conventional mercury electrodes. A variety of methods have been suggested for film preparation; Van der Leest[199] has suggested that a layer of platinum black assists the coating process on platinum electrodes, whereas Ariel and Koster[200] recommend nickel plating prior to film forma-

[188] W. Kemula, Z. Galus, and Z. Kublik, *Bull. Acad. polon. Sci., Sér. Sci. chim.*, 1959, 7, 732.
[189] G. D. Robbins and C. G. Enke, *J. Electroanalyt. Chem. Interfacial Electrochem.*, 1969, 23, 343.
[190] G. W. Tindall and S. Bruckenstein, *J. Electroanalyt. Chem. Interfacial Electrochem.*, 1969, 22, 367.
[191] F. Malatesti and G. Raspi, *J. Electroanalyt. Chem. Interfacial Electrochem.*, 1970, 27, 283.
[192] F. Malatesti and G. Raspi, *J. Electroanalyt. Chem. Interfacial Electrochem.*, 1970, 27, 295.
[193] U. Eisner and H. B. Mark, jun., *J. Electroanalyt. Chem. Interfacial Electrochem.*, 1970, 24, 345.
[194] E. A. Zakharova, Z. G. Kilina, and G. A. Rachmanina, *Elektrokhimiya*, 1969, 5, 1494.
[195] M. G. Tamba and N. Vantina, *J. Electroanalyt. Chem. Interfacial Electrochem.*, 1970, 25, 235.
[196] M. G. Tamba, N. Vantina, and S. Maneschi, *J. Electroanalyt. Chem. Interfacial Electrochem.*, 1971, 31, 193.
[197] N. Velghe and A. Claeys, *J. Electroanalyt. Chem. Interfacial Electrochem.*, 1972, 35, 229.
[198] M. D. Booth, Ph.D. Thesis, University of London, 1970.
[199] R. Van der Leest, *Analyt. Chim. Acta*, 1970, 52, 151.
[200] G. Koster and M. Ariel, *Israel J. Chem.*, 1969, 9, 21.

tion. The latter observation is in accordance with earlier results which favour nickel or carbon as a substrate rather than platinum. In an interesting paper, Florence[201] describes the use of a glassy carbon electrode, which is mercury-plated *in situ* by addition of mercuric nitrate to the test solution during the pre-electrolysis step. It is claimed that very thin films (10^{-7} cm) are produced by this method which give excellent resolution and sensitivity. Brainina[202] has also suggested simultaneous pre-electrolysis and mercury-film formation by adding mercury oxynitrate to the test solution.

In spite of the problems arising from the heterogeneity of metal electrode surfaces, various types of solid electrode still find application in stripping voltammetry, especially for the determination of the more electropositive metals. Thus platinum,[190, 203] silver,[204] and gold[205] electrodes have all found application under suitable conditions. The platinum ring-disc electrode system described by Tindall and Bruckenstein[190] for the stripping analysis of copper and silver offers interesting possibilities of selective monitoring of multiple-peak dissolution by suitable adjustment of the disc potential. Another interesting application of stripping at metal electrodes has been described by Dolezal *et al.*,[206, 207] who determined Mn^{II} and Pb^{II} after pre-electrolysis under anodic conditions to form MnO_2 and PbO_2 on the electrode, respectively. The cathodic stripping process then involves dissolution of the oxides.

Apart from the above examples the evidence would tend to suggest that one of the various forms of carbon is the most suitable solid electrode material. Vitreous or glassy carbon is becoming increasingly popular as an electrode material and the observations of Kopanica and Vydra[208] on the stripping of Cu and Ag showed slightly superior performance to the wax-impregnated graphite electrode. A rotating carbon-paste electrode has been described by Monien and Jacob,[209] and Chulkina *et al.*[210] have examined the influence of the composition of the paste on the efficiency of anodic stripping at the nanogram level.

Conventional cell design for stripping voltammetry employs a stationary electrode with the solution stirred by a synchronous stirrer. Alternatively, the electrode may be rotated[211] and there is evidence that the latter approach

[201] T. M. Florence, *J. Electroanalyt. Chem. Interfacial Electrochem.*, 1970, **27**, 273.
[202] E. M. Roizenblat and Kh.Z. Brainina, *Elektrokhimiya*, 1969, **5**, 396.
[203] D. P. Sandoz, R. M. Peekema, H. Freund, and C. F. Morrison, *J. Electroanalyt. Chem. Interfacial Electrochem.*, 1970, **24**, 165.
[204] D. J. Astley, J. A. Harrison, and H. R. Thirsk, *J. Electroanalyt. Chem. Interfacial Electrochem.*, 1968, **19**, 325.
[205] R. C. Propst, *J. Electroanalyt. Chem. Interfacial Electrochem.*, 1968, **16**, 319.
[206] E. Hrabankova, J. Dolezal, and V. Masin, *J. Electroanalyt. Chem. Interfacial Electrochem.*, 1969, **22**, 195.
[207] E. Hrabankova, J. Dolezal, and P. Beran, *J. Electroanalyt. Chem. Interfacial Electrochem.*, 1969, **22**, 203.
[208] M. Kopanica and F. Vydra, *J. Electroanalyt. Chem. Interfacial Electrochem.*, 1971, **31**, 175.
[209] H. Monien and P. Jacob, *Z. analyt. Chem.*, 1971, **255**, 33.
[210] L. S. Chulkina, S. I. Sinyakova, and E. K. Vul'fson, *Zhur. analit. Khim.*, 1970, **25**, 1268.
[211] M. J. D. Brand and B. Fleet, *J. Polarog. Soc.*, 1967, **13**, 77.

gives improved precision. Koster and Ariel[212] and Stulik et al.[213] have described flow cells for stripping voltammetry, and Brand et al.[214] have designed a fully automated system programmed by a cam-operated cycle timer. Kemula[215] has described a 'sluice' arrangement for carrying out stripping with automatic standard addition.

One approach to increasing sensitivity is the use of a micro cell. Huderova and Stulik[216] report on a cell design capable of handling solution volumes of 100 µl, while Eliseeva and Sinyakova[217] have designed a thin-layer cell which will permit stripping on as little as 2.5 µl. Whereas most modern polarographs include stripping voltammetry as only one of their many functions, an instrument has recently appeared[218] designed specifically for this technique. The instrument is capable of operating up to 12 separate cells and uses a composite mercury–graphite electrode.

Most applications of stripping voltammetry employ a linear potential sweep for the stripping step, but one way of improving sensitivity is to use one of the more sensitive polarographic variants for the stripping process.

Several groups of workers have employed chronopotentiometric stripping.[219-222] In one system[219] the potential of the galvanic cell element formed between the reference electrode and the hanging mercury drop electrode containing the deposited metals is monitored as the stripping process occurs.

The use of a.c. polarography for the stripping step[223, 224] offers a very useful approach since capacitance current and the effect of irreversibly reduced interferences are effectively eliminated. An increase in sensitivity over conventional linear-sweep methods of one order of magnitude has been demonstrated. Pulse stripping has also been described[225, 226] and has the advantage that it can be used with low concentrations of background supporting electrolyte. Monien et al.[225] have compared several techniques, including pulsed voltage sweep, for carrying out the stripping process.

Differential or substractive techniques have also been used in stripping voltammetry. Kemula[219] used a system of two similar electrodes in the test solution and carried out the pre-electrolysis at each for varying lengths of

[212] G. Koster and M. Ariel, *J. Electroanalyt. Chem. Interfacial Electrochem.*, 1971, 33, 339.
[213] L. Zieglerova, K. Stulik, and J. Dolezal, *Talanta*, 1971, 18, 603.
[214] M. D. Booth, M. J. D. Brand, and B. Fleet, *Talanta*, 1970, 17, 1059.
[215] W. Kemula, *Pure Appl. Chem.*, 1967, 15, 283.
[216] L. Huderova and K. Stulik, *Talanta*, 1972, 19, 1285.
[217] L. V. Eliseeva and S. I. Sinyakova, *Zhur. analit. Khim.*, 1971, 26, 1171.
[218] Multiple Anodic Stripping Analyser, Joyce Loebl Ltd., Gateshead.
[219] W. Kemula, *Pure Appl. Chem.*, 1970, 21, 449.
[220] A. Baranski and Z. Galus, *J. Electroanalyt. Chem. Interfacial Electrochem.*, 1971, 30, 219.
[221] M. S. Zacharov, V. V. Pnev, and V. I. Bakanov, *Zavod. Lab.*, 1970, 36, 643.
[222] M. S. Zacharov, V. V. Pnev, and V. I. Bakanov, *Elektrokhimiya*, 1971, 7, 611.
[223] T. Miwa, S. Oki, and A. Mizuike, *Bunseki Kagaku*, 1969, 18, 1406.
[224] A. M. Bond, *Analyt. Chem.*, 1970, 42, 1165.
[225] B. Lendermann, H. Monien, and H. Spekker, *Z. analyt. Chem.*, 1970, 250, 296.
[226] G. D. Christian, *J. Electroanalyt. Chem. Interfacial Electrochem.*, 1969, 23, 1.

time. By displaying the difference in stripping currents the effects of charging current and impurities were eliminated.

Applications.—The range of applications of stripping voltammetry is rapidly expanding. The traditional areas of application to the analysis of trace metallic impurities in pure chemicals, metals, semiconductors *etc.* (see refs. 177—183 for extensive bibliography) have been extended to cover problems in the broader field of environmental pollution monitoring. The determinations of lead in biological materials,[227] trace metals in sea water,[228] and marine samples[229] have recently been reported. Eisner and Mark[193] showed that stripping voltammetry was more sensitive than neutron activation analysis for the determination of silver at the 10^{-10} mol l^{-1} level in rain and snow samples from AgI 'seeded' clouds. The measurement of trace heavy metals in natural water and industrial effluent has been carried out by anodic stripping voltammetry using pulsed anodic stripping.[230] This method has several valuable features: firstly, the measurement does not require the additions of electrolyte or buffer to the sample; secondly, by the addition of controlled demasking agents the degree of complexation of metal pollutants can be studied;[231] finally, the technique is very economical in terms of equipment cost. Pulse polarographic equipment costing around £1200 is now available whereas the cost of equipment for spectroscopic techniques capable of achieving a similar sensitivity would be at least a factor of three higher.

In the technique of cathodic stripping voltammetry the electroactive species is deposited at a working anode either as an insoluble film[206, 207] or as an insoluble mercury complex. Brand and Fleet have described the determination of the thiocarbamate group of pesticides[232, 233] by stripping of the insoluble mercury derivative from a rotating mercury film electrode. Tetramethylthiuram disulphide was determined after chemical reduction to the dimethyldithiocarbamate anion,[232] and the multiple stripping peaks obtained for ethylenebisdithiocarbamate series were attributable to known impurities in the original sample.[233]

Stripping voltammetry has also been applied for the determination of organometallics. The reduction of triphenyltin derivatives, which are widely used as stabilizers and fungicides, proceeds *via* the formation of an organotin radical in the first step:[234]

[227] W. Oelschlaeger and R. Gilg, *Landwirtsch Forsch.*, 1969, **22**, 218.
[228] J. D. Smith and J. D. Redmond, *J. Electroanalyt. Chem. Interfacial Electrochem.*, 1971, **33**, 169.
[229] T. M. Florence, *J. Electroanalyt. Chem. Interfacial Electrochem.*, 1972, **35**, 237.
[230] M. E. Allen, W. R. Matson, and K. H. Mancy, *J. Water Pollut. Control Fed.*, 1970, **42**, 573.
[231] H. K. Hundley and E. C. Warren, *J. Assoc. Offic. Analyt. Chemists*, 1970, **53**, 705.
[232] M. J. D. Brand and B. Fleet, *Analyst*, 1970, **95**, 1023.
[233] M. J. D. Brand and B. Fleet, *Analyst*, 1970, **95**, 1136.
[234] M. D. Booth and B. Fleet, *Analyt. Chem.*, 1970, **42**, 825.

Q

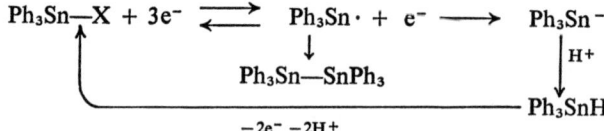

The radical, which is stabilized by adsorption at the mercury electrode, has a sufficient lifetime, even in a protogenic medium, to enable stripping voltammetry to be carried out. Triphenyltin residues in potato samples at the 0.001 p.p.m. level were detectable.

4 Coulometry

There are two main branches of coulometric techniques, controlled potential and controlled current. During the period under review, however, relatively few papers of significance in the field of controlled-potential coulometry have appeared so both techniques will be considered together.

A number of reviews covering constant-current coulometry,[235] coulometry,[236, 237] coulometric analysis of oil products,[238] and the coulometric determination of carbon, oxygen, chlorine, and sulphur in biological materials[239] have been written. Mann and Barnes[240] have compiled electrochemical data for a wide range of inorganic and organic substances in non-aqueous solvents which should be of great value.

Numerous designs for constant-current sources have been described.[241-246] A solid-state coulometric titrimeter for the semi-automatic determination of mercaptans has been designed by Devay et al.,[247] and Nebesar[248] has reviewed commercial instruments for the coulometric determination of sulphur. An instrument which provides a continuously variable current depending on the derivative of the indicator electrode potential has been described by McCracken et al.[249]

Cell designs for coulometry have been reported by Lindquist[250] and Clem

[235] F. P. Ijsseling, *Chem. Tech.*, 1970, **25**, 657.
[236] W. H. Lee, *Electrometric Methods*, 1969, 104.
[237] G. Muto, Y. Takata, *Kagaku No Ryoiki, Zoken*, 1969, 188.
[238] S. Miyake, *Bunseki Kagaku*, 1970, **19**, 1341.
[239] K. Hoshino and T. Ihara, *Rinsho Byori, Rinjizokan*, 1970, **17**, 49.
[240] C. K. Mann and K. K. Barnes, in 'Electrochemical Reaction in Non-aqueous Systems', Marcel Dekker, New York, 1970.
[241] M. Astruc, J. Bentata, J. Bonastre, A. Castelbon, and P. Grenier, *Chim. analyt.*, 1970, **52**, 534.
[242] D. M. Coulson, U. S. P. 3 563 873.
[243] Jungner Instrument AB, B. P. 1 212 890.
[244] J. Martin, Ger. Offen. 1 931 438.
[245] D. R. Rhodes, U. S. P. 3 580 832.
[246] J. T. Stock, *J. Chem. Educ.*, 1969, **46**, 246.
[247] J. Devay, T. Garai, J. Havas, and B. Juhasz, *Hung. Sci. Instr.*, 1972, **22**, 11.
[248] B. Nebesar, *J. Chem. Educ.*, 1972, **49**, A9–A10, A12–A17.
[249] J. E. McCracken, J. C. Guyon, W. D. Shults, and H. C. Jones, *Chem. Instr.*, 1972, **3**, 311.
[250] J. R. O. Lindquist, Ger. Offen. 2 156 669.

et al.[251, 252] The latter design used a rotating cell and permitted fast (6 min) determinations by controlled potential coulometry. A cell design for use as a detector in liquid chromatography has been patented by Muto et al.[253] The errors in coulometry arising from diffusion between different parts of coulometric cells are considered by Lindberg.[254]

A wide range of applications of coulometry has been noted. The absolute nature of the technique makes it ideally suited for calibration of primary standards while its sensitivity and ease of automation make it ideally suited for environmental analysis.

Marinenko and Foley[255] have redetermined the atomic weight of zinc by coulometry, a value of 65.377 being reported with the 95% confidence limits being ±0.003. High-precision coulometric assays of primary standards (potassium hydrogen phthalate, sodium chloride, and potassium dichromate) have been described by Yoshimori et al.[256, 257] The feasibility of using single crystals of sodium chloride as a primary standard was also investigated by these authors.[258] The high-precision determination of boric acid (s.d. 0.0033% for a single determination) has been described by Marinenko and Champion.[259]

Coulometric finishes to the elemental analysis of organic compounds appear to be growing in popularity. Methods for the determination of carbon and hydrogen have been reported,[260-262] the general procedure being to calcine the sample followed by coulometric estimation of the H_2O formed, e.g. with a Pt–P_2O_5 electrolytic cell. The CO_2 is determined either by titration with hydroxide ions or by reaction with lithium hydroxide to give water which may then be measured as above. An automatic coulometric method for carbon and hydrogen has been described by Anisimova et al.[263] These authors have also developed[264] a method for the simultaneous measurement of hydrogen and nitrogen in organic compounds. A Dumas combustion procedure was used, the nitrogen being measured volumetrically whereas hydrogen (as water) was determined coulometrically. Oxygen determination

[251] R. G. Clem, F. Jakob, D. H. Anderberg, and L. D. Ornealas, *Analyt. Chem*, 1971, **43**, 1398.
[252] R. G. Clem, *Analyt. Chem.*, 1971, **43**, 1853.
[253] G. Muto, J. Takata, and Y. Hamano, Ger. Offen. 2 024 008.
[254] I. Lindberg, *J. Electroanalyt. Chem. Interfacial Electrochem.*, 1972, **40**, 265.
[255] G. Marinenko and R. T. Foley, *J. Res. Nat. Bur. Stand.* (*A*), 1971, **75**, 561.
[256] T. Yoshimori and I. Matsubara, *Bull. Chem. Soc. Japan*, 1970, **43**, 2800.
[257] T. Yoshimori, I. Matsubara, K. Hirosawa, and T. Tanaka, *Bunseki Kagaku*, 1970, **19**, 681.
[258] T. Yoshimori and T. Tanaka, *Analyt. Chim. Acta*, 1971, **55**, 185.
[259] G. Marinenko and C. E. Champion, *J. Res. Nat. Bur. Stand.* (*A*), 1971, **75**, 421.
[260] G. F. Anisimova, V. A. Klimova, and I. A. Lavrov, *Otkrytiya, Izobret., Prom. Obraztsy, Tovarnye Znaki*, 1972, **49**, 142.
[261] K. Nakamura, K. Kuboyama, K. Ono, and K. Kawada, *Mikrochim. Acta*, 1972, 353.
[262] K. Nakamura, K. Ono, and K. Kawada, *Microchem. J.*, 1972, **17**, 338.
[263] G. F. Anisimova, V. A. Klimova, I. A. Lavrov, and P. M. Shishkin, *Izvest. Akad. Nauk S.S.S.R., Ser. khim.*, 1972, 700.
[264] G. F. Anisimova and V. A. Klimova, *Izvest. Akad. Nauk S.S.S.R., Ser. khim.*, 1972, 583.

in organic compounds by means of controlled-potential coulometry is described by Karrman and Karlsson.[265] The method was based on the classical chemical procedure except that the iodine produced from the $CO-I_2O_5$ reaction was passed into an electrolytic cell where it was reduced coulometrically at constant potential on a rotating platinum electrode. Coulometric methods for chlorine,[266] phosphorus,[267] manganese and chromium,[268] arsenic, antimony, and copper,[269] and active hydrogen[270] in organic compounds have also been detailed.

The determination of non-metals in metals by coulometry has received some attention. Methods for carbon in iron and steel by determination of the carbon dioxide formed in combustion have been described.[271-273] Sulphur in steel samples has been determined by Kajiyama and Hoshino,[274] and a method for oxygen in titanium is given by Ono et al.[275]

Toxic impurities in the atmosphere and in industrial gas streams can conveniently be monitored coulometrically. A method for sulphur dioxide in gas samples is described by Lahmann and Prescher[276] in which the sulphur dioxide is oxidized by iodine electrolytically generated at a carbon electrode. Interference from hydrogen sulphide was noted but not from ozone, chlorine, ammonia, nitrogen oxides, and carbon monoxide. Bailey and Bishop[277] have used electrolytically generated bromine for titrating sulphur dioxide at concentrations greater than 40 µmol l^{-1} in aqueous solutions using a differential electrolytic potentiometric end-point system. Austin and Creighton[278] have patented a method for the determination of small amounts (<2 p.p.m.) of hydrogen sulphide in gas mixtures, and Austin and Robison[279] have described a procedure for monitoring sulphide and total sulphur. A method for estimating nitrogen dioxide based on the liberation of bromine from bromide solutions and subsequent electroreduction of the bromine has been detailed.[280] An apparatus for continuously monitoring nitrogen dioxide and nitric oxide present in various atmospheres at concentrations below 1.0 p.p.m. has been described by Harman.[281] For the determination of nitric oxide it was

[265] K. J. Karrman and R. Karlsson, *Talanta*, 1972, **19**, 67.
[266] W. Krijgsman, G. de Groot, W. P. van Bennekom, and B. Griepink, *Mikrochim. Acta*, 1972, 364.
[267] R. F. Sympson, *Analyt. Chim. Acta*, 1972, **61**, 148.
[268] M. Bigois and M. Marchand, *Talanta*, 1972, **19**, 147.
[269] M. Bigois and M. Marchand, *Talanta*, 1972, **19**, 157.
[270] G. F. Anisimova and V. A. Klimova, *Izvest. Akad. Nauk S.S.S.R., Ser. khim.*, 1972, 581.
[271] A. I. Orzhekhovskaya, *Novye Metody Khim. analit. Materialy*, 1971, **2**, 45.
[272] W. Thomich, *Arch. Eisenhuettenwissen*, 1972, **43**, 239.
[273] B. Metters, B. G. Cooksey, and J. M. Ottaway, *Talanta*, 1972, **19**, 1605.
[274] R. Kajiyama and K. Hoshino, *Analyst*, 1971, **96**, 835.
[275] H. Ono, K. Hoshino, and T. Ihara, *Bunseki Kagaku*, 1972, **21**, 901.
[276] E. Lahmann and K. E. Prescher, *Wasser, Luft Betr.*, 1971, **15**, 366.
[277] P. L. Bailey and E. Bishop, *Analyst*, 1972, **97**, 311.
[278] R. T. Austin and D. M. Creighton, Ger. Offen., 2 157 632.
[279] R. R. Austin and J. R. Robison, *Amer. Gas Assoc., Oper. Sect. Proc.*, 1971, D28.
[280] V. Z. Al'perin, T. K. Khamrakulov, D. A. Il'yasov, Z. A. Pepelyaeva, and V. V. Chernyakov, *Zavod. Lab.*, 1972, **38**, 643.
[281] J. N. Harman, *Ann. ISA Conf. Proc.*, 1971, **26**, 554.

first converted into nitrogen dioxide using ozone, this being more efficient than chromium trioxide which is more generally used. Methods for chlorine[282] and dichloroacetylene[283] in gases have also been developed.

Karlsson[284] has described the determination of small amounts of water with coulometrically generated Karl Fischer reagent, and a similar method has been used by Sistig and Reinermann[285] to determine water in ethane and by Mika and Cadersky[286] for water in silicone compounds at the p.p.m. level. Water measurements in minerals[287] and olefins[288] have also been made.

Vajgand et al.[289] have used a non-aqueous system for the coulometric determination of mixtures of bases. Binary mixtures of primary or secondary amines with tertiary amines were determined in an acetic acid medium. The hydrogen-ion titrant was generated by oxidation of hydroquinone at a platinum anode. Microgram amounts of biogenic amines have been titrated with hydrogen ions in a acetone–sodium perchlorate medium containing less than 2% water.[290] Amine hydrochlorides could also be determined if mercuric acetate was first added. The applications of non-aqueous coulometry to the analysis of petroleum products has been considered by Miyake:[291] applications to the measurement of moisture, bromine value, and neutralization value are described.

A coulometric method for penicillins has been developed by Mondzhoyan[292] and compared with the conventional iodometric procedure. The coulometric procedure was completely automated and permitted analyses every three minutes. The activity of phosphatases in serum has been determined using a coulometric procedure, the method being to release phenol from the phenyl phosphate enzymatically and then titrate the phenol against electrogenerated bromine.[293] A method for amylase in serum has also been described.[294]

Examples of coulometry used for the determination of pharmaceuticals have appeared. Oliveri-Vigh et al.[295] have determined N-isopropyl-α-(2-methylhydrazino)-p-toluamide hydrochloride in capsules by oxidation to the azo-compound with electrogenerated iodine. Mercury(II) electrolytically

[282] W. H. Parth, *Analyt. Instr.*, 1972, **10**, 9.
[283] F. W. Williams, *Analyt. Chem.*, 1972, **44**, 1317.
[284] R. Karlsson, *Talanta*, 1972, **19**, 1639.
[285] E. Sistig and K. H. Reinermann, Ger. Offen. 2 103 089.
[286] V. Mika and I. Cadersky, *Z. analyt. Chem.*, 1972, **258**, 25.
[287] M. Cremer, H. N. Elsheimer, and E. E. Escher, *Analyt. Chim. Acta*, 1972, **60**, 183.
[288] G. Winkler, *Period. Polytech., Chem. Eng.*, 1972, **16**, 71.
[289] V. J. Vajgand, T. J. Pastor, and R. P. Mihajlovic, *Lucrarile 3rd Conf. Nat. Chim. Analyt.*, 1971, **1**, 45.
[290] J. Becker and K. Beyermann, *Z. analyt. Chem.*, 1972, **258**, 20.
[291] S. Miyake, *Sekiyu Gakkai Shi*, 1971, **14**, 975.
[292] A. L. Mondzhoyan, R. A. Kropivnitskaya, Yu. Z. Ter-Zakharyan, A. A. Sarkisyan, and K. S. Lusararyan, *Khim. Farm. Zhur.*, 1972, **6**, 5.
[293] M. A. Brooks and W. C. Purdy, *Clin. Chem.*, 1972, **18**, 503.
[294] J. R. Moody, *Diss. Abs. (B)*, 1970, **31**, 3227.
[295] S. Oliveri-Vigh, J. J. Donahue, J. E. Heveran, and B. Z. Senkowski, *J. Pharm. Sci.*, 1971, **60**, 1851.

generated has been used for the determination of barbiturates,[296] and manganese(III) has been applied to the determination of N-substituted phenothiazines.[297]

An interesting method for the titration of uranium(VI) with uranium(III) has been described by Farrington and Lingane.[298] Uranium(III), the most powerful reductimetric titrant used to date, can be generated in acidic solution with nearly 100% current efficiency, and 200 mg samples of uranium(VI) can be determined to $+0.09\%$. A controlled-potential coulometric determination of uranium(VI) in the presence of nitrate has been described by Sobkowska,[299] and McCracken et al.[300] have used feed-back controlled electrolysis current to determine uranium and iron in mixtures of these elements. A two-step flow-coulometry procedure for plutonium has been developed by Kihara et al.[301] The sample is first passed through a column where all the plutonium is reduced to the tervalent state and subsequently passed to a second column in which the plutonium is oxidized electrolytically to Pu^{IV}.

Methods for the determination of iron in non-stoicheiometric ferrous oxide,[302] in ferrites and ferric oxide,[303] in wustite,[304] and in electroplating baths[305] have been described. Differential coulometric procedures for manganese(VII), cerium(IV), and vanadium(V) in alloys have been determined with electrolytically generated iron(II),[306] and electrogenerated dichromate in phosphoric acid[307] has been used to titrate iron(II), manganese(II), vanadium (IV), and cerium(III). The titration of chromium(III) with electrogenerated ferricyanide in the presence of a 10^5-fold excess of chromium(VI) has been shown to be feasible[308] and applied to the measurement of chromium(III) impurities in chromium trioxide.[309]

Khamrakulov et al.[310-314] have described what they term substoicheio-

[296] J. R. Monforte and W. C. Purdy, Analyt. Chim. Acta, 1970, 52, 25.
[297] G. J. Patriarche and J. J. Lingane, Ann. pharm. franç., 1970, 28, 511.
[298] G. C. Farrington and J. J. Lingane, Analyt. Chim. Acta, 1972, 60, 175.
[299] A. Sobkowska, Radiochem. Radioanalyt. Letters, 1971, 8, 357.
[300] J. E. McCracken, J. C. Guyon, and W. D. Shults, Analyt. Chim. Acta, 1971, 57, 151.
[301] S. Kihara, T. Yamamoto, K. Motojima, and T. Fujinaga, Talanta, 1972, 19, 657.
[302] E. I. Pil'ko and V. A. Kozheurov, Sb. Nauch. Trudov, Chelyabinsk Politekh. Inst., 1970, 66, 67.
[303] S. N. Nikolaeva, O. N. Fedorov, and T. E. Komissarova, Zavod. Lab., 1971, 37, 913.
[304] E. I. Pil'ko, Sb. Nauch. Trudov, Chelyabinsk Politekh. Inst., 1970, 66, 72.
[305] G. De Kauilis, D. Merigot, and C. Pourcel, Chim. analyt., 1971, 53, 696.
[306] E. R. Nikolaeva, T. N. Lakeeva, and P. K. Agasyan, Zhur. analit. Khim., 1972, 27, 497.
[307] A. I. Kostromin, A. A. Akhmetov, and L. N. Burygina, Zhur. analit. Khim., 1972, 27, 315.
[308] V. M. Masalovich and T. Kh. Pirskaya, Zhur. analit. Khim., 1972, 27, 802.
[309] V. M. Masalovich and T. Kh. Pirskaya, Sb. Trudov Ural. Nauch.-Issled. Khim. Inst., 1971, 26, 24.
[310] T. K. Khamrakulov and V. V. Chervyakova, Trudy Samarkand. Univ., 1970, 180, 70.
[311] T. K. Khamrakulov, P. K. Agasyan, and V. V. Chervyakova, Zhur. analit. Khim., 1972, 27, 492.
[312] P. K. Agasyan, L. N. Smyshlyaeva, and T. K. Khamrakulov, Zhur. analit. Khim., 1972, 27, 257.

metric coulometry for the determination of various elements. The method consists of depositing a known or reproducible fraction of the material to be determined on an electrode and then coulometrically stripping it from the electrode in a manner similar to stripping voltammetry. Bismuth could be determined by this procedure over the concentration range 2×10^{-5}—10^{-3} mol l^{-1} with an error of less than 5%.[310] The optimum conditions for determining silver and bismuth in alloys,[311] zinc in electroplating baths,[312] and selenium(IV)[313, 314] have also been given.

A number of papers concerned with the electrolytic generation of various ions for titrimetric procedures have appeared. Of interest is the generation of fluoride ions from a europium-doped lanthanum fluoride crystal.[315] Kostromin et al.[316] have studied the generation, with an efficiency ca. 99%, of vanadium(IV) from a vanadium metal electrode in 5% K_2SO_4 (0.1N-H_2SO_4) medium. The use of externally generated vanadium(IV) for the coulometric titration of potassium ferricyanide in the presence of potassium ferrocyanide has been described by these authors.[317] The conditions for the electrolytic generation of titanium(III) and tin(II) on a lead cathode for use in determining iron(III), vanadyl, and copper(II) ions has also been given.[318]

Fuel cell electrodes, surprisingly, have received little attention analytically, presumably owing to the high cost and technical problems in their manufacture. In a news release from NASA[319] a low-temperature H_2–O_2 fuel cell is described as an oxygen detector. Fleet, Tenygl, and Ho have recently reported the use of a porous catalytic electrode, initially developed as a fuel cell cathode, for a range of oxygen determinations. The device works on a coulometric principle and has been applied to the determination of sodium hypochlorite and hydrogen peroxide[320] and also for monitoring various redox systems.[321] The latter principle has been utilized for a fully automated method for the determination of chemical oxygen demand.[322]

[313] T. K. Khamrakulov, P. K. Agasyan, and D. A. Il'yasov, *Zhur. analit. Khim.*, 1972, **27**, 399.
[314] T. K. Khamrakulov and D. A. Il'yasov, *Trudy Samarkand. Univ.*, 1970, **180**, 83.
[315] Y. K. Lee, K. J. Whang, K. Nozaki, and G. Muto, *Bunseki Kagaku*, 1971, **20**, 1441.
[316] A. I. Kostromin, V. N. Basov, and V. V. Mosolov, *Issled. Elektrokhim., Magnetokhim. Metod. Analyt.*, 1970, **3**, 43.
[317] P. K. Agasyan, V. N. Basov, and A. I. Kostromin, *Vestnik Moskov. Univ.*, 1972, **13**, 353.
[318] A. I. Kostromin, V. V. Mosolov, and I. D. Yuzhanina, *Zhur. analit. Khim.*, 1972, **27**, 1115.
[319] NASA Tech. Brief 65–10066 March 1965.
[320] B. Fleet, A. Y. W. Ho, and J. Tenygl, *Talanta*, 1972, **19**, 317.
[321] B. Fleet, A. Y. W. Ho, and J. Tenygl, *Analyt. Chem.*, 1972, **44**, 2156.
[322] B. Fleet, A. Y. W. Ho, and J. Tenygl, *Analyst*, 1972, **97**, 321.

Errata

Vol. 2, 1971

p. 49 The account of the work by Bailes and Leveson (the penultimate sentence on p. 49) should read:

'An intermediate corresponding to the first six-electron wave was established by controlled-potential electrolysis of p-bromo-ω-diazoacetophenone at the foot of the wave. p-Bromophenylglyoxal hydrazone was identified in the catholyte.'

Author Index

Abdullah, M. I., 220
Accascina, F., 28
Adams, R. N., 211
Aditya, S., 4
Afghan, B. K., 216
Agasyan, P. K., 236, 237
Agawa, T., 173
Aggarwal, I. P., 35
Aguiar, A., 158
Ahluwalia, J. C., 196, 198, 201
Ahmed, I. Y., 32
Aitken, H. W., 206
Akhmetov, A. A., 236
Akimoto, N., 9
Akiyama, A., 87
Albery, W. J., 54
Albisson, A., 125
Alder, J. F., 53, 211
Alei, M., 197, 205
Aleikina, S. M., 59, 61
Alfenaar, M., 5, 19
Allen, M. E., 231
Al'perin, V. Z., 234
Alting, I., 7
Amaya, T., 205
Amblard, J., 75
Ambrose, D., 188
Amis, E. S., 20, 28, 188
Ammann, D., 18
Ammar, I. A., 61, 80, 81, 90
Andalaft, E., 32
Anderberg, D. H., 233
Anderson, H. L., 196
Anderson, K. P., 192
Anderson, R. E., 214
Andreev, K. N., 115
Andreoli, R., 150
Andronati, S. A., 138
Andrusev, M. M., 47
Andruzzi, R., 138, 152
Angell, C. A., 23, 209
Anisimova, G. F., 233, 234
Anstine, W., 225
Antropov, L. I., 119
Aouanouk, F., 54
Aragon, P. J., 60
Arel, M., 24
Ariel, M., 3, 228, 230
Armand, A. J., 133, 137, 143, 148
Armstrong, R. D., 43, 57, 62, 69, 70, 76, 94
Arnett, E. M., 198, 203, 205
Arora, P. C., 167
Arrington, D. E., 31
Arshadi, M., 202

Artamonov, B. P., 23
Arvia, A. J., 31, 77
Asakura, S., 88, 89
Asawa, S., 65
Asplund, J., 216
Assikis, J. H., 211
Astley, D. J., 229
Astruc, M., 232
Atkinson, G., 22, 27, 209
Attia, A. H., 145
Aucouturier, M., 81
Austin, L. G., 79
Austin, R. R., 234
Austin, R. T., 234
Avaca, L. A., 153
Averjanov, V. I., 7
Avrutskaya, I. A., 177
Azim, A. A. A., 60
Azzerri, N., 88

Baba, Yu. I., 85
Bach, H., 9
Backmann, K. J., 52
Baes, C. F., 205
Bagger, C., 79
Bagotskaya, I. A., 71, 102
Bagotskii, V. S., 46, 77, 111
Bahe, L. W., 193
Bailey, A. R., 193
Bailey, P. L., 234
Bailey, P. C. A., 79
Baine, P., 205
Baizer, M. M., 130, 131, 143
Bakanov, V. I., 230
Baker, C. T., 14
Balashov, V. F., 103
Balashova, N. A., 77, 114
Balasubrahamanyan, K., 25
Baldwin, W. G., 5
Ballash, N. M., 207
Bangert, F. K., 194, 196
Bapat, M. R., 106
Barak, M., 4
Baranski, A., 230
Barbey, G., 175
Barbouth, N., 72
Barbré, H., 79
Barclay, D. J., 100
Bard, A. J., 131
Bardin, M. B., 54
Bardina, N. G., 71
Barendrecht, E., 227
Barger, H. J., 167
Barker, G. C., 221
Barlow, A. J., 209
Barmashenko, V. I., 57
Barnartt, S., 51, 88

Barnes, D., 125, 127
Barnes, K. K., 232
Bartak, D. E., 133, 156
Bartenev, V. Ya., 48
Barthel, J., 35, 209
Bartlett, E. S., 93
Barradas, R. G., 109, 175
Bascombe, K. N., 207
Basov, V. N., 237
Basson, A. J., 14
Bates, R. G., 13, 206
Batrakov, V. V., 90, 94
Batyaev, I. M., 5
Baucke, F. G. K., 4, 9
Bauer, H. H., 105, 106, 223, 224
Baum, G., 15
Baumann, E. W., 13, 19
Baumann, W. M., 7
Baumgartner, E., 23
Bax, D., 5, 19
Beacom, S. E., 73
Bech-Nielsen, G., 88
Bechtler, A., 209
Becker, J., 235
Behar, B., 97
Behl, W. K., 63
Behr, B., 103
Behret, H., 209
Belew, W. L., 212
Belinskaya, F. A., 14
Belkova, N. L., 5
Bell, R. P., 207
Belyaeva, M. E., 46
Benn, R. C., 66
Ben-Naim, A., 189
Benner, E. J., 217
Bennetto, H. P., 3
Bennion, D. N., 31, 51
Benoit, R. C., 206
Bentata, J., 232
Beran, P., 229
Berne, D. H., 19
Bernhardsson, E., 161
Beronius, P., 30
Berry, W. E., 93
Bertha, S. L., 23
Bertocci, C., 65
Bertram, J., 165
Bessonova, T. A., 61
Beutler, P., 68
Bewick, A., 120, 153
Beyermann, K., 235
Bezborouah, C. P., 195
Bezuglyi, V. D., 107, 153
Bhatnagar, O. N., 193
Bieder, A., 226
Biedermann, G., 195
Biegler, T., 112

239

Author Index

Bierly, T., 24, 208
Bignold, G. J., 205
Bigois, M., 234
Birke, R. L., 51
Bishop, E., 234
Blakeley, C. P., 216
Blandamer, M. J., 20, 188, 196
Bobbitt, J. M., 181
Bockris, J. O'M., 55, 83, 89, 187, 193
Boden, D. P., 7, 33
Boden, P. J., 73
Bodennec, G., 170
Bodyu, V. I., 227
Bolzan, J. A., 31
Bogatskii, A. V., 138
Bogdanovskaya, V. A., 112
Boháčková, V., 187
Boileau, S., 31
Boksay, Z., 8
Bolton, P. D., 206
Bolocofsky, D., 200
Bombara, G., 88
Bonastre, J., 232
Bond, A. M., 210, 215, 216, 223, 224, 225, 230
Bond, A. P., 88
Bondarevskaya, E. A., 13
Bonnemay, M., 77
Bonner, O. D., 192, 206
Bontempelli, G., 184
Booth, M. D., 216, 218, 224, 226, 228, 230, 231
Bordet, J., 54
Borello, A., 212
Borisov, A. I., 177
Bornong, B. J., 91
Borovaya, N. A., 106
Bostanov, V., 66
Bothwell, A., 203, 204
Bothwell, A. L. M., 5, 203, 204
Bottenberg, W. A., 201
Boudreau, C. F., 193
Boulares, L., 143
Bouquet, G., 8
Bower, V. E., 195
Boyd, G. E., 195, 199
Brabec, V., 221
Brafov, B. M., 54
Brainina, Kh. Z., 227, 229
Brand, M. J. D., 2, 8, 13, 229, 230, 236
Braun, J., 217
Braunsburger, S., 57
Braunstein, H., 195
Braunstein, J., 20, 195
Breiter, M. W., 72
Breitschwerdt, K. G., 209
Bressel, R. D., 25, 209
Breukel, J. S. M. C., 222
Brewer, A. D., 205
Brezlau, B. R., 197
Britton, W. E., 134
Britz, D., 105, 106, 223
Bro, P., 63
Broadhead, J., 3

Broadwater, T. L., 23, 26, 208
Brocke, W. A., 221
Brodsky, A. M., 55
Bronöel, G., 77
Brook, P. A., 59
Brooker, M. H., 208
Brookes, H. C., 30
Brooks, M. A., 235
Brophy, R. G., 216
Broszkiewicz, R. K., 219
Brown, O. R., 134
Brown, P. G. M., 190
Brownsword, R., 75
Bruckenstein, S., 67, 78, 117, 228
Bruk, B.-S., 222
Brummer, S. B., 37, 39
Brun, B., 40
Brunel, P., 226
Brusic, V., 89
Bryksin, I. E., 222
Brzostowska, M., 43
Bubelis, J., 66
Buchanan, E. B., jun., 226
Buchanan, T. S., 209
Buck, E., 51
Buck, R. P., 1, 8
Budewskii, E., 66, 78
Buisson, C., 206
Buhl, H., 100
Bukhtiarov, A. V., 183
Bulischeck, T. S., 192
Bull, G. C., 213
Bull, R. H., 213
Bulter, J. N., 205
Buncel, E., 206, 207
Bunting, J. W., 209
Burbank, J., 60
Burley, D. M., 191
Burshstein, R. Kh., 46, 99, 113, 114
Burton, R. E., 202
Bury, R., 21, 23
Burygina, L. N., 236
Busch, M., 23
Bushrod, C. J., 74
Butin, K. P., 132, 158
Butler, C. G., 24
Butler, E. A., 192
Butler, J. N., 16, 195
Butler, R. N., 208
Byallozor, S. G., 67

Cabana, A., 197, 199
Cachet, C., 110
Cachet, H., 110
Cadersky, I., 235
Cadle, S. H., 67, 117
Cahan, B., 213
Caille, G., 217
Caja, J., 100
Calandra, A. J., 112
Calmes-Perraud, F., 32
Campanella, L., 43
Campbell, A. N., 27, 193
Campbell, M. H., 224
Campbell, R., 10
Campion, J. J., 198

Canterford, D. R., 210
Canterford, J. H., 223, 224
Card, D. N., 190
Cardinali, M. E., 138, 152
Carelli, I., 152
Carmack, G. P., 11, 18
Carman, P., 21
Carpenter, A. K., 174
Carr, J. P., 46, 74
Carrozza, J. S. W., 77
Cassel, R. B., 201
Castell, J. F., 28
Castlebon, A., 232
Cattrall, R. W., 11
Caullet, C., 175
Cauquis, G., 173, 175, 184
Cemenov, V.-A., 135
Cezner, V., 50
Chakrabati, S., 4
Champion, C. E., 233
Champion, P. 45
Chang, B., 183
Chang, J., 121
Chantooni, M. K., 204
Chao, F., 72, 77
Chase, J. W., 199
Chattopadhay, P. K., 205
Chawla, B., 196, 201
Cheh, H. Y., 71, 72
Chelepin, I. V., 168
Chen, H., 205, 206
Chen, Y., 189
Chernega, L. P., 227
Chernomorskii, A. I., 79
Chernyakov, V. V., 234
Chervyakova, V. V., 236
Cheytanov, H., 50
Chidambaram, S., 151
Chikamori, K., 73
Chiku, T., 74
Childs, C. W., 195
Childs, W. V., 131, 183
Chin, D. T., 54, 72, 73
Chin, R. J., 91
Chizhov, A. V., 43
Chkir, M., 168
Cho, S. J., 189
Choppin, G. R., 23
Choux, G., 206
Christensen, J. J., 3, 17
Christenson, P. G., 195
Christian, G. D., 16, 230
Christie, I. R. A., 69
Christoffersen, M. R., 192
Chruma, J. L., 130
Chugunova, L. V., 68
Chulkina, L. S., 229
Chupina, L. A., 177
Chu-Xuan-Anh., 226
Clack, D. W., 202
Claeys, A., 228
Clavilier, J., 45
Cleghorn, H. P., 143
Clem, R. G., 214, 233
Clementi, E., 202
Clerc, J. T., 10
Coetzee, C. J., 14
Coetzee, J. F., 32, 197, 204
Coeuret, F., 54

Author Index

Cogley, D. R., 205
Cognard, J., 173
Cognet, G., 54
Cohen, B. J., 123
Cohen, E. G. D., 21
Cohen, E. M., 218
Colchester, J. E., 157
Coleman, J. P., 167
Collie, C. H., 209
Connery, J. G., 212
Convert, O., 148
Conway, B. E., 118, 128, 193, 199
Cook, R. O., 4, 209
Cooksey, B. G., 234
Cornish, D. C., 10
Corsi, R., 218
Cosgrove, R. E., 17
Costa, M., 72, 77
Cottrell, P. T., 155
Coulson, D. M., 232
Courtot-Coupez, J., 34
Cover, R. E., 212
Covington, A. K., 1, 5, 12, 187, 192, 195, 202, 206, 208
Crawford, R. A., 24
Creason, C., 214
Creighton, D. M., 234
Cremer, M., 235
Crespy, G., 45
Criss, C. M., 193, 198
Cros, J.-L., 173
Csakvari, B., 8
Csanyi, P. F., 27
Cunningham, G. P., 23, 26
Curthoys, G., 193
Cwiklinski, C., 163

Dagless, M. N., 2
Daguenet, M., 54
Dahlberg, D. B., 203
Dahlgren, G., 3
D'Alkaine, C. V., 97
Daly, J., 21
Daly, J. E., 202
Damaskin, B. B., 43, 94, 96, 98, 103, 104, 106, 108, 119
Damjanovic, A., 55
Danesi, P. R., 15
Danford, M. D., 196
D'Aprano, A., 23, 26, 208
Darbari, G. S., 196
Darchen, A., 150
Darlington, W. B., 24
Das, M. N., 204, 205
Das, N. N., 5
Dass, N., 189
Date, K., 11, 13
Date, Y., 132
Daver, A., 122, 133
Davies, A., 4, 209
Davies, D. E., 90
Davies, J., 208
Davies, J. E. W., 11
Davies, J. S., 197
Davis, D. D., 204
Davydov, A. D., 57
Day, M. C., 205

Day, R. A., jun., 123
De, A. L., 5
De Carolis, M., 212
Dechy, P., 45
Degner, D., 122
de Groot, G., 234
de Haas, K. S., 200
De Kauilis, G., 236
Delahaye, D., 175
Delesalle, G., 29
de Ligny, C. L., 5, 19
De Lisi, R., 28, 29
Demange-Guerin, G., 6
Demerie, W., 220
Demey, J. P., 29
Denessen, H.-J. M., 19
Denning, K. F., 198
Dennison, R. W., 168
de Paz, M., 202
Desideri, P. G., 172
Desiento, R. P., 30
De Smet, D. G., 89
Desnoyers, J. E., 24, 193, 198, 199
Despić, A. R., 56, 83
de Trobriand, A., 197, 199
Deurainne, P., 29
Deurenberg, H., 12
Devanathan, M. A. V., 44, 118
Devay, J., 52, 223, 232
de Visser, C., 203
Dexter, R., 189
Dhawle, S. W., 106
DiBari, G. A., 58
Diebler, H., 209
Diggens, A. A., 9
Dikusar, A. I., 54
Dill, J. F., 209
Dirkse, T. P., 83
Dirlam, J. P., 160
Dismanovic, D., 218
Ditter, W., 189
Dmitriev, V. A., 82
Dmitrievskaya, L. I., 153
Dobos, S., 8
Dobren'kov, G. A., 105
Dobson, J. V., 2
Dogonadze, R. R., 49
Dohner, R., 10
Dojlido, J., 103
Dolenko, A., 206
Dolezal, J., 227, 229, 230
Dolman, D., 207
Dolofer, K., 106
Donahue, J. J., 235
Doo, V. Y., 92
Dorfmann, G. G., 22
Doucet, Y., 32
Douglass, D. C., 189
Doupeux, H., 137
Dousek, F. P., 59
Doustin, D. R., 188
Downes, C. J., 195
Drury, J. S., 54
Dryhurst, G., 140, 142, 176, 218, 223
Dubini-Paglia, E., 14, 15, 192
Ducauze, C., 63

Duer, W. C., 13
Dumas, J., 6
Dunaeva, T. I., 78
Dunlop, P. J., 192, 201
Dunn, L. A., 26, 36, 194, 209
Dunning, J. S., 51
Dunsmore, H. S., 23, 208
Durou, C., 69
Durst, R. A., 10
Dvořák, A., 58
Dvořák, J., 187
Dwivedi, A. N., 71
D'yachenko, I. A., 141
Dyatkina, S. L., 119
Dzhaparidze, D. I., 104
Dzhaparidze, Sh. S., 104
Dzidic, I., 202

Eagland, D., 197
Eatough, D. J., 3
Ebeling, W., 22, 188
Eberil', V. I., 70
Ebersbach, U., 89
Eberson, L., 121, 160, 161, 167
Eckfeldt, E. L., 9
Egorov, L. Ya., 44
Ehrenson, S., 202
Eichkorn, G., 64
Eigen, M., 209
Eisner, U., 228
Elczov, E. U., 146
El'gashevich, A. M., 190
Eliášek, J., 85
Elina, L. M., 70
Eliseeva, L. V., 230
Elsaid, M., 60
El-Said Mahgaub, A., 27
Elsdale, R. N., 83
El Sheikh, F. M., 168
Elsheimer, H. N., 235
El-Sobki, K. M., 60
Elving, P. J., 3, 138
Engel, G., 35
Engell, H.-J., 88
Enke, C. G., 228
Entwisle, J. H., 157
Enyo, M., 48
Epelboin, I., 75, 91
Erdey Gruz, T., 27
Erlich, R. H., 205
Ershler, A. B., 43, 47, 121, 158
Escalante, E., 87
Escher, E. E., 235
Esikova, I. A., 158
Essig, T. R., 207
Evans, C. E., 207, 209
Evans, D. F., 23, 26, 29, 35, 197, 208
Evans, D. H., 125
Evans, J. M., 73
Evans, U. R., 86
Ewald, A. H., 40
Eyal, E., 18

Faita, G., 121
Falck, J. R., 184
Falkenhagen, H., 22, 188

Author Index

Fano, V., 213
Farkas, J., 54, 65
Farr, J. P. G., 75
Farren, G. M., 13
Farrimond, M. S., 202
Farrington, G. C., 236
Farsang, Gy., 53
Fasman, A. B., 59
Fateev, S. A., 102
Fay, D. P., 209
Feakins, D., 204
Fedash, P. M., 68
Fedoroňko, M., 129
Fedorov, O. N., 236
Fedorova, L. A., 132
Fedoseev, D. V., 57, 60
Fedoseeva, T. A., 60
Feillolay, A., 201
Feller, H. G., 76
Feltham, A. M., 2, 78
Feng, E., 121
Ferles, M., 145
Fernandez-Prini, R., 21, 23
Fidelus, J., 217
Filinovskii, V. Yu., 54
Finch, A., 193
Fioshin, M. Ya., 177
Fischer, H., 64
Fisher, D. J., 212
Fisher, F. H., 209
Flato, J. B., 210
Flatt, R. K., 59
Fleet, B., 53, 211, 216, 218, 223, 224, 226, 229, 230, 231, 237
Fleischmann, M., 54, 134, 165, 178, 185, 186
Fleury, M.-B., 122, 184
Flid, R. M., 158
Flinn, D. R., 46, 111
Flora, H. B., 206
Florence, T. M., 229, 231
Florin, A. E., 197
Førland, T., 19
Foley, R. T., 91, 233
Forchhammer, P., 88
Forcier, G. A., 32
Forsén, S., 197
Forster, T. E., 211
Fossdal, K., 218
Fouad, M. G., 69
Fox, H. M., 183
Fox, M. F., 188
Franck, E. U., 36, 37
Frank, H. S., 19
Frankel, L. S., 205
Frankenthall, R. P., 89
Frant, M. S., 16
Fratiello, A., 23, 204
Frazer, J. W., 214
Freeman, J. G., 206
Freidlin, G. N., 168
Freiser, H., 11, 15, 18
Frensdorff, H. K., 17
Freund, H., 229
Fricke, G. H., 14
Fried, I., 4
Friedman, H. L., 24, 190, 194, 197, 203

Friedman, L., 202
Fritsch, K., 209
Frumkin, A. N., 94, 98, 110, 113, 121
Fry, A. J., 134
Fülleova, E., 129
Fueno, T., 183
Fujinaga, T., 159, 236
Fujishima, A., 54
Fujiwara, S., 222
Fukui, K., 219
Fuoss, R. M., 21, 26, 27, 208
Furuta, S., 6
Furuta, T., 163
Furutani, S., 227
Froment, M., 75, 86

Gabe, D. R., 66
Gagnon, E. G., 79
Gal, I. J., 209
Galatin, A. F., 138
Galinker, V. S., 67
Galli, R., 14, 15, 192
Galus, Z., 228, 230
Galushko, V. P., 68
Gambaretto, G., 183
Gamboa, J. M., 91
Gamburg, E. M., 66
Gammann, K., 10
Gancy, A. B., 37, 39
Gandhi, A. H., 206
Ganshina, I. M., 103
Garai, T., 223, 232
Gardam, P., 29
Gardner, A. W., 190, 191
Gardner, P. J., 193
Garrera, H. A., 77
Garz, I., 92
Gaskin, J. E., 143
Gaunitz, U., 109
Gavach, C., 10
Gavar, P. A., 148
Gavioli, G. B., 150
Gedansky, L. M., 71
Genieś, M., 173, 175
Genshaw, M. A., 70, 89
Gerasimenko, M. A., 119
Gerasimenko, Yu. S., 119
Gerovich, V. M., 104
Giammario, P., 2, 5
Gibbard, H. F., 194
Gibbons, J. J., 5, 203, 204
Gibofsky, A., 19
Giese, K., 209
Gieskes, J. M., 195
Gildseth, W. A., 189
Gileadi, E., 5, 43, 123
Gilg, R., 231
Gilkerson, W. R., 36, 209
Gill, D. S., 34
Gillen, K. T., 189
Giller, S. A., 148
Gillibrand, M. I., 79
Gilman, S., 117
Ginsburg, V. A., 151
Giordano, M. C., 175
Giuliani, L., 87
Gladkova, L. K., 157
Glasser, F. D., 51

Gleim, V. G., 24
Glover, D. E., 222
Glueckauf, E., 190, 191, 209
Gnusin, N. P., 56
Goffman, Ya. A., 58
Goffredi, M., 28, 29, 198
Gokhshtein, A. Ya., 42, 46, 111
Gokman, N. S., 9
Goldberg, R. N., 19
Gol'dfarb, Ya. L., 141
Goldman, S., 206
Goldstein, G., 6
Golsworthy, W. W., 214
Golubev, V. N., 67
Gonzalez, E. R., 97
Goodfellow, G. I., 9
Goodfriend, M. J., 3
Gopal, R., 34
Gorbunova, K. M., 65
Gordievskii, A. V., 13, 14
Goulden, P. D., 216
Govrilova, V. I., 7
Grachev, D. K., 64
Grafor, B. M., 54
Grahame, D. C., 98, 102
Grand, R., 96
Grandi, G., 150
Granovskii, Y. V., 227
Gravchev, D. K., 62
Greco, P. W., 30
Green, R. D., 204
Greenspan, L., 78
Grekovich, A. L., 14
Grenier, P., 232
Greyson, J., 196, 203
Griepink, B., 234
Griffiths, G. M., 11
Grigor'ev, N. B., 102, 107
Grilikhes, M. S., 23
Grimshaw, J., 121
Grimshaw, J. T., 121
Griswold, E., 31
Grochowska, Z., 217
Grofcsik, M. N., 211
Gromyko, V. A., 112
Gronwall, T. H., 191
Grunwald, E., 205
Gubbins, K. E., 200
Gubskaya, V. P., 82
Gudin, N. V., 66
Gürtler, O., 226
Guggenheim, E. A., 22, 208
Guidelli, R., 211
Guidotti, G. R., 212
Guilbault, G. G., 12
Guillanton, G. L., 122, 129, 133
Guitton, J., 75
Gultyai, V. P., 138, 154
Gundersen, N., 224
Gupta, S. L., 151
Gurevich, L. I., 66
Gurevich, Yu. Ya., 55
Guseva, L. T., 105
Gutknecht, W. F., 213, 214
Guyon, J. C., 232, 236

Author Index

Haas, Y., 204
Haase, R., 25
Habenschuss, A., 189
Haberman, J. P., 216
Hackerman, N., 45, 51, 90, 91
Häfke, U., 92
Hafez, A. M., 29
Haggis, G. H., 209
Haihizume, M., 27
Hair, M. L., 7
Hall, D. G., 204
Hallcher, R. C., 181
Haluk, J. P., 129, 130
Halvorsen, S., 217
Hamann, S. D., 37
Hamano, Y., 233
Hamelin, A., 45, 47
Hampson, N. A., 46, 62, 64, 69, 74, 83
Hand, R., 174
Hanna, E. M., 21, 208
Hansen, B. H., 142, 176
Hansen, E. H., 11
Harman, J. N., 234
Harrar, J. E., 214
Harrison, J. A., 57, 154, 229
Harrison, M., 222
Hart, A. C., 92
Hartman, D., 37
Haruyama, S., 65, 72, 89
Hasebe, K., 221
Hashinaga, F., 227
Hashizume, G., 90
Haskett, P. R., 216
Hassall, M. L., 208
Hasted, J. B., 209
Havas, J., 232
Hawley, M. D., 133, 156
Hayakawa, Y., 93, 219
Haynes, R., 213
Haynes, W. M., 15
Hazard, R., 147
Hearn, B., 205
Heasman, M. J., 217
Heckner, K. H., 115
Hefter, G., 223
Heimler, D., 172
Hemmes, P., 29
Henderson, M., 69
Henry, R. P., 6
Henry, W D., 86
Hepfinger, N. F., 3
Heplcr, L. G., 71, 205, 206
Hermann, R. B., 196
Hermolin, J., 123, 152
Herrmann, C. C., 52
Hertz, H. G., 188, 197
Heumann, T., 80
Heveran, J. E., 235
Heyrovsky, J., 210
Heyrovsky, M., 150, 225
Hibbs, C., 28
Hickman, H. J., 19
Higashiyama, K., 13
Hill, J. O., 17
Hill, R. J., 194
Hills, G. J., 20, 36, 96
Hine, F., 25

Hinton, J. F., 20, 188
Hirata, H., 11, 13
Hirosawa, K., 233
Hitchman, M. L., 54
Ho, A. Y. W., 237
Ho, F. C., 77
Hoar, T. P., 59
Hoare, J. P., 73
Hoch, M. J. R., 189
Hochary, J. M., 29
Hoft, E., 189
Hoffman, A. K., 141, 163
Hojman, J., 157
Holland, F. S., 168
Holleck, L., 150
Holly, S. N., 46
Holzaple, W. B., 37
Homolka, J., 211
Honda, K., 54
Honda, M., 219
Hoogland, J. G., 77
Hopkins, H. P., 207
Horanyi, G., 77, 115, 116, 117, 119
Horita, K., 83
Horner, L., 122
Hoshino, K., 232, 234
Hosono, K., 67
Hostomsky, J., 50
Hotz, M. C. B., 30
House, H. O., 121
Howard, L. O., 206
Hozumi, K., 9
Hrabankova, E., 229
Hranilovic, J., 150
Hrubá, H., 145
Hsai, K. L., 208
Hsu, E. T., 163
Hsueh, L., 50
Huang, S. J., 163
Huber, J. F. K., 219
Huderova, L., 230
Huebert, B. J., 222
Hull, M. N., 74, 82
Humphrey, R. E., 215
Humphries, W. T., 192
Hundley, H. K., 231
Hung, J. H., 200
Hunter, J. L., 209
Huong, N. V., 45
Hurkot, D. G., 205
Huston, J. L., 215
Huston, R., 16, 195
Hutson, V. C. L., 191
Hyne, J. B., 199

Ihara, T., 232, 234
Ijsseling, F. P., 232
Ikeda, A., 164
Iketani, H., 54
Il'Yasov, A. V., 66
Il'yasov, D. A., 234, 237
Imoto, E., 145
Inaba, Y., 75
Indelli, A., 24, 194
Inui, T., 67
Inuta, S., 25
Iofa, Z. A., 90
Irish, D. E., 192, 205, 206, 208

Ishibashi, N., 15
Ismail, M. I., 69
Ito, S., 73
Ivanov, V. F., 103
Ivanov, V. T., 52
Ivanova, R. V., 103
Ivanovskaya, I. S., 7
Iversen, P. E., 151, 219
Ives, D. J. G., 4
Ives, M. B., 92
Iwachide, T., 209
Iwakura, C., 173
Iwamoto, R. T., 4
Iwasaki, T., 155
Izatt, R. M., 3, 17
Izutsu, K., 159

Jackopin, L. G., 209
Jacob, P., 229
Jacobsen, E., 217, 218, 224, 225
Jacobsen, T. V., 217
Jacquot, D., 63, 68
Jakli, G., 188, 192
Jakob, F., 233
Jakuszewski, B., 119
Jalota, S. K., 23, 208
James, H. J., 11, 18
James, W. J., 60, 82, 163
Jamk-Czachor, M., 89
Jana, D., 204, 205
Jansco, G., 188
Janssen, L. J. J., 55
Jansta, J., 59
Janz, G. J., 25, 32
Jarrousseau, J. C., 25
Jaselskis, B., 215
Jastrzebska, J., 102
Jeanne, A., 80
Jee, R. D., 223, 226
Jehring, H., 107
Jeminet, G., 156
Jenevein, R. M., 158
Jenkins, C. M., 175
Jenkins, D. A., 70
Jennings, V. J., 211
Jensen, E. T., 138
Jezkova, G., 221
Johansson, G., 6, 8, 15
Johari, G. P., 35
Johnson, A. M., 70
Johnson, C. A., 51
Johnson, J. S., 194, 195
Johnson, J. W., 60, 81, 82, 163
Johnson, K. W., 86
Jolicoeur, C., 197, 199
Jones, A. L., 192
Jones, D. A., 84
Jones, D. O., 213
Jones, H. C., 212, 232
Jones, P. C., 83
Joshi, K. M., 106
Jouve, G., 81
Jovanovic, R., 218
Joynes, P. L., 219
Jozefowicz, M., 65
Juhasz, B., 232
Juillard, J., 6

243

Juodkazis, K., 50
Justice, J. C., 21, 27
Justice, M. C., 21

Kaatze, U., 209
Kabanov, B. N., 57, 60, 79
Kabasawa, Y., 226
Kaganovich, R. I., 103, 104, 108
Kahr, G., 10
Kajiyama, R., 234
Kakihana, H., 205
Kalinowski, M. K., 159
Kalish, T. V., 99
Kalvoda, R., 185, 225
Kamal, I., 80
Kambara, T., 221
Kamenev, A. I., 227
Kanakam, R., 151
Kane, P. O., 53
Kang, H. Y., 63
Kano, G., 83
Karbainov, Y. A., 227
Karbainova, S. N., 227
Kariv, E., 123, 152
Karlberg, B., 6
Karlberg, G., 89
Karlsson, R., 234, 235
Karpenko, G. V., 85
Karrman, K. J., 234
Kartzmark, E. M., 208
Kashin, A. N., 158
Katayama, A., 27
Kato, H., 93
Kato, M., 76
Kato, S., 145
Katz, S., 207
Kawada, K., 233
Kawai, S., 93
Kay, R. L., 23, 26, 32, 197
Kazarinov, V. E., 113, 115, 116
Kebarle, P., 202
Kelbg, G., 22
Keller, H. E., 220
Kelley, M. T., 212
Kemula, W., 219, 228, 230
Kenny, N. C., 13
Kepnev, G. P., 9
Kesselman, W., 54
Kessler, Ya. M., 34
Kesten, M., 76, 87
Keszthely, C. P., 131
Khairyi, F. M., 27
Khalil, F. Y., 29
Khalil, M. W., 61, 90
Khalturina, T. I., 71
Khamrakulov, T. K., 234, 236, 237
Khavalits, T. V., 226
Khazova, O. A., 77
Kheifets, L. Ya., 153
Kheifets, V. L., 43
Khentov, V. Ya., 24
Khlopotina, L. V., 68
Kho, B. T., 221
Khonina, V. F., 103
Khoo, K. H., 5, 204
Khoo, S. W., 59
Khudair, A. I., 168

Kihara, S., 236
Kilina, Z. G., 228
Kim, C. P., 69
Kimura, Y., 164
King, D. L., 206
Kirowa-Eisner, E., 5
Kir'yanov, V. A., 43, 101
Kiseleva, I. G., 60
Kishore, N., 151
Kiss, L., 54, 65
Kitaev, Yu. P., 142
Kitagawa, T., 211, 219
Kivalo, P., 98
Klapka, V., 78
Klemm, L. H., 154
Kliever, L. B., 24
Klimova, V. A., 132, 233, 234
Kludt, J. R., 197
Klygul, T. A., 138
Knerr, M., 35
Knots, L. L., 46, 111
Knunyants, I. L., 183
Ko, H. C., 205
Koch, H. R., 134
Kochkin, D. A., 226
Koczorowski, Z., 103
Koehl, B. G., 93
Koen, J. G., 219
Kogan, N. B., 221
Kohara, H., 15
Kojima, H., 222
Kokoulina, D. V., 112
Koloskova, N. H., 168
Kolthoff, I. M., 20, 204
Komissarova, T. E., 236
Kon'kova, A. V., 227
Konnik, E. I., 153
Kono, T., 226
Konstantinov, P. A., 168
Kopanica, M., 229
Kopanskaya, L. S., 225
Kor, S. K., 209
Korinek, K., 178
Korobkov, U. I., 29
Korshikov, L. A., 107
Korshunova, K. S., 43
Koryta, J., 187
Koryushin, A. P., 65
Koshcheev, V. D., 57
Kostelitz, M., 72
Koster, G., 228, 230
Kostromin, A. I., 236, 237
Kotaka, M., 29
Kouton, V. N., 68
Kovac, Z., 59
Kovalskaya, V. D., 46
Kovarskii, N. Ya., 67
Kousman, L. P., 168
Kozheurov, V. A., 236
Kozlowska, H. A., 118
Kozowoski, Z., 119
Kraeft, W. D., 22
Krasovitskaya, Yu. I., 112
Krastskov, B. S., 43
Kratochvil, B., 3, 32, 33
Krause, M. S., jun., 223
Kravstov, V. I., 77
Krawczyk, A. R., 145
Kray, A., 214

Krayanskii, O.-B., 181
Kreis, R. W., 192
Kremp, D., 22
Kreuzer, F., 12
Kreysa, G., 89
Krijgsman, W., 234
Krishnan, C. V., 203
Krishtalik, L. I., 49, 112
Kron, A. K., 190
Kroon, D. J., 83
Kropivnitskaya, R. A., 235
Kruger, J., 87
Kruglikov, S. S., 75
Krull, I. H., 17
Krumgalz, B. S., 22
Kruse, I., 217
Krylov, V. S., 43, 101, 107
Kryszozynska, H., 159
Kryuchkova, E. I., 162
Kublik, Z., 228
Kubo, V., 204
Kubota, T., 153
Kuboyama, K., 233
Kudish, A. I., 189
Kudo, K., 89
Kudryavtsev, N. T., 80
Kuduyavtseva, I. V., 23
Kugatova-Shemyakina, G. P., 129
Kugler, K., 27
Kuhn, A., 121
Kulezneva, M. I., 77, 114
Kulkarni, A. G., 24
Kundu, K. K., 5, 204, 205
Kurihara, Sh., 60
Kuroki, N., 27
Kuznetsov, A. M., 49
Kwee, S., 138
Kwong, G. Y. W., 197

LaBoda, M. A., 73
Lacombe, P., 81
Ladańyi, L., 151
Laforge-Kantzer, D., 80
Lagrou, A., 220
Lahmann, E., 234
Lai, S. C., 163
Laird, C. E., 215
Lakeeva, T. N., 236
Lakomov, V. I., 55
Lakshmanan, S., 194
Lakshminarayan, G. R., 25
Lal, S., 16
Laliberté, L. H., 199
La Mar, V. K., 191
Lamberts, L., 30
Lamm, C. G., 11
Lamy, C., 52
Landolt, D., 68
Langford, C. H., 205
Langrish, L., 79
Lanier, R. D., 10
Lantzke, I. R., 208
Lapatin, V. A., 54
Large, R. F., 121
Larsen, D. W., 197
Larson, J. W., 193, 205, 208
Lassigne, C., 205
Latham, R. J., 62, 64

Author Index

Laurent, A., 170, 185
Laurent, E., 170
Laviron, E., 125
Lavrov, I. A., 233
Lawrence, J., 95, 118, 199
Lawrence, K. G., 204
Lawrenson, I. J., 188
Lecoeur, J., 45
Leduc, P. A., 24, 199
Lee, D. G., 207
Lee, J. B., 69
Lee, L. H., 92
Lee, R. E., 204
Lee, T. P., 200
Lee, W. H., 232
Lee, Y. K., 237
Le Goff, P., 54
Le Gorrec, B., 75
Leibson, V. N., 153
Leifer, L., 199
Leikis, D. I., 48
Leja, J., 74
Lelandais, D., 168
Lemley, A. T. G., 208
Lendermann, B., 230
Leng, T. T., 207
Lengyel, B., 52
Lenzi, F., 207
Lepri, L., 172
Leskovšek, D., 67
Lestrade, J. C., 110
Leung, C. S., 205
Leuschke, W., 4
Levay, B., 27
Leverkusen, P. V., 183
Levich, V. G., 54
Levin, E. D., 45
Levine, S., 100
Levins, R. J., 18
Levinson, I. M., 158
Levitskaya, N. K., 34
Lewis, M. C., 170
Leyendekkers, J. V., 194, 195
L'Her, M., 34
Libert, M., 175
Liberti, A., 13
Libus, W., 32
Lietzke, M. H., 192, 196
Light, T. S., 3
Lilley, T. H., 187, 200, 202, 206, 207
Lin, C. K., 193
Lind, J. E., 25
Lindauer, R., 33
Lindberg, I., 233
Lindenbaum, S., 193, 195, 196, 199
Lindman, B., 197
Lindquist, J. R. O., 232
Lindsay, W. T., 192
Lindström, M., 98
Linek, K., 129
Lingane, J. J., 214, 236
Linge, H. G., 192
Linton, M., 37
Liotta, C. L., 207
Lippold, B., 226
Litchman, W. M., 197

Littlehailes, J. D., 3
Liu, C., 192
Lizlovs, L. A., 88
Lizogub, A. V., 146
Llenado, R., 12
Llopis, J., 91
Lloyd, D., 143
Lobachev, V. A., 76
Lockwood, D. J., 208
Lomax, G. R., 79
Longchamp. S., 175
Longhi, P., 2, 5
Lopatin, V. A., 54
Lorenz, W., 100, 109
Lorenz, W. J., 68
Losev, V. V., 57, 113
Loshkarev, Yu. M., 68
LoSurdo, A., 197, 199
Loutfy, R. O., 66
Lovrecek, B., 150
Lowe, B. M., 198
Lown, D. A., 39, 207
Loyd, R. J., 214
Luborsky, F. E., 72
Lucas, M., 197, 199, 200, 201
Luck. W. A. P., 189
Lukashenkova, N. V., 221
Lukovstev, P. D., 73
Luk'yanycheva, V. I., 46, 111
Lund, H., 123, 138, 219
Lusararyan, K. S., 235
L'Vova, L. A., 62
Lyalikov, Yu. S., 225
Lyubimova, N. A., 71

Maahn, E., 79
McAllister, D. L., 140, 223
McAllister, R. A., 10
Macaskill, J. B., 26
Macau, J., 30
McBryde, W. A. E., 6
McCracken, J. E., 232, 236
MacDonald, D. D., 199
MacDonald, J. R., 51
MacDonald, K. I., 69
McDonald, R. L., 197
Macdougall, B., 118
Machavariani, D. N., 60
Mache, H.-R., 64
McKelvey, D. R., 203
McKenzie, I. D., 27
McKeon, M. G., 134
McKinney, W. J., 205
McMaster, C. H., 215
Maeda, M., 205
Maggs, R. J., 219
Magno, R. P., 184
Magnusson, C., 160
Mahendran, K., 206
Mairanovskii, S. G., 129, 141, 153, 154, 157
Makenc, J., 69
Maksimov, Yu. M., 114
Malaterre, P., 52
Malatesti, F., 228
Malodov, A. I., 57
Maloy, J. T., 131

Malysheva, Zh. N., 114, 115, 116
Mamajek, R. C., 30
Mamina, F. A., 135
Mamontov, E. A., 56
Manahan, S. E., 1
Mancy, K. H., 231
Mandell, L., 123
Maneschi, S., 228
Mangold, K., 36
Mann, C. K., 155, 232
Manning, C. W., 4
Manohar, G., 107
Manousek, O., 143
Mansfeld, F., 84, 87
Månsson, M., 206
Mantella, L., 15
Mao, G. W., 86
Mao, K.-W., 73
Maranville, L. F., 205
Marchand, M., 234
Marchesini, L., 183
Marciaq-Rousselot, M. M., 197
Marcoux, L., 162
Marinenko, G., 233
Marinkowsky, A. E., 192
Marinsky, J. A., 192, 207
Mark, H. B., jun., 212, 228
Markovic, S. V., 216
Markushina, I. A., 183
Marsh, K. N., 201
Marshakov, I. K., 59, 61
Marshall, W. L., 26, 36, 37, 190, 196, 209
Martin, J., 232
Martin, P., 91
Martinet, P., 137
Masalovich, V. M., 236
Mascini, M., 13
Masciopinto, D., 31
Masin, V., 229
Mask, C. A., 17
Maslova, G. V., 48
Maslova, L. I., 142
Masterton, W. L., 24, 200, 208
Mastroianni, I., 193, 198
Masui, M., 175
Materova, E. A., 14
Matesich, M. A., 29, 35
Mathieson, J. G., 193
Matson, W. R., 231
Matsubara, I., 233
Matsubara, Y., 83
Matsuda, H., 54
Matsuda, K., 215
Matsuda, Y., 173
Matsui, M., 15
Matsui, Y., 132
Matsumoto, K., 155, 227
Matsumura, Y., 165, 226
Matsuoka, M., 155
Matsushi, H., 6
Matulis, J., 66
Mauger, R., 6
Maurin, G., 75
Mayeda, E. A., 181
Mayell, J. S., 184
Mayer, G. E., 25

Means, D. K., 212
Meatherall, R. C., 206
Mecarelli, E., 218
Medvedev, A. N., 151
Medvedev, G. I., 80
Meklati, M. H., 54
Melicharek, M., 174
Menéndez, V., 91
Mennereau, G., 148
Merigot, D., 236
Mesmer, R. E., 205
Mészaros, L., 52, 223
Metche, M., 129, 130
Metters, B., 234
Meunier, J.-M., 125
Mihajlovic, R. P., 235
Mijajima, K., 195
Mika, V., 235
Mikhailov, V. S., 154
Mikhilev, A. D., 29
Mikolajek, A., 217
Mikuni, F., 62, 77
Miller, D., 123
Miller, I. F., 197
Miller, J. E., 207
Miller, L. L., 181, 184
Millero, F. J., 188, 189, 194, 199
Mil'man, V. I., 181
Milovzorov, V. P., 67
Minasz, R. J., 205
Minc, S., 102
Mironov, V. E., 5
Mishra, A., 27
Mitani, M., 126
Mitra, S. K., 189
Mitsuish, M., 27
Miwa, I., 211, 217
Miwa, T., 230
Miyake, S., 232, 235
Miyaoka, Y., 163
Miyashita, H., 60
Miyazaki, H., 153
Mizuike, A., 226, 230
Mockle, J. A., 217
Modena, C., 87
Mohan, M. S., 14
Mohanty, R. K., 198
Mohilner, D. M., 95, 106, 118
Mohr, N. J., 5
Moiseeva, Z. I., 71
Moldaver, T. I., 58
Molenat, J., 25, 40
Mondzhoyan, A. L., 235
Monforte, J. R., 236
Monheim, H. N., 183
Monica, M. D., 31, 196
Monien, H., 227, 229, 230
Monk, C. B., 207, 209
Montrose, C. J., 209
Moody, G J., 10, 11
Moody, J. R., 235
Moore, J. M., 2, 7
Moore, J. T., 192
Morcos, I., 42, 119
Morel, P., 91
Morgan, C. E., 3
Morrison, C. F., 229
Mosaad, A. I., 27

Mosolov, V. V., 237
Mostkova, R. I., 34
Motojima, K., 236
Motomizu, S., 209
Moynihan, C. T., 25, 209
Mueller, T. R., 212
Mukherjee, L. M., 7, 33, 207
Mukhovikov, V. V., 14
Muller, H., 191
Muller, L., 115
Muller, R. H., 68
Murakami, Y., 171
Muravich-Alexander,H.L., 146
Murphy, J. J., 21
Mussini, T., 2, 4, 5, 14, 15, 192
Muto, G., 232, 233, 237
Myers, R. L., 213

Nadas, J. A., 28, 35
Nadjo, L., 127
Nagamori, N., 168
Nagasaki, K., 65, 72, 89
Nagy, F., 77, 115, 117, 119
Nakagawa, S., 90
Nakagawa, Y., 165
Nakahara, M., 40
Nakajima, K., 74
Nakamura, K., 233
Nakamura, S., 138
Nakanishi, N., 183
Nakashima, M., 227
Nakaya, J., 145
Nakayama, F. S., 209
Nakayama, H., 193
Nanis, L., 54
Narula, S. P., 34
Navon, G., 204
Neal, E. G., 20
Nebesar, B., 232
Nechaev, E. A., 80
Neeb, R., 212, 227
Nekrasov, L. N., 150
Nelson, D. L., 208
Nelson, R. F., 174
Nemirovsky, Y., 3
Nesterov, B. P., 53
Newman, J., 50, 51, 70
Newman, K. E., 202
Nicholson, R. S., 227
Nielsen, N. A., 85
Nigretto, J. M., 65
Niki, E., 214
Niki, H., 83
Niki, K., 51
Nikiforova, Yu. A., 90
Nikitskaya, E. P., 34
Nikolaeva, E. R., 236
Nikolaeva, G. M., 129
Nikolaeva, S. N., 236
Nilsson, A., 30
Nishikida, K., 153
Nobe, K., 69, 88, 89, 91
Noguchi, I., 181
Nomura, T., 159
Nonaka, T., 132
Nord, H., 88
Nothnagel, K. H., 198

Novosel, B., 226
Novosel'skii, I. M., 44
Nozaki, K., 237
Nyberg, K., 159, 160, 161

Oae, S., 169
O'Brien, J. F., 205
Odier, M., 225
Odobesku, N. S., 225
O'Donnell, T. A., 215, 216, 225
Oelschlaeger, W., 231
Østwald, T., 19
Ogle, J., 214
Ogo, H., 181
O'Grady, W. E., 55
Ogura, K., 72
Ohnesorge, W. E., 160
Ohnishi, M., 17
Ohshiro, Y., 173
Ohtaki, H., 195, 206
Okazaki, M., 83
Oki, S., 230
Okochi, H., 227
Okuda, G., 211, 217
Okuda, J., 211, 217
Okuda, Y., 51
Okumura, K., 155
Okuyama, M., 89
Olander, J. A., 205
Oldfield, J. W., 54
Oldham, K. B., 84, 220
Oliver, B. G., 25, 27
Oliveri-Vigh, S., 235
Olivier, A., 18
Olofsson, B., 160
Olofsson, G., 207
Olson, D. R., 154
Olver, J. W., 32
Ong, P. T., 216
Ono, H., 234
Ono, K., 233
Onsager, L., 20
Ord, J. L., 77, 89
O'Reilly, J. E., 138
Ormonroyd, S., 197
Ornealas, L. D., 233
Orttung, W. H., 189
Orzhekhovskaya, A. I., 234
Osajima, Y., 227
Oshe, A. I., 76
Osipchuk, N. Yu., 78
Osman, A., 168
Osten, W., 37
Osteryoung, J., 53
Osteryoung, R. A., 53, 220
Osugi, J., 30
O'Sullivan, T. D., 200
Otsuka, A., 195
Ott, W. T., 9
Ottaway, J. M., 234
Outhwaite, C. W., 191
Ovchinnikova, R. A., 153
Ovsyannikov, N. N., 43
Ozaki, K., 67

Paabo, M., 206
Padova, J., 9
Pagella, A., 4, 192
Pakhmurskii, V. I., 85

Author Index

Palagyi-Fenyes, B., 223
Palecek, E., 221
Paligoric, I., 209
Palm, U., 44
Panckhurst, M. H., 26, 193
Pangarov, N. A., 56
Pariaud, J. C., 6
Parker, A. J., 188
Parker, K., 9
Parker, V. D., 160, 179
Parry, E. P., 87, 220
Parsons, G. H., 205
Parsons, R., 50, 97, 102, 104
Parth, M. H., 235
Partyka, S., 119
Pascal, Y. L., 125
Past, V., 44
Pastor, T. J., 235
Patel, S. R., 206
Paterson, P., 23
Paterson, R., 208
Pathy, M. S. V., 151
Patriarche, G. J., 236
Patterson, C. S., 192
Paul, R. C., 34
Pavlikova, G. P., 158
Payne, R., 103, 111
Pchel'nikov, I. J., 30
Peak, S., 204
Pearce, P. J., 39
Pecei, G., 35, 209
Pechan, Z., 221
Pechova, H., 5
Pedersen, C. J., 17
Pedler, A. E., 175
Peekema, R. M., 229
Peet, N. P., 121
Peleg, M., 208
Pepela, C. N., 192, 201
Pepelyaeva, Z. A., 234
Périchon, J., 163
Perjéssy, A., 129
Perone, S. P., 213, 214
Perron, G., 198, 199
Pesterbe, G., 77
Pethybridge, A. D., 21, 191, 208
Petree, L. A., 196
Petrii, O. A., 46, 94, 110, 113, 114, 115, 116
Petrovich, J. P., 131
Petrucci, S., 29, 196, 209
Pfeifer, S., 218
Phaermark, V. Z., 151
Philip, P. R., 199
Piccardi, G., 211
Pierre, G., 184
Pietrzyk, D. J., 121
Pilac, I., 4
Pil'ko, E. I., 236
Pilla, A. A., 51, 52
Pilling, G., 197
Pilloni, G., 184
Pinson, J., 133, 137, 143
Pioda, L. A. R., 10, 16
Pirskaya, T. Kh., 236
Pistoia, G., 35, 209
Pitts, E., 21
Pitzer, K. S., 191

Plambeck, J. A., 198
Plane, R. A., 208
Plank, W. M., 218
Platford, R. F., 195
Plesch, P. H., 157
Pleskov, Yu. V., 55
Pletcher, D., 120, 134, 165, 178, 185, 186
Plichon, V., 225
Plock, C. E., 215, 216
Pnev, V. V., 230
Podesta, J. J., 59
Podgaetskii, É. M., 54
Podlibner, B. G., 150
Podolyanko, V. A., 30
Polev, V. V., 46
Polyanovskaya, N. S., 98
Pomosov, A. V., 66
Popkie, H., 202
Popov, A. I., 205
Popov, K. I., 83
Popov, V. I., 59
Popova, T. I., 71
Popovych, O., 19, 188
Popp, G., 121
Porai-Koshits, E. A., 7
Porthouse, C. A., 202
Portnoy, N. A., 158
Potsepkina, R. N., 13
Pottel, R., 209
Pottkamp, F., 215
Pourcel, C., 236
Povarov, Yu. M., 73
Prasad, D., 4
Prejean, G. W., 158
Premaswarup, D., 209
Prescher, K. E., 234
Pretsch, E., 10, 18
Pribadi, K., 23
Privat, M., 96
Proctor, W. E., 9
Pronkina, I. I., 14
Propst, R. C., 229
Prue, D. G., 221
Prue, J. E., 6, 21, 190, 191, 192, 196, 208
Pruppacher, H. R., 189
Przhilgovskaya, N. M., 153
Przybylowicz, Z., 219
Pshenechnikov, A. G., 46
Psrasnyski, M., 119
Pullerits, R., 44
Pungor, E., 10
Pupezin, J., 108
Purdie, N., 209
Purdy, W. C., 4, 235, 236
Pyshnogreeva, I. I., 77
Pyzhkov, E. M., 64

Qazi, M. A., 74
Quint, J., 24
Quist, A. S., 36, 37

Race, W. P., 43, 70, 94
Rachmanina, G. A., 228
Rafinski, A., 186
Raghavan, P. S., 151
Rahman, A., 189
Rajagopalan, S., 106

Rajie, S. R., 216
Rald, K., 11
Ramaley, L., 213, 223
Ramananurty, M. V., 27
Ramanathan, P. S., 24, 194
Rand, D. A. J., 77, 112
Randin, J.-P., 45
Randles, J. E. B., 97
Rangarajan, S. K., 51, 194
Rao, P., 86
Rao, P. S.-K. M., 209
Rasaiah, J. C., 190
Rashid, A., 185
Rason, H. P., 216
Raspi, G., 228
Rastogi, P. P., 34, 198
Rea, E. J. F., 121
Read, A. J., 196
Read, H. J., 58
Rechnitz, G. A., 2, 8, 10, 12, 13, 14, 18
Reddy, A. N., 187
Reddy, M., 192
Reddy, T. B., 141
Redmond, J. D., 231
Reeves, R. M., 96
Reid, W. E., 88
Reihman, G. O., 148
Reilly, P. J., 20, 188, 194, 195, 196
Reinermann, K. H., 235
Reishakhrit, L. S., 153
Rendall, H. M., 198
Renkert, H., 37
Rhodes, D. R., 232
Rhodes, K. A., 207
Rhotenbacher, P., 66
Richards, J. A., 125
Riddell, J. D., 208
Rietz, B., 161
Rifi, M., 134
Ritson, D. M., 209
Rizzardi, G., 24
Robbins, G. D., 20, 228
Robertson, E. B., 207
Robic, G., 148
Robinson, D. J., 66
Robinson, R. A., 13, 191, 194, 195, 201
Roblson, J. R., 234
Robson, D., 157
Rochester, C. H., 205, 206, 207
Rock, P. A., 1
Rodewald, L. B., 170
Roessler, N., 25
Roffia, S., 211
Rogers, J. W., 142, 150
Rogers, R. S., 207
Roizenblat, E. M., 229
Rojahn, T., 225
Romanova, I. L., 74
Romanovski, S., 119
Romero, G., 196
Rontan, A., 179, 180
Rosen, M., 46, 111, 113
Rosenko, S. G., 30
Ross, J. W., 16
Rossall, B., 206

Author Index

Rosseinsky, D. R., 188, 194
Rossotti, F. J. C., 6
Rotenberg, Z. A., 55
Rothwell, G. P., 58, 59
Rotinyan, A. L., 43, 45
Roughton, C. L., 222
Roušar, I., 50
Roy, R. N., 5, 203, 204
Royle, L. G., 220
Royon, J., 45
Roze, V. P., 13
Rozhkov, I. N., 183
Rozsondai, B., 53
Rubinstein, I., 123, 152
Rudd, E. J., 128
Ruehlen, F. N., 183
Rush, R. M., 194, 195
Ruzicka, J., 11
Rybakov, B. N., 48
Rybalka, K. V., 46
Ryvolova-Kejharova, A., 123

Sadek, H., 29
Sadovnichaya, L. P., 30
Saeki, T., 183
Sageman, D. R., 25
Sagner, P., 5, 206
Sainsbury, M., 163
Saji, T., 87
Sako, H., 163
Sakuma, M., 209
Salikhdzhanova, R. M., 222
Salim, R., 81
Salomon, M., 203
Saluja, P. P. S., 193
Salvinien, J., 40
Sandig, R., 22
Sandoz, D. P., 229
Sandved, K., 191
Sano, K., 209
Sard, R., 71
Sarkar, A., 192
Sarkisyan, A. A., 235
Sarma, T. S., 198, 201
Sathyanarayana, S., 107
Sato, N., 89
Sato, Y., 83
Satrikova, N. N., 66
Savéant, J. M., 127
Savenko, P. V., 67
Savvin, N. I., 13, 14
Sayo, H., 175
Scalvini, M., 213
Scatchard, G., 194
Schäfer, H., 121
Schertel, M. E., 217
Schiffrin, D. J., 94, 98, 109
Schmidt, E., 68
Schmithals, F., 209
Schmulbach, C. D., 32
Schneider, O., 3
Schoene, K., 215
Scholer, R. P., 10, 16
Schrier, E. E., 10, 200
Schroeder, T. D., 226
Schuldiner, S., 46, 111, 113
Schultze, J. W., 72
Schuster, R. E., 204

Schwabe, K., 4, 89
Schwartz, L. M., 206
Schweider, H., 25
Schwing, J. P., 6
Schwitzgebel, G., 35
Scibona, G., 15
Scoggin, D. I., 174
Scudder, J. A., 40
Sedlak, J. M., 109
Seiber, J. N., 134
Sekine, T., 132, 168
Selekhova, N. P., 71
Selitskii, I. A., 74
Selvaratnam, M., 191
Semenova, V. N., 34
Senkevich, V. V., 224
Senkowski, B. Z., 235
Senne, J. K., 3
Sergeeva, I. A., 34
Sergeivskii, V. V., 14
Serve, D., 173
Sevast'yanov, É. S., 48
Shain, I., 213
Shakunthala, A. P., 151
Shalaby, L. A., 60
Shallal, A. K., 105, 106, 224
Shamin, M., 191
Shapiro, S. A., 207
Shapnik, M. S., 66
Sharma, S. K., 224
Sharp, M., 15
Sharpe, W. R., 197, 204
Sheberstov, S. V., 55
Sheffield, W. H., 175
Sheppard, A. J., 217
Sheppard, N., 204
Shibasaki, Y., 75
Shibata, S., 77
Shields, T. M., 156
Shiozaki, K., 219
Shirai, H., 214
Shishkin, P. M., 233
Shkodin, A. M., 30, 34
Shkorbatova, T. L., 226
Shoesmith, D. W., 154
Shono, A., 173
Shono, T., 126, 164, 165
Shoor, S. K., 200
Shterman, V. S., 13
Shults, M. M., 7, 9
Shults, W. D., 232, 236
Shulyakovskaya, N. V., 183
Simkovic, J., 67
Simon, W., 7, 10, 16, 18
Simonet, J., 125, 133, 137, 156
Simonov, V. D., 135
Simonova, N. A., 71
Simpson, R. J., 10
Singh, D., 27, 35, 71
Singh, K., 34
Singla, J. P., 34
Sinyagovskaya, L. A., 48
Sinyakova, S. I., 229, 230
Sioda, R. E., 50
Sirak, L. D., 226
Sirohi, R. S., 70
Sistig, E., 235
Skalozubov, M. F., 78

Skrebkova, I. M., 142
Slaiman, Q. J. M., 90
Slusher, R., 190
Sluyters, J. H., 222
Sluyters-Rehbach, M., 222
Smirnov, V. A., 181
Smit, W., 72
Smith, D. E., 196, 211, 214, 219, 222
Smith, D. F., 207
Smith, G. L., 142
Smith, H. M., 205
Smith, J. D., 231
Smith, N. O., 200
Smith, T., 87
Smolianizkaia, V. V., 151
Smyshlyaeva, L. N., 236
Snell, H., 196, 203
Sobkowska, A., 236
Socrates, G., 207
Soga, T., 132
Sokolowski, M. D., 207
Sokol'skii, D. V., 77
Šolc, M., 48
Solomin, A. V., 162
Solov'eva, L. I., 168
Solt, J., 77, 115, 116, 119
Somsen, G., 203
Souchay, P., 143
Spalek, O., 53
Spalthoff, W., 197
Spedding, F. H., 189
Spekker, H., 230
Spink, M. Y., 10, 200
Spiro, M., 2, 23, 78, 191
Spong, A. H., 30
Springer, C. H., 32
Srinivasan, K. V., 5
Stankova, V., 16
Stanković, B., 157
Stanley, E. M., 189
Staples, B. R., 27
Starchenko, A. A., 59
Staroshitskii, P. Y., 22
Stasko, A., 157
Staveley, L. A. K., 4, 209
Stead, J. A., 24
Steadman, C. J., 193
Steckel, F., 189
Stefanac, Z., 10
Stefanović, A., 157
Steigman, J., 196, 200
Stelzner, R. D., 212
Stengle, T. R., 205
Stephens, F. B., 214
Stermitz, F. R., 184
Stern, J. H., 200, 201
Sternberg, B. M., 222
Stewart, R., 207
Stickley, J. D., 216
Stillinger, F. H., 189
Stock, J. T., 232
Stocker, J. H., 158
Stockton, A., 104
Stokes, R. H., 191, 195, 201
Stolica, N., 80
Stotz, R., 174
Stoughton, R. W., 192, 196
Stoyanovich, F. M., 157

Author Index

Strachan, A. N., 83
Stradins, J., 214
Stradyn, Ya. P., 148
Straka, P., 185
Strauss, W., 39
Strehblow, H. H., 89
Strochkova, E. M., 46, 111
Stroka, J., 103
Stromberg, A. G., 227
Struck, B. D., 3
Strzelecki, H., 32
Stuart, J. D., 160
Stulik, K., 227, 230
Štverák, B., 50
Subramanian, G. S., 151
Subramanian, S., 198, 201
Sudo, E., 227
Sueliam, A. A., 168
Sugii, A., 226
Sukava, A. J., 66
Sukhatin, A. M., 64
Sumino, M. P., 77
Sun, Y. C., 82
Sunder, S., 198
Sundholm, G., 98, 177, 185
Sunner, S., 206
Suranova, M. A., 108
Surfleet, B., 222
Surikova, N. V., 129
Sutzkover, E., 3
Swanson, J. A., 193
Swearingen, J. T., 200
Sweeton, F. H., 205
Syalikov, Yu. S., 227
Sybrandt, L. B., 213
Sykes, J. M., 58
Symons, E. A., 207
Symons, M. C. R., 197, 208
Sympson, R. F., 234
Syrchenkov, A. Ya., 13, 14
Szilagyi, E. G., 27
Szklarska-Smialowska, Z., 46, 84, 92

Tabor, B. E., 21
Tackett, S. L., 216
Tadjeddine, A., 72, 77
Takagi, M., 211
Takahashi, K., 213
Takamura, K., 219
Takamura, T., 62, 77, 83
Takata, J., 232, 233
Takeda, A., 171
Takehara, Z., 76
Takemori, Y., 219
Takenouti, H., 91
Tallec, A., 147, 148
Tamamushi, R., 215
Tamba, A., 87, 88
Tamba, M. G., 228
Tamm, K., 209
Tammilehto, S., 217
Tamura, H., 173
Tanaka, H., 163, 181
Tanaka, N., 206
Tanaka, S., 221
Tanaka, T., 233
Tandberg, G., 225
Tanenaka, T., 206

Tarasevich, M. R., 112, 113, 114
Tarasyants, R. R., 13
Tardivel, R., 185
Tarkhanov, G. A., 168
Tatlow, J. C., 175
Taylor, A. H., 86
Taylor, H., 24
Taylor, R., 46, 74
Taylor, R. J., 216
Tedoradze, G. A., 47, 132
Temkin, O. N., 158
Temmerman, E., 220
Tennakoon, L., 54
Tenygl, J., 237
Tersoc, G., 31
Ter-Zakharyan, Yu. Z., 235
Tessari, G., 31
Thain, J. M., 12
Thangavelu, S., 151
Thé, N. D., 197, 199
Thirsk, H. R., 2, 39, 57, 70, 134, 207, 229
Thoma, K., 226
Thomas, H. G., 169
Thomas, J., 29, 35
Thomas, J. D. R., 10, 11
Thomich, W., 234
Thomsen, A. D., 123
Thong, K. M., 209
Thornton, B., 134
Thulin, L. V., 19
Tiepel, E. W., 200
Tilak, B. V. K. S. A., 44, 118
Timofeeva, E. N., 146
Tindall, G. W., 78, 228
Titova, V. B., 107
Tjell, C. J., 11
Tkachenko, L. I., 34
Tkachik, Z. A., 65
Tobias, C. W., 68
Todres, Z. V., 109
Tôei, K., 209
Tohier, J., 122, 184
Toktabaeva, F. M., 77
Tokuda, T., 92
Tomat, R., 186
Tomcsányi, T., 53
Tomilov, A. P., 132, 151, 158
Tomkins, R. P. T., 3, 32
Tomonari, T., 51
Tomsons, P., 214
Toni, J. E., 63, 82
Tonomura, M., 226
Topolev, V. V., 113
Toren, P. E., 216
Torigoe, Y., 88, 89
Torii, S., 163, 169, 181, 184
Toth, K., 10
Tourky, A. R., 60
Trabenelli, G., 87
Trachtenburg, I., 14
Traore, I., 125
Trasatti, S., 41, 95, 99
Traubenberg, S. E., 91
Trazza, A., 138, 152
Treiner, C., 27

Trevisan, R., 183
Triolo, R., 198
Trukhan, A. M., 73
Truter, M. R., 17
Tsai, K.-C., 90
Tschernikovski, N., 43
Tsentovskii, V. M., 22
Tsuji, K., 213
Tsuru, T., 89
Tsveniashvili, V. Sh., 109
Turner, D. J., 206
Turner, J. C., 125
Turner, P. J., 3
Tutani, I., 214
Tvarusko, A., 55
Tyagai, U. A., 56
Tyrrell, H. J. V., 20
Tyssee, D. A., 143
Tyzack, C., 85

Uden, P. C., 127
Udupa, H. V. K., 63, 151
Uedaira, H., 201
Ugai, Ya. A., 61
Uhlig, H. H., 87
Ulbricht, H., 22
Ullman, E., 226
Umland, F., 215
Uneyama, K., 163, 169, 184
Unz, M., 84
Urushadze, Z. D., 49
Usachev, D. N., 69
Usanovich, M. I., 162
Utley, J. H. P., 167
Uvarov, L. A., 57, 60, 76

Vagramyan, A. T., 60, 76
Vaidyanathan, H., 45, 91
Vajda, M., 151
Vajgand, V. J., 235
Vajtner, Z., 150
Vakhidov, R. S., 59
Valcher, S., 53
Valeev, A. Sh., 68
Valensi, G., 25
Valentukeviciute-Sliesaraviciené, L., 66
Valette, G., 47
Valleau, J. P., 190
Vamos, Gy., 151
Van Bennekom, W. P., 234
Vance, C. J., 134
Van der Leest, R., 228
Vanderzee, C. E., 193, 206
Vanel, P., 95
Van Hook, W. A., 188, 192
Van Kempen, L. H., 12
Van Kerchove, C., 211, 218
Vantina, N., 228
Varenko, E. S., 68
Varshneya, N. C., 189
Vasak, M., 10
Vasil'ev, Yu. B., 77
Vasileva, L. N., 221
Vasina, S. Ya., 113
Vasquez, J., 215, 216
Vasudeva Rao, P. V., 63
Vavresyuk, I. V., 59, 61

Vavrička, S., 150
Vecchi, E., 186
Vegramyan, A. T., 57
Velghe, N., 228
Verbeek, F., 220
Verdier, E., 95
Vermilyea, D. A., 84
Vernon, W., 5, 203, 204
Verrall, R. E., 193, 199
Vértes, G., 77, 116
Veselovskii, V. I., 78
Vesheva, L. V., 153
Vetter, K. J., 72, 89
Viallard, A., 24
Vianello, E., 211
Vidulich, G. A., 23
Vigdorovich, V. I., 30
Vignaud, C., 86
Vijayavalli, R., 63
Vijh, A. K., 85, 179
Vikhliaev, I., 138
Vilinskaya, V. S., 113, 114
Vinet, D., 63
Vinogradova, E. N., 227
Vishnevetskaya, A. N., 24
Visomirskis, R., 50
Vituccio, T., 197
Voice, P. J., 204
Volki, J., 218
Volkov, V. A., 75
Vondrák, J., 53
Vorlickova, M., 221
Vorontsov-Velmaginov, P. N., 190
Vorotyntser, M. A., 49
Vosta, J., 85
Vrobel, L., 58
Vul'fson, E. K., 229
Vydra, F., 229

Wachholz, E., 218
Wachter, R., 35
Wada, G., 206
Wada, S., 171
Waddington, D., 196
Wadsö, I., 206
Wagenknecht, J. H., 15
Wahl, K., 225
Wales, C. P., 78
Walker, G. W., 167
Walker, R., 66
Wallace, A. J., 73
Walrafen, G. E., 189
Wanninen, E., 216
Ward, F. B., 15
Ward, J. J. B., 69
Warner, C. R., 221
Warren, E. C., 231
Wasa, T., 211
Watanabe, N., 183
Waters, J. A., 171
Watkins, L. S., 55
Watson, B., 23
Watson, W. H. H., 150
Waugh, A. B., 216, 225
Wawzonek, S., 121
Wear, J. O., 28

Webber, H. M., 9
Weber, J., 61
Weeda, C. L., 203
Weedon, B. C. L., 167
Weedon, C. J., 70
Wehrli, P. A., 179
Weinberg, N. L., 141, 157, 163
Weisgraber, K. H., 181
Weiss, A., 198
Welch, G., 206, 207
Wen, W.-Y., 195, 200, 201
Wennerström, H., 197
Wernet, J., 225
West, G, D., 62
West, T. S., 216
Wetzel, R., 115
Whang, K. J., 237
Whelwell, R. J., 6
White, D. V., 154
Whitehouse, D. R., 110
Whitfield, M., 7, 10, 195
Whitnack, G. C., 216
Wiart, R., 75
Wicker, R. K., 192
Widmer, H. M., 204
Wieczorek, G., 46
Wiemann, J., 125
Wikander, G., 30
Wikby, A., 8
Wilcox, F., 200
Wilde, B. E., 86
Willems, G., 212
Willgallis, A., 89
Williams, F. W., 235
Williams, J., 211
Williams, L. R., 216
Williams, P. N., 93
Willmott, A. R., 3, 204
Wilson, F. G., 89
Wilson, I. R., 192
Wilson, R. D., 196
Win, S., 216
Winkler, G., 235
Winnick, J., 189
Wipf, H. K., 18
Wirth, H. E., 194, 196, 197, 199
Witkop, B., 171
Witton, R. W., 206
Wolf, D., 189
Wolf, J. F., 181
Wolff, C. M., 6
Wong, M. K., 205
Wood, A. J., 217
Wood, G. C., 59
Wood, R. H., 20, 188, 192, 194, 196
Wood, R.-W., 59
Woodall, B. J., 3
Woods, R., 60, 77, 112
Woodson, A. L., 219
Woolford, R. G., 167
Woolley, E. M., 192, 205, 206
Wranglen, G., 89
Wright, G. A., 79

Wróbel, J. T., 145
Wroblowa, H., 89
Wu, C. L., 81
Wu, Y. C., 194
Wuhrmann, H. R., 10
Wulff, C. A., 193
Wyatt, P. A. H., 207
Wynne-Jones, Lord, W. K. F., 39

Yadav, R. C., 27
Yagi, H., 181
Yahalom, J., 79
Yakovleva, A. A., 78
Yamada, J., 54
Yamadagni, R., 202
Yamamoto, T., 236
Yamane, M., 209
Yamasita, S., 181
Yamate, N., 226
Yanase, H., 183
Yao, N. P., 31
Yaroshko, N. M., 52
Yasuda, M., 25
Yates, K., 206, 207
Yavich, A. A., 76
Yeager, E., 45, 193, 209
Yeager, H. L., 32, 33
Yeatts, L. B., 36, 196, 209
Yokoyama, T., 48
York, R. J., 167
Yoshida, K., 183
Yoshida, T., 138
Yoshimori, T., 233
Yoshizawa, S., 76
Young, T. F., 205
Yuzhanina, I. D., 237

Zacharov, M. S., 230
Zacharova, E. A., 227
Zacharová-Kalavská, D., 129
Zagórska, I., 102, 103
Zahavi, J., 79
Zakharova, E. A., 228
Zakumbaeva, C. D., 77
Zamastynik, I. E., 85
Zamboni, R., 194
Zana, R., 193
Zawoyski, C., 197
Zein, F. N., 69
Zelenskii, M. I., 77
Zelensky, I., 129
Zhamagortsyan, M. A., 76
Zhdanov, S. I., 109
Zheleztsov, A. V., 222, 224
Zhilina, Z. I., 138
Zieglerova, L., 227, 230
Zietek, M., 217
Zimmer, J. P., 125
Zubov, M. S., 57
Zucchi, F., 87
Zuman, P., 123, 125, 127, 157
Zverev, V. V., 142
Zwanzig, R., 22

QD
551
E53
v.3
1971

DEC 20 1973